Mathematical Models of the Cell and Cell Associated Objects

This is volume 206 in
MATHEMATICS IN SCIENCE AND ENGINEERING
Edited by C.K. Chui, *Stanford University*

A list of recent titles in this series appears at the end of this volume.

Mathematical Models of the Cell and Cell Associated Objects

VIKTOR V. IVANOV AND NATALYA V. IVANOVA
UNIVERSITY OF FLORIDA
GAINESVILLE
USA

ELSEVIER

Amsterdam • Boston • Heidelberg • London • New York • Oxford
Paris • San Diego • San Francisco • Singapore • Sydney • Tokyo

Elsevier
Radarweg 29, PO Box 211, 1000 AE Amsterdam, The Netherlands
The Boulevard, Langford Lane, Kidlington, Oxford OX5 1GB, UK

First edition 2006

Library of Congress Cataloging-in-Publication Data
A catalog record for this book is available from the Library of Congress

British Library Cataloguing in Publication Data
A catalogue record for this book is available from the British Library

ISBN-13: 978-0-444-52714-1
ISBN-10: 0-444-52714-1
ISSN: 0076-5392

Printed and bound in The Netherlands

06 07 08 09 10 10 9 8 7 6 5 4 3 2 1

И в мире, где всё граница,
Всё только предел и преграда,
Бездонная бесконечность, -
Ты мне лишь одна отрада!

(And in the world, where all are boundaries,
All are only limits and frustration,
Fathomless limitless is
My unique consolation!)

E. Vinokurov

Contents

Chapter 8: **MM of the Immune Network** **151**

Chapter 9: **MM of Some Other CAO** **169**

Part 3: **INTRODUCTION TO APPLICATIONS** **177**

Chapter 10: **AIDS** **179**

Preface

"Long ago it became evident that the key to every biological problem must finally be sought in the cell, for every living organism is, or at some time has been, a cell." (The Cell in Development and Heredity, Macmillan, Inc., 1925, by Edmund B. Wilson).

However, the cell itself and various cell-associated objects (CAO) are products of evolution and are still in a state of development as any other so-called evolutionary systems (ES).

So, it is natural to use mathematical models (MM) or mathematics of development to better understand the cell and CAO.

This volume is devoted to mathematical modeling of the cell and cell processes based on novel MM of development. These MM are Volterra-like non-linear dynamic models that allow us to model problems of analysis and synthesis for a wide class of ES, in particular, the cell and CAO. CAO relate to separate parts of the cell and various systems of cells including tissues, organs, subsystems of an organism, and whole organisms.

The main features of these MM may offer descriptions of the cell and CAO creation and restoration processes, of dynamics of elimination or reservation of obsolete "technology" in them, of dynamics of resources distribution for fulfillment of internal and external functions of the cell and CAO, etc.

There are potentially unlimited uses of these MM due to the unlimited quantity of different biological systems that can be integrated at various levels of their consideration, including at the molecular or Human Genome levels.

At the same time, the monograph contains a survey of well known MM of life sciences (Part I of the book), and also a comparison of wide-accepted and novel models (in various parts of the book). Of course, this novel approach does not belittle other approaches, but rather complements and hopefully enriches them.

One of the reviewers believes that this book is essential because the subject has great potential and is not well covered in the literature.

The audience of this book includes researchers in life sciences and students of applied mathematics, especially those with an interest in mathematical applications to cell biology, immunology, medicine, and health care.

The book consists of both the general theory of modeling and various applications.

The base MM of development and their various complications as well as examples of applications of these MM to many problems of the cell and CAO can be found in Part II of the book.

One of the main complications for successful modeling of the cell and CAO consists of proper description of objects and of proper identification of the respective MM. Therefore, special parts of the book are devoted first to biological background of the objects and then to the problems and the methods of identification of initial data of the models.

In particular, the book contains descriptions of the main known facts of cell biology and immunology with the point of view of a mathematician who wants to carry over those facts onto the language of MM.

In the cell and CAO, genes of the entire genome often give necessary characteristics for MM. Therefore, serious attention is devoted to the exploration of various genes and proteins.

The authors feel that the results obtained from statements above confirm the importance and the reliability of the novel technique as well as add to the rapid growth of possibilities for mathematical modeling of progressively more complicated cell processes as time passes and as new information is gathered.

Recently, in particular, many important biological experimental results are related to the problems of regulation of cell processes (i.e., their dynamics). Some of them have been described and investigated here with the help of MM under consideration.

The general mathematical theory represented in the Appendix is mainly devoted to the existence and the uniqueness of solutions for the systems of relations of MM and for respective optimization problems. Considerable attention is also devoted to asymptotic properties of the solutions and to the qualitative investigation of the solution structure for the optimization problems.

In addition, a specific part of the Appendix is focused on respective optimal numerical methods and software. The solution of Volterra-type integral equations with pre-assigned accuracy by the optimal methods is stated and investigated. Reduction of many other problems to the solution of Volterra-type equations is analyzed.

Introduction to applications is devoted mainly to AIDS, cancers, and life longevity problems (Part III of the book).

The monograph is a further development of the techniques for modeling of ES, represented for the first time in the book *Model Development and Optimization*, KAP, 1999, by V. V. Ivanov.

Essential differences consist of detailed investigation of such complicated objects as the cell and CAO, and more complete results of the general mathematical theory of ES modeling are presented.

The main purpose of any investigation in life sciences is finding new effective ways for disease-targeted research and for cure of life-threatening diseases. We attempt to reach this purpose.

However, the experience of application of MM in trying to reach this goal shows that progress here depends on power of connection among specialists in various fields.

As a rule, mathematicians attempt to understand the results of biologists, but not the other way around. Many biologists believe that so far only methods of mathematical statistics may be useful to them.

In this regard, we would like to emphasize special capability of mathematicians to separate essential factors from unessential due to mathematicians' long training in abstract thinking. It means that mathematicians can be very useful for different complicated problems (including biological), and, of course, for formalization of these problems by means of MM. After this, experienced mathematicians can analyze those MM and provide useful feedback to biologists.

One of the special concerns of any scientific endeavor is the effort to acquire accurate measurement in a form that can be understood and verified by others. We would like to emphasize the New International System of Units (S.I.) that recognizes seven basic units: mass (Kilogram - kg), length (Meter - m), time (Second - s or sec), electric current (Ampere - A), temperature (Kelvin - K), luminous intensity (Candela – cd), and quantity of a substance (Mole – mole). The recommended use is the joule (J) rather than the calorie (cal) as the standard unit of energy (one J = 1 kgm^2/s^2 = 0.239 cal). In physics, energy is expressed usually in electron volts, ergs, or inverse centimeters. The relations between these units are 1 eV = 1.602 10^{-12} erg = 8066.1sm^{-1} = 23.05 kcal/mole.

Finally, the authors would like to express thanks to reviewers for fruitful criticism, to Dr. Keith Jones, Publisher of Elsevier, and to Andy Deelen, Administrative Editor Physics, Mathematics, Computer Science and Astronomy Department, for their help and concern about the fate of this book.

Viktor Ivanov, Natalya Ivanova

St. Petersburg, Tampa, Gainesville, FL and Boston, MA USA December, 2005

List of Abbreviations

A subsystem of CAO for fulfillment of its internal functions
AC active center (s), enzyme-substrate complex (es)
ADP adenosine di- phosphate
Ag antigen (s)
AIDS acquired immune deficiency syndrome
ATP adenosine tri-phosphate, source of energy
B subsystem of CAO for fulfillment of its external functions
C cancer (s)
CAO cell associated objects
Ch. chapter (s)
DNA deoxyribonucleic acid (basic constituent of the gene)
E extraction subsystem of CAO
ES evolutionary system (s)
G genome subsystem of CAO
HIV human immune-deficiency virus (es)
ICC immune competent cells
Id-Aid idio-type anti-idio-typical interaction
Ig immune-globulin
iff if and only if
IN immune network
IS immune system
ISO immune status of an organism
Lc lymphocyte (s)
MM mathematical model (s)
OS organism subsystem (s)
R recreation subsystem of CAO
RPO remained part of an organism
s. section (s)
SR specific rate
Tef T effectors of hyper-sensitivity and T killers Lc
Th T helpers Lc
Ts T supressors Lc
V. i. e. Volterra integral equation (s)

List of Notations

$|*|$ - absolute value or measure of *.
$\approx *$ - approximately equal to *.
$\sim *$ or $\Theta(*)$ - asymptotically equal to * in order.
<~ *, >~ *- asymptotically less, greater than *.
« *, » * - by far less, greater than *.
$|$ - condition sign.
$\times, \oplus$ - direct product, sum.
$\varnothing$ - empty set.
 - end of the proof.
$\succ *, \prec *$ - follows (succeeds), precedes *.
∞ - infinity sign.
$\cap, \cup$ - intersection, union.
$\boldsymbol{M}(f,R), \boldsymbol{m}(f,R)$ - max, min f $(t \in R)$.
$x = o(y), O(y)$ - mean respectively $x/y \to 0$, $x/y < C$, where C is a positive constant, when $y \to 0$ or $y \to \infty$.
$x \cong y$ - means $x - y < C|y|/t$, $t \to \infty$, where C is a positive constant.
$\|x\|$ - norm of x in an abstract space.
$\in, \notin$ - member, not a member.
R^n - n-dimensional Euclidean space
∂ - partial derivative.
$\prod, \sum$ - product, summation.
$\subset, \supset$ - proper subset, superset.
f^{-1} - reciprocal function or operator for any function or operator f.
(x, y) - scalar product of elements x and y from an abstract space.
$\to$ - tend to.
$C(R)$ - the metric space of continuous functions on R to R.
$C(S, X)$ - the metric space of continuous functions on S to X, $S \to X$.
$C^l(R)$ - the metric space of continuously differentiable functions on R to R.
$C(R)^*$ - the space of functionals l_r on $C(R)$.
$L_p(R)$ - the space of functions on R to R absolute integrable to the power p, $p > 0$.
$\mathrm{B}(T, \Sigma, \mu, S; X)$ [or $\mathrm{B}(T, S; X)$ or B] - the vector space of functions $\varphi: T \times S \to X$ such that, for all $(t, s) \in T \times S$, $\varphi(t,.) \in C(S, X)$, $\varphi(.,s)$ is μ-measurable and there exists a μ-integrable $\psi_\varphi: T \to R$ with $|\varphi(t,.)|_{\sup} \le \psi_\varphi(t)$.

For simplicity, we denote sometimes $\sum f_i$ $(i \in A)$ and $\int g(s)ds$ $(s \in B)$ instead of the notations $\sum_{i \in A} f_i$ and $\int_B g(s)ds$, respectively, and we denote usually $\sum_1^n f_i$ instead of $\sum f_i$ $(i = 1, \ldots, n)$ and $\int_a^b g(s)ds$ instead of $\int g(s)ds$ $(s \in [a, b])$ or $\int_{[a, b]} g(s)ds$.

To save place and setting, we denote everywhere $g(x,y)$, where x, y are arguments of any function g.

Introduction

EVOLUTIONARY SYSTEMS AND DEVELOPMENT MODELING

To better understand the objects investigated various examples of *evolutionary systems* (ES) and the respective base mathematical models (MM) are presented.

There are three classes of ES: *Artificial ES* (AES) that are being created or already have been created by human beings and are functioning with their participation; *Natural ES* (NES), in particular, the cell and cell associated objects (CAO); and *Joined ES* (JES). The principal point of this introduction is to recognize the similarities between ES and the differences between AES and NES. In this introduction, we mainly describe structure-functional properties of ES that are independent of specific character of ES (such as separate businesses, different species of plants, or diversity of cells).

The base MM of development and general qualitative properties of MM for ES modeling are also described. In addition, certain principal differences between traditional MM and novel MM of this book are established.

1. Evolutionary Systems (ES)

1.1. ARTIFICIAL ES

Some examples of AES include industry, education, science, art, health services, etc.

The main elements of AES are *work places* (WP). *A work place* is usually localized in the time and space aggregate of labor functions of the respective ware: material, energy, and information, which should be fulfilled by a respective specialist. The results of WP functioning are various goods and services or products.

The main characteristics or indices of the WP functioning are *efficiency* (e.g., the quantities of products produced per unit expenditure and per unit time). Different WP and AES relate to different kinds of human activities. Apparently, diversity of WP and their indices are in continuous development.

Let us consider in more detail AES in industry, education, science, art, and health care. Each separate enterprise or corporations and industry as a whole can be considered as ES. Of course, the specific character of WP in each case depends on the aim and the result of their functioning.

There are two important classes of WP: one that reproduces external goods with respect to ES, and the second that reproduces or creates new more effective WP for the ES itself. Theoretically, it is possible that there could be WP creating new WP, which, in turn, create new WP, and so on. However, in practice this recurrence has to be interrupted.

While restricting the combination of creation and self-creation functions on some WP, *self-creation* or self-perfection *functions* become inevitable. It is also possible that some WP create copies of WP, which are similar to the process of cell proliferation. Other WP create new special WP, which are similar to the process of cell differentiation. The necessities of new products and greater efficiency of WP stimulate creation of new WP. But what does the *new WP* mean? If at least one of the labor functions and/or any element of the respective "wares", including personnel-ware, are changed, there will be a *new WP*.

The branch of industry in which ES creates new WP is called a *subsystem A* of ES. The branch of industry that creates or produces goods and services that are external with respect to ES is called a subsystem B of ES. Clearly, the subsystem A creates new WP for B and for itself. Therefore, A is the subsystem of creation or reproduction of ES and B is the subsystem of production of goods and services external to ES.

The distribution of WP between the subsystems A and B is very important. A share can be used for controlling ES. It is not clear in advance what consequences arise from increasing or decreasing the share. The problem of the best distribution has arisen.

In the course of time, *functioning WP becomes obsolete* (i.e., at least one of the WP elements becomes obsolete). So, these obsolete WP should be eliminated, reserved, or reconstructed. The temporal boundaries for obsolete WP and/or the *intensities* of the employment of WP are also significant. These boundaries and/or intensities can also be used to control ES. In general, there may be two kinds of boundaries: one for WP in the subsystem A and another for WP in the subsystem B.

The next important feature of ES is the existence of the external resources that enter ES. It means that ES are *open dynamic systems*. We can use these external and internal resources to create balance equations for ES.

To conclude with industry as ES we may consider two cases:

1. t_0 is the moment of ES origin and the starting point for modeling, and
2. $[t_0, t^\wedge]$ is the given prehistory of ES, $t^\wedge > t_0$, and $t^\wedge$ is the starting point.

Any educational center, including school, college, university, etc., and education as a whole can be considered as ES.

All the features of ES in industry (including WP as the main elements of ES, existence of the subsystem A for realization of the internal functions of the system development and the subsystem B for realization of the external functions, allocation of the system's resources between its internal and external functions, the out of date or obsolete WP, the inflow of resources from the outside, and existence of the moment of ES origin or ES prehistory) are also suitable for the case of education. An essential difference between education and industry as ES consists of their WP and their products.

The *educational product* is the number of specialists of determined quality, and the index of efficiency of WP is the number of new specialists of determined quality per unit of WP in the subsystem B per unit time. The quality of specialists can be designated by different rules. To avoid the question: “And who are the judges?” different tests are used and their results are calculated with the help of computers.

The subsystem B in education as ES is a set of WP, the product of which is the number of the specialists. The subsystem A in education as ES is a set of WP, the products of which are new, more effective WP in both A and B subsystems of education. Creation of these WP also means training of the respective, more effective specialists in the field of education. Any science-methodical centers in the field of education, of public educational office, and administration of any educational center will belong to the subsystem A if their functions are the creation of new WP. Actually, any educator belongs to A while he/she is increasing his/her own qualification. Thus, the primary indication of whether WP belongs to the subsystem A or B is not place, time, or person, but rather the kind of labor function fulfilled.

The *science product* is a new, more effective technology. Since new technology results in higher-level values of indices of WP efficiency, we can assume that the science external products are these indices of WP efficiency for different ES. The internal product of the subsystem A in science as ES is a new technology, which is new WP efficiency indices in A that further create new technology, including creation or restoration of A itself. The external product of the subsystem B in science is a new technology for other branches of human activities. So, we can conclude that indices of WP efficiency in the subsystem A of science and the main products of WP are actually the same.

Any scientific center, any branch of science, and science as a whole can be considered as ES. At all events, the subsystem A of these ES is the set of WP. The product of this set of WP is new, more effective WP, which create new technology for themselves and for external ES. Any scientist, while his/her labor function is perfection of his/her own WP, belongs to the subsystem A of science as ES.

Thus, it is not difficult to see that the essence of development of science (and other AES) is the presence of the *subsystem self-development* or perfection.

Considering art as ES, we only would like to emphasize that a writer, an artist or a musician also has his/her own work place and his/her own products. Any organization in art and in art as a whole can be considered as ES by separating the subsystems A and B depending on WP functions. The modern means of technology can considerably enlarge efficiency of labor functions in art.

The problem of how to best distribute these means between the subsystems A and B and the problem of how to determine the best temporal boundary for obsolete WP are also important for art.

We dwell on health care as ES in more detail since it is a vital issue and since important parts of its applications relate to health services or health care (HC).

The HC external product is the cure. In our definition the *cure* is the number of people who have improved their health to such an extent that they do not need providers of HC or the number of people who have prolonged their active life for more than a year after curing.

Unfortunately, United States has a difficult HC status: "There are thousands of billing codes in U. S. insurance for health care procedures; there is not one single billing code for a cure." (See G. Halvorson [75], p. 3).

To improve the situation in HC, a systematic approach is considered in [75], in particular, Continuous Quality Improvement that "means that the outcome of a process and the process itself are under continuous review, with a formal approach used to continuously improve both process and results."

In the ES language it means that we need to improve indices of efficiency of functioning for both subsystems A and B in HC. The subsystem A in HC is the subsystem that creates new (more effective) WP for itself and for the subsystem B in HC. B in HC, on the other hand, is an aggregate or a team of WP, having curing as the main labor function. The subsystem A comprises not only the doctors, who are organizers of HC, but also any doctor who improves in qualification and any person who helps to perfect HC as ES.

A special feature of HC as ES is the fact that in WP of the subsystem B "raw" materials that should be cured are living organisms or their subsystems and organs that in turn can be considered as NES. Therefore, we have to deal with interaction between AES and NES. Later we will consider different types of interactions between two ES. We would like to emphasize that, according to the theory of NES, the most difficult cases of diseases and their primary reasons are related to a defect in the organism subsystem, namely A-type.

For successful development of HC (under the same resources inflow), it is important that we distribute its internal resources between its subsystems A and B in a reasonable way, that we determine in the same way the temporal boundary for obsolete WP in the subsystems A and B, and that we have a competitive market of interaction between HC and its consumers, etc.

The respective formalized mathematical apparatus allows us to consider and investigate such kinds of problems both on the qualitative and quantitative level.

1.2. NATURAL ES

The examples of NES are the cell and CAO, a separate plant, a separate organism, a population of animals, the biosphere, etc.

The main elements of NES are reaction, *active centers* (AC) or sub-units (biological analogy of WP). An *active center* is usually localized in the time and space aggregate of chemical reactions (or appropriate physiological functions) with the respective ware: material, energetic, and information. A medium of the respective ware is usually biomass. The results of AC functioning are some products and/or biomass.

The main characteristics of AC are the indices of efficiency of AC functioning (e. g., the quantities of products and/or biomass created per unit of the respective ware expenditure and per unit of time). Sometimes, it is convenient to assume that AC are cells of an organism or, in other cases, parts of a cell (e.g., proteins, organelles, macromolecules or ionic particles).

Let us consider in more detail NES such as a separate plant, a separate animal, a population of animals, and the biosphere. There are two important classes of AC of a plant as ES: AC in the subsystems B in a plant that creates some product fulfilling a plant's interaction with its environment and AC in the subsystem A in a plant that creates new AC. In a cell's case as AC, proliferation and differentiation are processes of creation of new AC in the subsystem A.

In addition, share or ratio of AC distributed between the subsystems A and B is also very important in the case of plants. This ratio can be used to control productivity of plants. The problem of the best share has also arisen.

In time, earlier created AC become obsolete ones. These obsolete AC should be eliminated or reserved or renovated. Temporal boundaries for obsolete AC also seem to be very important and useful for controlling NES. Of course, a plant as ES cannot exist without external resources, which a plant consumes. We can use external and internal resources to create balance equations for NES. Using the comparatively simple balance of the process of photosynthesis [150], we can obtain the additional equation for new biomass or better phytomass of a plant.

It should be noted that there are essentially two different phases of plant development:

1. From the instant of germination of a seed to the instant of appearance of leaves when the process of photosynthesis does not take place, and
2. After the appearance of leaves.

In each of these phases we can consider two cases:

i) t_0 is the moment of ES origin and starting point for modeling, and

ii) $[t_0, t^\wedge]$ is the given prehistory of ES, $t^\wedge > t_0$, and $t^\wedge$ is the starting point for modeling.

All features of ES in a plant [including AC as the main elements of ES and the existence of the subsystem A for realization of the internal functions of the ES development and the subsystem B for realization of the external functions of interaction between ES and its environment, allocation of ES resources between its internal and external functions, the obsolete AC, the temporal bounds for the obsolete AC, consumption of the external resources, presence of two different phases of development (1. from the instant of conception to the instant of birth and 2. after birth) and existence of the moment of ES origin and/or ES prehistory] also take place in an animal as ES. Indices of efficiency for AC are usually by far higher for animals than for plants, and, of course, interpretation and composition of AC are different for different NES. We dwell here on differences between an animal and a plant and between NES and AES. Some differences between NES and AES are the following (see H. Curtis [37], V. Vernadsky [194]:

i) There are different compositions of AC and WP ware. "All life forms employ two categories of chemical, one that stores information and a second that acts, based upon that information, to duplicate the organism." (A. Levine, [120], p. 1).

ii) Indices of efficiency for AC of NES are considerably higher than for WP of AES. For example, some cells recreate cells during minutes with a negligible outlay, while WP create new similar WP during years with great expenditure. It means that efficiency of AC can be approximately a billion times higher than WP. A reason may be the very prolonged period of evolution.

"Any living cell carries with it the experience of a billion years of experimentation by its ancestors" (Max Debrick, 1944). But the rate of growth of WP indices is usually very high. Presupposing their growth of 2% on average each year, it is easy to obtain the increase of efficiency indices on 8-10 orders during about 10^3 years. Therefore, AES can probably achieve that level of efficiency in about a thousand years for which NES required a billion years.

iii) The life spans for some NES may be considerably shorter or considerably longer than for most AES. It seems advantageous for nature to provide the best survival rate of the population at the expense of the comparatively high death rate of its species. The life span of a population as ES may be longer, but for some species much shorter than for most AES.

According to H. Curtis [37], living things: i) are highly organized, ii) are homeostatic, iii) reproduce themselves, iv) grow and develop, v) take energy from the environment and change it, vi) respond to stimuli, and vii) are adaptable.

In the case of population as ES, the internal product is the birthrate of the species. The external product is the result of interaction between the population and the environment. The subsystem A in population is that of reproduction of species, and the index of efficiency of AC in A is the number of new species that appeared per unit of resource expenditure for the unit of time. Therefore, obsolete AC are dead species. The subsystem B in population is that of production of an external product.

The moment of ES origin is a time when mutation and recombination of the genetic blueprint results in a new population. The important meaning for the development and control of population as ES is the distribution of its internal resources between its subsystems A and B and is a temporal bound for obsolete AC.

The natural hypothesis is that a temporal bound is mainly determined by the genetic information, whereas distribution of resources depends mainly on the external effects from the environment. If a population has severe environmental conditions, its reproduction will decrease. The theory of development permits us to evaluate the internal and external effects more precisely.

According to V. Vernadsky [194] the subsystem A in the biosphere as ES is the subsystem of recreation of the living substance of the planet, mainly of the phytomass by way of photosynthesis. The subsystem B in the biosphere is the subsystem of creation of the so-called inorganic substance (mainly oxygen). The obsolete product is the dead or bio-boned substance, which is approximately equal by mass to the humus.

The important exception of the biosphere as ES is the composition of its external resources. In the biosphere, there is the *bio-geo-chemical cycle*, which results in the biosphere "consuming" a part of its own product: inorganic substance and the obsolete product. The other parts of its external resources are derived from the atmosphere, the hydrosphere, solar energy, and cosmic radiation.

It appears that the moment of the biosphere's origin is the time when the conditions on our planet were sufficient for self-creation of the living substances.

1.3. JOINED ES

AES are the results of human activities. The size of these activities is now so great that land, air, water and the whole biosphere have begun to suffer from them.

Destiny of human beings is now inseparable from the fate of their environment. So, academician V.I. Vernadsky [194] came to consider human activities and their environment as a unit and named this unit the *neosphere*. Thus, the *neosphere* includes the atmosphere, the human, the geo, the hydro and the biosphere. In other words, the *neosphere* is "Managing Planet Earth" [126].

We can consider the *neosphere* as one of NES in which the *subsystem A* recreates the living and bio-boned substance of our planet and in which the *subsystem B* creates the inorganic resources (parts of which leave the planet). The external resources are parts of NES's own products, solar energy, and cosmic radiation. We can also consider the *neosphere* as JES that is the combination of two ES, one of which is human activity as AES (ES1) and another of which is the other part of our planet as NES (ES2).

The principal meaning for interactions between ES1 and ES2 is the composition of their external resources. The external resources of ES1 are parts of the products of ES2 and parts of all the external resources of the biosphere. The external resources of ES2 are the remaining parts of the external resources of the biosphere and a portion of the obsolete products of ES1. The critical value is the size of the parts and the composition of the obsolete products leaving ES1. Starting from the determined size of the parts, the processes of recreation in ES2 can become irreversible, and even a small quantity of special obsolete products of human activities can be the reason for environmental disaster. Thus, we need to create strategies with finite resources for sustainable human development on our planet.

2. Generalized Structure of ES

According to the description above, the generalized *structure of ES* has the form of Fig.1. Here **f** are the external resources of ES; **m** are the new internal products (resources) of the subsystem **A**; **c** are the new external products of the subsystem **B**; **y** are relative shares of distribution of the internal resources between **A** and **B**, $0 \leq \mathbf{y} \leq 1$; **g** are the obsolete products of ES; $\boldsymbol{\lambda}$ is the intensity of the subsystem **A** functioning along the channels **ym--m**, $0 \leq \boldsymbol{\lambda} \leq 1$; $\boldsymbol{\alpha}$ are the indices of efficiency of functioning along the channels **ym--m**; $\boldsymbol{\mu}$ are the intensities of the subsystem **B** functioning along the channels **(1–y)m--c**, $0 \leq \boldsymbol{\mu} \leq 1$; $\boldsymbol{\beta}$ are the indices of efficiency of functioning along the channels **(1–y)m—c**, **1** is the unit matrix.

New technology or the respective matrices $\boldsymbol{\alpha}$ and $\boldsymbol{\beta}$ can be created outside ES (the exogenous case) or inside ES (the endogenous case).

It should be emphasized that the subsystems **A** and/or **B** can be, in turn, ES with the same structure as initial ES, and so on.

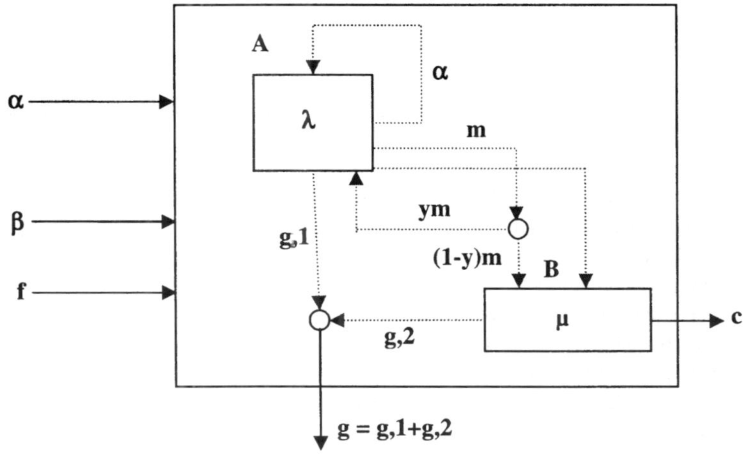

Figure 1. Generalized structure of ES

3. Generalized Structure of Two ES Interactions

We already considered two ES interactions in the case of the neosphere.

Another example of interaction in JES is the interaction between health care as AES (ES1) and patients as NES (ES2).

The main functions of ES1 are perfection of their own WP and curing of patients. As the response to that curing, ES2 gives to ES1 a certain part of its external products.

This part is also an important additional control function that can be determined by estimating the maximum number of cured patients.

More examples include interactions between directing ES in economy (ES1) and executing ES (ES2), interactions between two independent countries as ES1 and ES2, and so on. In these cases, we have, as a rule, mutually beneficial interactions of Fig.2.

Of course, we can consider both kinds of ES as one ES, but this is not the case when considered in more detail.

The main functions of ES1 are the creation of new WP for itself and for ES2. So, external products of ES1 are the new additional WP for ES2, including a new "personnel-ware" for ES2 as a result of training persons with new professions for ES2. In response, ES2 gives to ES1 a certain part of its external products. This part is an important additional control function, of which the best value can be found, for example, from analysis of joined maximal profit as ES1 and ES2.

In the case of the interaction between two independent countries as ES1 and ES2, determined parts of all kinds of products of ES1, including new technologies (which are products of its subsystem A1) and new goods and services (which are products of its subsystem B1), are subject to interchange with determined parts of all kinds of products of ES2 and vice versa. These parts are the important additional control functions.

MM of ES allow us to select the appropriate values, in a certain sense, of these parts. One of the criteria of optimization is the idea of coordinated maximization of profits for ES1 and ES2.

The classical example of interactions between two populations, specifically between beasts or birds of prey (ES1) and their victims (ES2), was investigated by V. Volterra [198, 199]. Here the subsystems A1 and A2 are subsystems of reproduction of plunderers and victims, respectively. The subsystem B1 consumes the maximal possible part of the victims, and the subsystem B2 minimizes that part.

So, we have the mini-max problem under the condition of finite external resources. The solution of this problem gives determined values to the distribution in each of ES: between A1 and B1 in ES1 and between A2 and B2 in ES2.

According to the description in this section, the generalized *structure of two ES interactions* has the form shown in Fig. 2 and 3.

Values **fi**, **mi**, **ci**, **yi**, and **gi**, **i = 1** for ES1 and **i = 2** for ES2, have the same sense as values **f, m, c, y,** and **g** in Fig. 1.

Fig. 2 is related best to the above-mentioned example of peaceful interactions between two independent countries or amicable coexistence of two different NES.

Unfortunately, interaction between human activities and the rest of our planet (the biosphere) is related more closely to Fig. 3, which depicts interactions between plunderers and victims or cancerous cells and normal cells.

Note that in the case of Fig. 1 as well as in Fig. 2 and 3, the bold notations mean that in general all considered magnitudes could be respective vectors and matrices. Therefore, the inequalities mean the same inequalities for their respective components.

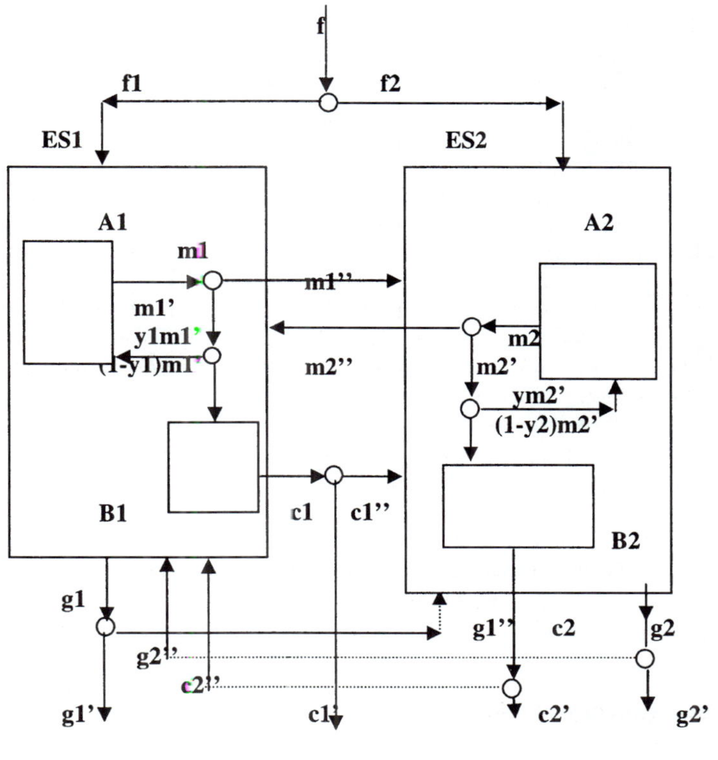

Figure 2. Generalized structure of two ES interactions (Peaceful coexistence)

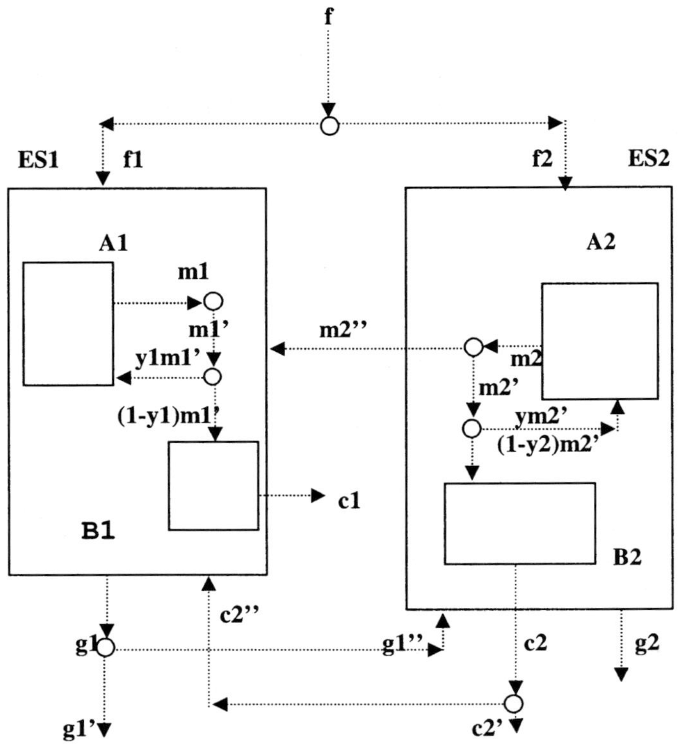

Figure 3. Generalized structure of two ES interactions (Plunderers and victims)

4. Generalized Structure of CAO

According to the well-known biological background, especially self-producing aspects in the cases of enzyme complexes, mitochondria, stem cells, and immune cells, the generalized *structure of* the typical model of CAO also has a form of Fig. 1.

Here **f** are the external resources of CAO; **m** are the new internal products (resources) of its subsystem A; **c** are the new external products of its subsystem B; **y** are relative shares of distribution of the internal resources between A and B, $0 \leq \mathbf{y} \leq 1$; **g** are the obsolete products of CAO; $\boldsymbol{\lambda}$ are the intensities of the subsystem A functioning along the channels **ym--m**, $0 \leq \boldsymbol{\lambda} \leq 1$; $\boldsymbol{\alpha}$ are the indices of efficiency of functioning along the channels **ym--m**; $\boldsymbol{\mu}$ are the intensities of the subsystem B functioning along the channels **(1–y)m--c**, $0 \leq \boldsymbol{\mu} \leq 1$; $\boldsymbol{\beta}$ are the indices of efficiency of functioning along the channels **(1–y)m--c**; new technology or the respective matrices $\boldsymbol{\alpha}$ and $\boldsymbol{\beta}$ can be created outside CAO (the exogenous case) or inside CAO (the endogenous case). It should be emphasized that the subsystems A and/or B can be, in turn, CAO with the same structure as initial CAO, and so on (see later in the Ch. 7 the more complicated generalized structure of the cell).

Generalized *structure of two CAO interactions* is represented by Fig. 2 and 3.

The main functions of CAO1 are self-perfecting and interacting with CAO2. As a result of that interaction, CAO1 gives to CAO2 a certain part of its external products and vice versa. We can consider both CAO1 and CAO2 as one CAO, but this is not the case when considered in more detail. Another case of the main functions of CAO1 is the creation of new active centers (AC) for itself and for CAO2. The external products of CAO1 are the new additional AC for CAO2. In response, CAO2 gives to CAO1 a certain part of its external products. This part is an important additional control function, of which the best value can be found, for example, from analyzing joined maximal usefulness or maximization of the fatigue function for both CAO1 and CAO2.

In one more case, when determined parts of all kind of products of CAO1, including new technologies (which are products of its subsystem A1) and new external resources (which are products of its subsystem B1), are subject to interchange with determined parts of all kinds of products of CAO2 and vice versa. These parts are important additional control functions. MM of CAO allows us to select the appropriate values, in a certain sense, of these parts. One of the criteria of optimization is the idea of coordinated maximization of advantages for CAO1 and CAO2. This means that under a certain variation of the control functions it is possible to increase both advantages simultaneously until we achieve the so-called *Pareto set* on which an increase of both advantages is impossible. So, we should first try to find the best values of the control functions outside the Pareto set.

Let us consider briefly the example of interactions between cancer cells (CAO1) and their victims, normal cells (CAO2). Here the subsystems A1 and A2 are subsystems of reproduction of cancer cells and normal cells, respectively. The subsystem B1 "consumes" or kills the maximal possible part of normal cells, and the subsystem B2 minimizes the damaged part. So, we have the mini-max problem under the condition of finite external resources. The solution to this problem gives determined values to the distribution in each of CAO: between A1 and B1 in CAO1 and between A2 and B2 in CAO2.

Thus, according to the description above, the generalized *structure of two CAO interactions* also has the form shown in Fig. 2 and 3. Values **fi, mi, ci, yi,** and **gi, i = 1** for CAO1 and **i = 2** for CAO2, have the same sense as values **f, m, c, y,** and **g** in Fig. 1.

Fig. 2 is related best to the above-mentioned example of "peaceful" interactions between two amicable CAO.

Fig. 3 is related to the interaction between cancer and normal cells.

5. On Development Modeling

A survey of well-known MM for modeling of various processes in the cell and cell-to cell interactions can be seen in Part I of this book. Those MM deal as a rule with the subsystem A or B of CAO or with the whole CAO without their representation in the form of two or more subsystems A- and B-types.

Here, we outline the main qualitative features of the novel class of MM.

5.1. BASE SIMPLEST MM OF DEVELOPMENT

The *base, minimal or simplest MM of development* has the form

$$m(t) = \int_{a(t)}^{t} \alpha(t,s)y(s)ds,\ 0 \le y \le 1,\ a(t) \le t,\ \alpha \ge 0;$$

$$c(t) = \int_{a(t)}^{t} \beta(t,s)z(s)ds,\ z(s) = 1 - y(s),\ \beta \ge 0;$$

$$R(t) = \int_{a(t)}^{t} m(s)ds,$$

$$M(t) = \int_{0}^{t} m(s)ds,$$

$$G(t) = M(t) - R(t),$$

$$f(t) = m(t) + c(t),\ t \ge t^{\wedge} \ge 0. \qquad (1)$$

where $m(t)$ is the rate of creation of the first kind *new generalized product* (resource) quantity at the time instant t, which provides the fulfillment of the internal functions of ES, that is, restoration of itself and creation of the second kind product; $y(t)m(t)$ is a share of $m(t)$ for fulfillment of internal functions in the *subsystem* A of restoration and perfection of the system as a whole; $\alpha(t,s)$ is the *efficiency index* for functioning of the subsystem A along the channel $\alpha(t,s)y(s)m(s)$ -- $m(t)$, i.e., the number of units of $m(t)$ created in the unit of time starting from the instant t per one unit of $y(s)m(s)$; $a(t)$ is a special temporal bound: the new product creating before $a(t)$ is never used at the instant t, but created after $a(t)$ is used entirely; $c(t)$ is the rate of creation of the second kind *new generalized product* quantity at the instant t, which provides the realization of the external functions of ES; $[1-y(s)]m(s)$ and $\beta(t,s)$ are similar to ym and α respectively but for the *subsystem* B of creation of the second kind product; $R(t)$ is the total quantity of the first kind product functioning at the instant t; $M(t)$ is the total quantity of the first kind product to be created during the time $t - 0$; $G(t)$ is the total quantity of the *obsolete product* at the instant t; $f(t)$ is the rate of the resource inflow from the outside ($m(t)$ and $c(t)$ are measured in the units of $f(t)$); $t^{\wedge}$ is the starting point for modeling; $[0, t^{\wedge}]$ is the *prehistory of ES*, for which all the functions are given (their values will be noted by the same symbols but with the sign "^," e g, $m(t) = m^{\wedge}(t),\ t \in [0, t^{\wedge}]$).

It is obvious that all the relations (1) are faithful representations by definition. In a general case, the indices α and β depend on *m, c, a, y, R, M, G, f.*

Thus, (1) consists of 7 equalities and 7 inequalities connecting 14 values, namely: *m, c*, α, *y*, β, *z, a, R, M, G, t, t^, f,* 0, all of which are nonnegative. Usually, α, β, *y, f* are given, and the others are found.

As can be seen, even in the simplified formulation, MM (1) is the system of nonlinear functional relations, in which along with the nonlinear integral equation of the unusual form (the lower bound *a(t)* can be unknown function) we have the system of functional inequalities. Note that we have a particular case here where the intensities of ES functioning (see Fig. 1) are

$$\lambda(t,s) = \mu(t,s) = 0,\ 0 < s < a(t);\ \lambda(t,s) = \mu(t,s) = 1,\ a(t) < s < t. \qquad (2)$$

We consider later in Ch. 5 and 14 the case of arbitrary λ and μ.

It is not hard to introduce different generalizations of MM (1).

The *n-product MM*, $n > 2$, can be formally written in the same form (1), where *m, a*, and *c* are the vector functions, and α, *y*, β, and *z* are the respective matrices (where the inequalities for the vectors and matrices are the same inequality for their appropriate components).

The continuous MM can be described in the same form considering *t* and *s* as many-dimensional variables and examining the appropriate integrals as multivariate ones.

The stochastic MM can be obtained by considering α, β, and *f* as functions of a random factor ω.

The discrete MM can be represented in the same form if the integrals in (1) are understood in the sense of Stieltjes.

One of the important *typical optimization problems for ES* is maximization of the functional

$$I(y) = \int_{t^\wedge}^{t} c(t)dt = \int_{t^\wedge}^{t} \{\int_{a(t)}^{t} \beta(t,s)[1-y(s)]ds\}dt \qquad (3)$$

over *y* with regard to MM (1). The first essential result on the properties of solutions of the problem (3) (see details in Ch. 15) has been obtained by V. Glushkov, V. Ivanov [62].

The result consisted qualitatively in that for "small" $T - t^\wedge$ the desired *y(t)* is minimally possible, but for "large" $T - t^\wedge$ the desired *y(t)* may differ from the minimally possible on the larger initial part of the segment [*t*^, *T*]. Only on the smaller final part of [*t*^; *T*] the desired *y(t)* is minimally possible.

The notions "small" and "large" depend on the values of the functions α and β; namely, the greater the functions in question, the closer to *t*^ is the boundary between "small" and "large" segments.

The result has obtained, in sequel, an important qualitative general interpretation:

The record of an external function for any evolutionary system (ES) can be obtained only under the conditions of its sufficiently comfortable guarantee, that is, under the significant fraction of resources sent to internal needs of ES.

5.2. COMPARISON WITH CLASSICAL MM

We dwell on the comparisons between MM in question and the classical MM. Detailed comparison with many well-known MM is considered more than once in different parts of the book.

Everybody is familiar with the approach of the so-called "black box", when only input $X = (x_1, \ldots, x_n)$ and output $Y = (y_1, \ldots, y_m)$ of a dynamic system are given. We have, in the linear approximation,

$$Y(t) = \int_{t-T}^{t} K(t,s)X(s)ds. \qquad (4)$$

Here, T is the upper bound for all the transient end times, $K(t,u)$ is the matrix of the pulse transition functions $k_{ij}(t,u)$ that are the response functions of the system for $x_j(s) = \delta(u-s)$, $t > u > t-T$, where δ is Dirac δ-function, and all the other $x_k(s) = 0$. A nonlinear dynamic system can be also represented as (4), but then K depends on X.

The MM (1) deals with the so-called "gray box" when the structure of a dynamic system is partly revealed. Indeed, it is possible to say that in (1) the matrix K in the case of the subsystem A has been factored into two parts $\boldsymbol{\alpha}$ and $\boldsymbol{\lambda}$ (in our particular case, $\boldsymbol{\lambda} = 1$, $a(t) \le s \le t$; $\boldsymbol{\lambda} = 0$, $a(t) \le s \le t$) such that each of them has its own applied sense.

In addition, several outputs of the system in (1) have served as its inputs. At last, due to the functions $c(t)$, $G(t)$, and $f(t)$ we deal with the so-called non-autonomous or open dynamic system.

If we consider the case, when in (4) $X = f$ and $Y = (c, g)$, $g(t) = G'(t) = m[a(t)]a'(t)$ (supposing that derivative $a'(t)$ *of* $a(t)$ exists), which is in correspondence to the generalized structure of ES (Fig. 1), then for determination of the respective K, we need to solve rather complicated system of nonlinear equations (see Appendix, Ch. 14). As a result, K will be a rather sophisticated non-linear vector-functional of f. The essential differences are that all the values in (1) are nonnegative by definition and a diminution of the output values is regulated not by a sign, but, for example, by the rate of $a(t)$ growth.

We can find certain analogies of MM (1)-(2) among MM in economics (so-called *macroeconomics MM*) L. Kantorovich, L. Gor'kov [108]; J. Malcomson [125]; G. Silverberg [171]; A. Petrov, I. Pospelov [148] and in population modeling R. Poluectov, Yu. Pykh, I. Shvytov [152]; F. Sharpe, A. Lotka [169].

Detailed comparison of MM (1) with some *integral dynamic MM* is published by V. Ivanov, Yu. Yatsenko, U. Galiev [104] and by N. Hritonenko, Yu. Yatsenko [82, 83].

MM (1) were described in detail in the monograph by V. Ivanov [92] with applications to ES (artificial, natural, and joined ES). That monograph also contains comparison of a novel MM with the well-known ones, mainly in economy, ecology, and immunology.

However, the applications from above do not include detailed consideration of the cell (and many other CAO) or the problems of immune and gene therapy against AIDS and cancers. Besides, we have many new theoretical results, relating to the global stability of the solution and to the so-called functional genomics and its mathematical challenges [R. Karp, 110].

Thus, this monograph is a further development of the previous one with the emphasis on the cell and CAO.

We would like to emphasize a principal place of MM of development from the point of view of the general theory of mathematical modeling in biology.

In the period of the flowering of mathematical biophysics (1970s), A. Bartholomay [18] noted that a *biological system* was summarized by the symbolic expression B (**C**, **S**, **E**, **F**, t), in which **C** refers to its composition, **S** to its structural aspects, **E** to its environment, **F** to the set of its biological, chemical or physical activity, and t to the time parameter.

MM is μ: B $\rightarrow$ M, where M is a mathematical domain associated with the various components (either **C**, **S**, **E**, etc.).

In the case of MM (1)--(3)-types, **C** is $\boldsymbol{m}, \boldsymbol{c}, \boldsymbol{\alpha}, \boldsymbol{\delta}, \boldsymbol{\beta}, \boldsymbol{\gamma}, \boldsymbol{f}$; where $\boldsymbol{\alpha}$ and $\boldsymbol{\beta}$ are indices of the active sites of B functioning; **S** is $\boldsymbol{y}$, $\boldsymbol{z}$ and the structure of Fig. 1; **E** is the other CAO of the types of Fig. 2, 3; **F** is the example of activity of (2), (3).

The main problem is how to determine $\boldsymbol{\alpha}$ and $\boldsymbol{\beta}$. In our opinion, these indices most likely can be found as a result of synthesis of known physical and chemical laws with the information contained in the Human Genome.

Indeed, the every step of biological activity, which cannot be realized spontaneously due to the existing natural laws, must probably be realized with the help of control from the genome.

Summarizing the implication of nonlinear dynamics to biological systems (1990s), J. Walleczek [200, 201] noted that a new epistemology for biology appears probably in such features as:

nonlinear sensitivities to weak external influences, whether they are chemical, mechanical or electromagnetic in nature;

spontaneous pattern formation;

self-organization; and so on.

All are connected with nonlinear bio-dynamics.

We show in Ch. 5, 6, and 14 that all those features are the intrinsic properties of MM of (1)--(3)-types.

Results of weak stimuli (**f**) cannot be, for example, uniqueness of the solutions of the system of the relation's (1)-type, which can be interpreted as such biological activity as proliferation and/or differentiation. And what is more, the origin itself of bio-systems can be modeled with the help of MM (1)-type (see details in S. Girlin [56] and in Ch. 6).

One of the interpretations of the solutions for MM (1)-type can be external and internal bio-fields of the bio-object considered, and the case is possible when rather weak and distant external electromagnetic stimuli can essentially influence the functioning of the bio-object [V. Glushkov, 61].

As to self-organized biological activities, this feature is an inherent property of any ES, and it is the main purpose of further investigation of MM of development with regard to and emphasis on the well-known properties of the cell and CAO and their applications to health-oriented modeling.

Bearing in mind that a great number of important scientific and applied works are devoted to mathematical modeling of various CAO, the monograph contains a survey of MM in those works, especially if they have connections with MM of development.

Detailed comparison of a novel class of MM with the well-known MM is presented.

5.3. A HISTORY, IN SHORT

MM of (1)-type were introduced in a paper by academician V. Glushkov [58], Director of Kiev Center of Cybernetics (KCC) of the Academy of Science of the Ukraine. Shortly after, similar models were also introduced by V. Glushkov and B. Pshenichnyj [73]. The problem of investigation of the functional (2) with regard to MM (1) was formulated by V. M. Glushkov as a competitive problem for all related scientists of KCC. V. M. Glushkov had foreseen the properties of possible solutions, but a rigorous proof for them was absent. A winner of that competition was Professor V. V. Ivanov, Head of KCC Computation Theory Department. The respective results were published in V. Glushkov, V. Ivanov [62] in 1977.

Since then, successors of V. M. Glushkov continue his work. Hundreds of scientific works some of which, summarized in monographs by V. Glushkov, V. Ivanov, V. Yanenko [69], Yu. Yatsenko [213], N. Hritonenko, Yu. Yatsenko [82, 83], and V. Ivanov [92], have been published.

Dr. Yu. P. Yatsenko conducted later on scientific works focused on the structure of the solution of optimization problems with regard to MM under consideration and applications focused mainly on economy and ecology. V. V. Ivanov conducted research with applications mainly to life sciences and cell biology.

V. V. Ivanov reported the general theory of this technique to the World Congress of Nonlinear Analysts in Tampa, FL, August 1992, and to the Meeting of the AMS, Atlanta, GA, October 1997. The participants agreed that the respective direction of the research was new for the USA and the Western countries. But it seems no one wants to participate in this kind of research.

The applications to modeling of the immune network were reported by V. V. Ivanov to the Conference on Mathematical Models in Medical and Health Sciences, Nashville, TN, May 1997.

Several proposals to National Science Foundation (NSF), National Research Council (NRC), and National Institutes of Health (NIH) of the USA, 1995 –1998, to conduct further development of this research were not acknowledge.

However, NIH introduced a program on "Quantitative Approaches to the Analysis of Complex Biological Systems," 1988. And NSF, in connection with the problem of "Bio-complexity 2000," gave away $40 million to stimulate the creation and the development of actually the same research.

Part I:

SURVEY OF MM OF CAO

The goal of this survey is to demonstrate the main methods and results of the cell and cell associated objects (CAO) mathematical modeling that are widely accepted, in contrast to the novel approach considered in the part II.

Ch. 1 is devoted to the general approach of introducing and analyzing different MM, including quantum mechanics and spatial MM. This chapter also shows examples of modeling of CAO such as cell cycles and super-cycles, cell excitable membrane, traffic through membranes, protein folding, tumor angiogenesis, and mesenchumal morphogenesis.

Ch. 2 considers general techniques of mathematical modeling of biochemical enzyme reactions with application to multi-enzyme systems and energetic metabolism.

Ch. 3 is mainly devoted to the kinetic cellular theory by N. Bellomo and L. Preziosi [20].

Ch. 4 contains many other MM, including delay differential equations, MM of population biology, physical models of DNA, some models of computational biology, the theory catastrophes, and information on the respective software.

Comparatively new MM, including delay differential equations, integral relations of the kinetic cellular theory and age-structured relations of the population theory, are in immediate proximity to MM of the next part. These new MM, especially the integral relations, are considered and compared here and later in more detail.

Some of the scientific works such as MM of the immune systems and MM of disease progression that we use later in other parts of the book are not considered in part I.

Of course, the list of references here is not exhaustive. There are many more related scientific works than we mention. We considered only some of the existing works, which give certain new tools for mathematical modeling in the field under inspection.

Chapter 1

GENERAL METHODS OF INLET AND ANALYSIS OF MM

1. Methods for Simplification of the Systems of Kinetic Equations

All processes in living organisms develop in time and space. It is widely accepted that the most exact MM of these processes are systems of equations with partial derivatives. However, in many cases, one may assume that the processes considered are synchronous in all parts of biological objects, and therefore dependence on space coordinates is absent. In these cases we usually deal with the systems of ODE, i.e. *ordinary differential equations*. Among them is the most common so-called *autonomous system of ODE* order n of the form

$$dx_i/dt = F_i(x_1, x_2, \ldots, x_n),\ i = 1, 2, \ldots, n, \qquad (1)$$

where F_i are nonlinear functions of dynamic variables x_i. As a rule, they are composed of positive terms describing increase of components x_i and composed of negative ones describing their decrease.

For construction of the respective MM (1)-types, it is sufficient to know the rates of inflow and outflow of each component and their dependence on variables x_i.

Simplification consists of a decrease in the number of equations and, at the same time, in the number of parameters determining behavior of the system. After certain transformations, the system (1) can be rewritten in the form

$$\varepsilon dx_i/dt = F_i(x_1, x_2, \ldots, x_n),\ i = 1, \ldots, r, \qquad (2')$$

$$dx_j/dt = F_j(x_1, x_2, \ldots, x_n),\ j = r + 1, \ldots, n, \qquad (2'')$$

where ε is a small parameter, $\varepsilon \geq 0$. If we are interested in the events of the system (2') during the short time $T \approx \varepsilon$, then we can assume that the values x_j are constant, from which there is a possibility to decrease the order of the system (2). On the other hand, there is a connection between the solutions of the whole system (2) and its subsystems (2') and (2'') for all values of variables $x_1, x_2, \ldots, x_n$.

Theorem (A. Tikhonov [183]; A. Tikhonov, A. Vasil'eva, A. Svechnikov [185]). *Let* (2') *be the adjoined system and* (2) *under* $\varepsilon = 0$ *the confluent system. The solution of the whole system tends to the solution of the confluent system under* $\varepsilon \to 0$ *if*

i) *the solution* $x^\wedge_1 = f_1(x_{r+1}, \ldots, x_n), \ldots, x^\wedge_r = f_r(x_{r+1}, \ldots, x_n)$ *is an isolated root of the system*

$$F_i(x_1, x_2, \ldots, x_n) = 0,\ i = 1, \ldots, r; \qquad (3)$$

ii) $x^\wedge_1, \ldots, x^\wedge_r$ *is the stable critical point of the adjoined system under all values* $x_{r+1}, \ldots, x_n$;

iii) *the initial values* $x^0_1, x^0_2, \ldots, x^0_n$ *belong to the domain of influence of the critical point*;

iv) *the solutions of the whole and adjoined system are unique, and the given functions are continuous.*

The conditions i), iii), and iv), but not the condition ii), usually take place in biological systems. The condition ii) is not valid, for example, in the case when an attracting isocline (see the next s. 2) has a so-called S-shaped form so that stationary states along its intermediate branch are unstable.

In that case, the reduction by Tikhonov is possible only in certain restricted domains of the phase-space. Note that the theorem affirms that the result (under $\varepsilon \to 0$) does not depend on initial conditions for the variable of the adjoined system.

Let us consider an example of the system

$$dx/dt = y,\ dy/dt = -x + d(1-z)y,\ \varepsilon dz/dt = -z + x^2. \tag{4}$$

Under $\varepsilon = 0$, from (4) follows the classical *Van-der-Pol system* that describes the auto-oscillation of variable x (or y) or the so-called *limiting cycle* on phase-plane (x, y) (see s. 2.1). Let $\varepsilon \neq 0$, but $\varepsilon \ll 1$. It means that the relaxation time for the variable z is much less than for the other variables. The stationary solution for the adjoined equation in (4) is the parabola $z = x^2$ for all x. In three-dimensional space the limiting cycle will be located on a parabolic cylinder. If a depicted point (x, y, z) is outside of the limiting cycle, then instantly, in practice (due to small ε), it goes again to the stable state. Hence, practically, instead of investigating the depicted point motion in three-dimensional space, we can investigate the motion only on the plane (x, y).

Small parameter ε arises naturally due to the specifics of various biophysical systems. However, even in the confluent system, one can select a certain domain of variables and parameters such that there will be additional possibility to decrease the order of the system. To consider the respective theorem of reduction we need to introduce the following so-called *characteristic equation* for the system of (1)-type or, more exactly, for its linear approximation:

$$\det\{a_{ij} - pk_{ij}\} = 0,\ a_{ij} = [\partial F_i/\partial x_j]^\wedge;\ k_{ij} = 0,\ i \neq j,\ k_{ij} = 1,\ i = j;\ i, j = 1, 2, \ldots, n, \tag{5}$$

where det means determinant and [*]^ means that the functions * are calculated in a stationary point $x^\wedge = (x^\wedge_1, \ldots, x^\wedge_n)$, which is a root of the system

$$F_i(x_1, x_2, \ldots, x_n) = 0,\ i = 1, 2, \ldots, n. \tag{6}$$

Thus, (5) is the algebraic equation of nth power with respect to p.

Under variation of parameters, let n - 1 roots or characteristic numbers of (5) not have small negative real parts, but the real root $p_1 = \varepsilon$ ($|\varepsilon| \ll 1$) with a sign change. Then system (1) can be rewritten in the form

$$dx'_1/dt' = a' - x'_1 + 1/\varepsilon(\{x'^2\}_1 + \{x'^3\}_1 + \dots), \; t' = \varepsilon t, \tag{7'}$$

$$\varepsilon dx'_i/dt' = \Sigma\, a_{ij} x'_j \; (j = 2, \dots, n) + \{x'^2\}_i + \dots, \; x'_i = x_i - x\hat{}_i, \; i = 2, 3, \dots, n, \tag{7''}$$

where x'_1 is deviation of x_1 from a certain value that may be different from $x\hat{}_1$ on the value ε in order, $\{x'^2\}$ means an algebraic expression of the type $\Sigma\, b_{ij}\, x'_i x'_j\, (i, j = 1, \dots, n)$, similar for $\{x'^3\}$, and so on, a' is a parameter in order ε.

The system (7'') is the adjoined system with the stable stationary point $x'_i = 0$, $i = 2, 3, \dots, n$. According to the theorem by Tikhonov, instead of the system (7') and (7'') we have only one equation

$$dx'_1/dt = a - \varepsilon x'_1 + bx'^2_1 + cx'^3_1 + O(x'^4_1), \tag{8}$$

where $O(x'^4_1)$ means all terms of the order not less than 4. Now, to investigate qualitatively the whole system, it is sufficient to investigate (8), depending on the four parameters: ε, a, b, and c.

If $a_{12} = - a_{21} = a$, and $|a_{11}| = |a_{22}| = \varepsilon$, then the roots of the equation (5) $p_{1,2} = -\varepsilon \pm ia$, and the system (6) on the strength of the theorem by Tikhonov reduces to two equations of the form

$$dx'_1/dt = a_{11}x'_1 + a_{12}x'_2 + \{x'^2\}_1 + \{x'^3\}_1 + \dots,$$

$$dx'_2/dt = a_{21}x'_1 + a_{22}x'_2 + \{x'^2\}_2 + \{x'^3\}_2 + \dots. \tag{9}$$

More complicated cases of the roots of the equation (5) and the system (6) can be reduced to 3, 4, etc. equations.

2. On the Qualitative Theory for the Systems of the Second Order

The most well known MM in biophysics are the systems of ODE of the second order or the systems that reduce to ODE of the second order. Let us consider the system of the form

$$dx/dt = P(x,y), \; dy/dt = Q(x,y), \tag{10}$$

where P and Q are continuous functions. For qualitative investigation of (10) it is convenient to consider x and y as coordinates of a depicted point on *phase-plane* (x, y). The collection of the solutions $x(t)$ and $y(t)$ of the system (10) or the phase curves corresponding various initial conditions is called *phase-portrait* of the system.

For the construction of phase-portrait, one should find a collection of integral curves of the equation

$$dy/dx = Q(x,y)/P(x,y). \tag{11}$$

On the strength of Cauchy theorem, there is only one integral curve passing through each point of phase-plane except for the so-called *singular points* for which $P(x,y) = 0$ and for *critical points* or *stationary points* for which simultaneously

$$P(x,y) = 0,\ Q(x,y) = 0. \tag{12}$$

The solutions or roots of (12) $x = x$^ = constant, $y = y$^ = constant is called stationary values of x and y. The phase-portrait is usually constructed qualitatively by the *method ISO-clines*, i.e., the lines that cross by integral curves under the same angle; their equations dy/dx = constant. Special interest is the so-called principal iso-clines; that is, iso-clines of vertical and horizontal tangents for which respectively $dy/dx = \infty$ or $P(x,y) = 0$ and $dy/dx = 0$ or $Q(x,y) = 0$. The horizontal iso-cline is also called the null-cline. The critical points are intersections of principal iso-clines. Qualitatively, one can obtain a good estimate of phase-portrait on the basis of disposition of principal iso-clines and character of stability of critical points.

Stability of critical points is usually determined with the help of linearization in the vicinity of the stationary values x^ and y^ equations of the form

$$dx'/dt = ax' + by',\ dy'/dt = cx' + dy',\ x' = x - x\text{^},\ y' = y - y\text{^},$$

$$a = [\partial P(x,y)/\partial x]\text{^},\ b = [\partial P(x,y)/\partial y]\text{^},$$

$$c = [\partial Q(x,y)/\partial x]\text{^},\ d = [\partial Q(x,y)/\partial y]\text{^}, \tag{13}$$

where [*]^ means that the respective function * calculates under $x = x$^ and $y = y$^. The linear system (13) has the non-trivial solution

$$x' = x'_0 e^{pt},\ y' = y'_0 e^{pt}, \tag{14}$$

where p is the root of the *characteristic equation*

$$p^2 - (a + d)p + ad - bc = 0. \tag{15}$$

Let p_1 and p_2 be two roots of the equation (15). Then if the discriminant of the characteristic equation $D = 4bc + (a - d)^2 \geq 0$ and $p_1,\ p_2 < 0$, the solutions (14) are decreasing exponents, i.e., the system brought out of the state of equilibrium tends to this state again. The respective critical point (x^, y^) is called *stable node.* If p_1 , $p_2 < 0$, we have *unstable node* (under any initial deviations the system is removing from the state of equilibrium). If the roots have different signs, the critical point is unstable and is called saddle-point. Only two integral curves are passing through this point. These curves are called *separatrixes*. If the discriminant D < 0, then p_1, $p_2 = \delta \pm \mathbf{i}\omega$, $\mathbf{i} = (-1)^{1/2}$. If $\delta < 0$, then damped oscillation is the result in the system. On phase-plane, it corresponds to a collection of spirals, tending to the critical point. This point is called *stable focus*.

When $\delta > 0$, we have an *unstable focus*. And when $\delta = 0$, we have the so-called *center*. In the system, undamaged oscillation is the result. On phase-plane, it corresponds to a collection of enclosed ellipses. The value of $\delta = 0$ is *bifurcation* since transition from positive δ to negative follows by a change of the phase-portrait (from the unstable focus to the stable one). In the cases when one of the roots is 0 or when the linear terms in expansion of P and Q by x' and y' are absent, the analysis of the system is more complicated (see, for example, G. Ivanitsky, V. Krinsky, E. Sel'kov [105]; G. Korn, T. Korn [115]).

The phase trajectories tend either to the critical stable points, to infinity, or to the so-called stable limiting cycles. The last one corresponds to the stable oscillations of the system. If both critical point and infinity are unstable, then there is at least one stable limiting cycle. A rather frequent case is when the system in the positive quadrant has three critical points, two of which are stable. Such system is called *trigger* or flip-flop.

The *separatrix* divides phase-plane into two zones of attraction, which lead to two stable points. The important property of this system is that it can be in one of the stable states as long as needed until a large external perturbation would transfer depicted point to the other side of the separatrix. Collection of separatrixes and of a limiting cycle divides phase-plane in elementary parts where behavior of all trajectories is similar.

Thus, the qualitative construction of phase-portrait is reduced to the construction of critical points, limiting cycles, and separatrixes.

Depending on values of parameters, the phase-portrait of the system can be qualitatively changed. The most important bounds of *parametric portrait* on the plane of usually two parameters α, β are

i) the line of neutrality or the bound of stability of the stationary state can be described by the system algebraic equations

$$P(x,y;\ \alpha,\beta) = 0,\ Q(x,y;\ \alpha,\beta) = 0,\ Sp(x,y;\ \alpha,\beta) = a + d = 0; \tag{16}$$

ii) the line of mono-stationary state or the bound of domain for the existence of the single stationary state (beyond this domain there exist three stationary states: one saddle and two nodes or focuses) is determined by the equations

$$P(x,y;\ \alpha,\beta) = 0,\ Q(x,y;\ \alpha,\beta) = 0,\ \Delta(x,y;\ \alpha,\beta) = ad - bc = 0; \tag{17}$$

iii) the line of multiplicity of roots of the characteristic equation (15) or the bound of domain for the existence of the stationary state of the focus-type (beyond this domain the stationary state is a node) is determined by the equations

$$P(x,y;\ \alpha,\beta) = 0,\ Q(x,y;\ \alpha,\beta) = 0,\ D(x,y;\ \alpha,\beta) = 0. \tag{18}$$

The parametric portrait constructed with the help of these bounds does not cover all possible cases. There exist qualitative changes of phase-portrait far from the stationary points. Such non-local changes are caused by fusion of separatrixes, fusion or fission of limiting cycles, fusion of separatrixe with limiting cycle, and so on. Numerical methods usually used for estimate of non-local branching or of *bifurcation*, as well as for estimation of a stable and an unstable limiting cycle and the so-called *strange attractors*.

The *strange attractor* is different from simple attracts (stable critical point and limiting cycles) because all its trajectories are unstable and mixed. The trajectories, however, remain in certain domain.

In the system of third or higher order, behavior of phase trajectories can be much more complicated. First of all, the critical points of saddle-focus or saddle-node types are possible. If, for example, two roots of the respective characteristic equation have negative real parts while the third root is positive, then phase trajectories located on certain surface, containing the critical point, tend to this point. The respective motion will be absolutely unstable since all trajectories outside the surface are absolutely divergent.

Sometimes, it is more convenient not to use a phase-plane, but rather another system coordinate. For example, MM of open enzyme reactions [105; D. Roberts 161] has the form

$$dx/dt = v_{xs} - k_x x - v(x,y),\ \varepsilon dy/dt = v_{ys} - k_y y + v(x,y), \tag{19}$$

where $v(x,y)$ is a rate of a certain reaction X → Y, v_{xs} and v_{ys} are rates of formation of substrates X and Y, and terms $k_x x$ and $k_y y$ regard expenditure of X and Y in other reactions.

When $\varepsilon \ll 1$, MM (19) allows a simple graphic investigation if coordinates (concentration, rate) are used. One can find quasi-stationary values $y = y^\wedge$ and $v = v(x, y^\wedge)$ as points of intersection of the curve $v = v(x,y)$ and the line $v = - v_{ys} + k_y y$ on the plane (y, v) (in this regard, concentration x is regarded as a constant). Changing x and determining for each x the value v, one can construct dependence $v(x)$, i.e., input or output quasi-stationary characteristic of the reaction. The points of intersection of the line $v = v_{xs} - k_x x$ with $v(x)$ on the plane (x, v) are the stationary points of the model (19).

Such analysis is equivalent to consideration of the confluent system

$$dx/dt = v_{xs} - k_x x - v(x,y),\ 0 = v_{ys} - k_y y + v(x,y), \tag{20}$$

which can be obtained from (18) under $\varepsilon \to 0$ based on the theorem by Tikhonov (see s. 1).

2.1. MM OF CELL CYCLES, I

The *cell cycle* is the cellular process from the cell origin to its next division. Often, this process is accompanied by *differentiation*; that is, a daughter cell differs from an initial cell (for example, in blood-formation). In many cases, however, daughter cells are identical to mother cells.

Usually, the cell cycle is divided into the following phases: G_1, phase S (DNA synthesis), G_2, and M (mitosis). Division of cells is called *proliferation*. Besides, there exist the resting state R_1, in which a cell can remain for a long time, and the resting state R_2, in which a cell can be blocked by the phase G_2. Switching from the resting state to proliferation and vise versa is under the influence of different factors that can sometimes act spontaneously. Mechanism of this switching is the main object of modeling.

According to Y. Romanovsky, N. Stepanova, D. Chernavsky [162], we accept the hypothesis that the regulator of the cell cycle is situated in the cell membrane. The problem is to find a model for the origin of cell membrane signals that influence transition from one phase of the cell cycle to another.

There are two types of processes in a cell membrane: chemical and physical. Let us consider these two processes.

Relating to the chemical processes is:

1. Oxidation of lipids with participation of free radicals. There exists an intermediate state in which lipids themselves become radicals, and this process is auto-catalytic. Characteristic concentrations of radicals are 10^{-6} mole/liter and the half-renovation time is 1 sec.
2. Removing from membrane lipids that are oxidized too much so that they stop becoming radicals.
3. Introducing new lipids into a membrane with the help of special protein-carriers. Characteristic half-renovation time of lipids is 10 hours.
4. Weak radicals and anti-oxidizers regulate the process of lipid. Characteristic half-renovation time of anti-oxidizers in a membrane is 1 minute.

Thus, MM of the process of oxidation of lipids in a membrane has the form (dimensionless) [162]

$$\tau_S dS/dt = \nu - \alpha SR - DS,$$

$$\tau_R dR/dt = \kappa + \alpha SR - RR^\wedge - R^2 - \delta_1 R,$$

$$\tau_{R^\wedge} dR^\wedge/dt = \gamma - RR^\wedge - \delta_2 R^\wedge, \tag{21}$$

where S, R, and $R^\wedge$ are concentrations of lipids, radicals of super-oxide-type, and anti-oxidizers.

Characteristic values of parameters in order are $\tau_S = 10^4$ sec, $\tau_R = 1$ sec, $\tau_{R^\wedge} = 10^2$ sec, the others have order 1.

The fastest process cannot be a stationary one because the null cline $dR/dt = 0$ contains unstable parts.

The intermediate in the speed process is always stable and we can put $dR^\wedge/dt = 0$, i.e., $R^\wedge = \gamma/(R + \delta_2)$. The reduced system takes the form

$$\tau_S dS/dt = \nu - \alpha SR - DS,$$

$$\tau_R dR/dt = \kappa + \alpha SR - \gamma R/(R + \delta_2) - R^2 - \delta_1 R. \tag{22}$$

The null-cline of this system is the attract that under $\gamma < \kappa$ has N-shaped character. The phase portrait of the system shows the following possibilities:

1. Stationary stable behavior under small R and large S. This behavior coordinates with the cell cycle resting state R_1.
2. Stable state under small S and large R. This case coordinates with the resting state R_2.
3. Auto-oscillating behavior under the condition that the point of intersection of the clines $dS/dt = dR/dt = 0$ is situated on the dropping line of the attractor. The fast phases of the attractor are accompanied with harsh changes of S and R. We can consider jumping to large S as a signal for the start of the DNK synthesis and jumping to large R as a signal for the start of mitosis.
4. Trigger-type behavior under the condition that the cline $dS/dt = 0$ intersects the attractor three times. Here we have two stable states, R_1 and R_2, and transfer from each to another is possible under a finite perturbation. In addition, transfer from R_2 to R_1 is accompanied by cell division. Switching the resting state to proliferation can be the result of hanging of parameters. So, increase of γ and ν and decrease of κ promote transfer to the resting state R_1.

However, this model is insufficiently crude (i.e., under small perturbations of parameters all its various behaviors can overlap), and it does not include the state of cell membrane. This model better suits the properties of the cancer cell cycle than of the normal cell cycle.

The minimal model regarding physical processes in a cell membrane has the form

$$\tau_S dS/dt = \nu - \alpha SR - DS,\ \tau_R dR/dt = \kappa + \alpha SR - \gamma R/(R + \delta_2) - R^2 - \delta_1 R,$$

$$\tau_\alpha d\alpha/dt = F(\alpha,S), \tag{23}$$

where τ_α is the time for phase equilibrium in a membrane and $F(\alpha,S)$ is a hysteresis-type function characterizing viscosity of the compound lipid layer of a membrane. Viscosity of a membrane in the "fluid" and "solid" state of membrane can differ.

We cannot reduce the model (23) to the case of two equations. But the whole process can be represented on the phase plane (S, R) projecting on it the motion of depicted point on two planes $\alpha = \alpha_1$ and $\alpha = \alpha_2$, which correspond to the certain values $S = S_1$ and $S = S_2$.

As a result, the model (23) has the same four main properties as the model (22). In addition, the model (23) is sufficiently crude.

There are some other MM of the cell cycle (see below and [1, 51, 146]).

Very complicated models of the system of partial differential equations are constructed and investigated by B. Sendov [168], assuming the hypothesis that the regulator of the cell cycle is located in the cell nucleus.

According to [51], mitosis and cell division are controlled by a protein called MPF (*M-phase promoting factor*). MPF, which consists of two subunits (cdc2 and cyclin), can be active or inactive. Active MPF is known to promote its own activation. An analysis of the mechanism of interactions between cdc2 and cyclin leads to a pair of ODEs similar to a well-known *MM* of the *oregonator*

$$dx/dt = x(1-x) - fz(x-q)/(x+q),\ dz/dt = x-z, \tag{24}$$

where x = active MPF, z = total cyclin, and where f and q are parameters. Originally, the model (24) arose to describe the well-known *Belousov-Zhabotinskiy (BZ) reaction* (A. Zhabotinskiy [215]). One of the important properties of this reaction is oscillation when $0.5 < f < 2.414$. Similarly to the BZ reaction, the MPF control circuit can be spontaneously oscillatory (as in early embryonic division) or excitable (as in somatic cell division). As expected, waves of nuclear division can propagate through the somatic tissue. These results were developed further in [146], where MM and phase-plane portraits of the G1, G2, and M-phase checkpoints of the cell cycle have been described.

2.2. MM OF HYPER-CYCLES

The hyper-cycle MM is developed by Eigen and Schuster (see, for example, G. Rowe [163]). The *hyper-cycle* is suggested by the cyclic property of the DNA-protein synthesis reactions in modern organisms. DNA contains the information for constructing proteins, which are in turn necessary to construct more DNA, and so on.

We take the simple step of closing the loop by requiring the last products as a catalyst for the production of the first.

We suppose that we have n different species of 'information molecules' I_i (such as RNA or DNA sequences), each of which contains the instructions for manufacturing one protein E_i. This protein is assumed to be a 'replicase,' whose function is to recognize and translate the next information molecule I_{i+1} into its replicase E_{i+1}.

A general hyper-cyclic system will have n species in the loop, with p species participating in each reaction. The rate term in such a system will have the general form

$$R_i = k_i\, x_i\, x_j \ldots x_r,\ i = 1, \ldots, n, \tag{25}$$

where p - 1 indices $j, \ldots, r$ denote all the other species participating in the reaction that produces a copy of species i.

The respective system of ODE can be written in the form

$$x_i' = R_i + S_i,\ i = 1, \ldots, n, \tag{26}$$

where S_i arises from flows into or out of the system. We assume that

$$S_i = S(t)x_i(t)/C(t), \tag{27}$$

where S is some overall flux and C is the total concentration of all species:

$$C = \sum x_i\,(i = 1, \ldots, n). \tag{28}$$

If we impose the constraint of constant concentration, $C = C_0$, then the system ODEs obeyed by the species in a hyper-cycle becomes

$$x_i' = k_i\, x_i\, x_j \ldots x_r - x_i/C_0 \sum k_r\, x_r\, x_s \ldots x_t, \tag{29}$$

where the sum is taken over all sets of indices $\{r, s, ..., t\}$ such that the resulting product of factors is one of the rate terms of R_i.

The general stationary point analysis for autonomous systems in the case of the equal rate constants results in the following (G. Rowe [163]).

When $n = 2$, 3, or 4, there is one stationary point with all species present, and this point is asymptotically stable. For $n > 4$, a stable limit cycle appears in which all species oscillate in quantity, but still participate all the time.

The general conclusion is that the hyper-cycle equations provide a chemical system in which all the information present in a set of molecules can be maintained over long periods.

"One can envisage hyper-cycles encompassing more and more information by inserting extra links in the chain as more information-carrying molecules evolve" [163], p. 95.

For Eigen and Schuster's application of the hyper-cycle MM to the origin of life, refer to [163].

2.3. MM OF CELL EXCITABLE MEMBRANES

Let $I(t)$ be the positive direction for the membrane current outwards from the axon. This current is due to individual ions, which pass through the membrane, and to the contribution from the time variation in the transmembrane potential (the membrane capacitance contribution). Thus we have (J. Murray [143])

$$I(t) = CdV/dt + I_i, \tag{30}$$

where C is the capacitance and I_i is the current contribution from the ion movement across the membrane. Based on experimental observation Hodgkin and Huxley (1952) took

$$I_i = I_{Na} + I_K + I_L = g_{Na}m^3h(V - V_{Na}) + g_K n^4(V - V_K) + g_L(V - V_L), \tag{31}$$

where V is the potential and I_{Na}, I_K, and I_L are respectively the sodium, potassium and 'leakage' currents. I_L is a contribution from all the other ions that contribute to the current.

The g_s are constant conductances with, for example, $g_{Na}m^3h$ sodium conductance, and V_{Na}, V_K, and V_L, which are constant equilibrium potentials.

The m, n, and h, bounded by 0 and 1, are variables that are determined by the differential equations

$$dm/dt = \alpha_m(V)(1-m) - \beta_m(V)m,$$

$$dn/dt = \alpha_n(V)(1-n) - \beta_n(V)n,$$

$$dm/dt = \alpha_h(V)(1-h) - \beta_h(V)h, \tag{32}$$

where the α and β are given functions of V (again empirically determined by fitting the results to the data); $\alpha_m(V)$ and $\alpha_n(V)$ are qualitatively like (1+ tanh V)/2 while $\alpha_h(V)$ is qualitatively like (1 - tanh V)/2, which is a 'turn-off' switch, when V is moderately large.

If an applied current $I_a(t)$ is imposed, then the governing equation using (29) becomes

$$CdV/dt = -g_{Na} m^3 h(V - V_{Na}) - g_K n^4(V - V_K) - g_L(V - V_L) + I_a \tag{33}$$

The system (33) with (32) constitutes the 4-variable *Hodgkin and Huxley MM*, which was solved numerically.

If $I_a = 0$, the rest state of the model (32) and (33) is linearly stable but excitable.

That is, if the perturbation from the steady state is sufficiently large, there is a large excursion of the variables in their phase space before returning to the steady state.

If $I_a \neq 0$, there is a range of values where regular repetitive firing occurs. The mechanism displays limited cycle characteristics. Both types of phenomena have been observed experimentally.

Because of the complexity of the equation system, various simpler mathematical models, which capture the key features of the full system, have been proposed ([134, 143, and 146]). At the same time, various generalizations and specifications have been worked out ([134, 146, and 163]. Furthermore, different neuro-simulators [134] and numerical methods for neuronal modeling (M. Mascagni, A. Sherman [130]) were developed.

It should be noted that A. Sherman [170] considers the models of electrical activity with the typical Z-shaped phase portraits with application to the pancreatic β-cell. The cell energetic metabolism is determined by the electrical activity and production of insulin. The time scales for m, n, and h in (32) are not the same order. The time scale for m is much faster than for the others. So we can put $dm/dt = 0$.

If we set h = constant, the system still retains many features that are experimentally observed.

The resulting 2-variable model in V and n can be qualitatively approximated by the dimensionless system

$$dv/dt = f(v) - w + I_a\,,\; dw/dt = bv - \gamma w,\; f(v) = v(a-v)(v-1), \tag{34}$$

where $0 < a < 1$ and b and γ are positive constants. Here v is like the membrane potential V, and w plays the role in all three variables m, n and h.

With $I_a = 0$ or with a constant, the system (34) is simply a two-variable phase plane system, the phase portrait for which can be easily constructed. There can be, for example, one or three steady states.

The excitability characteristic, a key feature in the Hodgkin-Huxley system, is now evident. That is, a perturbation, for example, from 0 to a point on the v-axis with $v > a$, undergoes a large phase trajectory excursion before returning to 0.

3. MM of Spatial Systems

Spatial Systems naturally arise from the general conservation law of mathematical physics (see J. Murray [143]; A. Samarskii, A. Mikhailov [164]; V. Vladimirov [195]).

Let S be an arbitrary surface enclosing a volume V. The general *conservation law* states that the rate of change of the amount of material in V is equal to the rate of flow of material across S into V plus the rate of material created in V, i.e.,

$$\partial/\partial t \int_V c(\mathbf{x},t)dv = -\int_S \mathbf{J}\cdot\mathbf{ds} + \int_V f\,dv\,, \tag{35}$$

where the vector $\mathbf{J} = (J_x, J_y, J_z)$ is the velocity or flux of material and f is the source of material that may be a function of density c, $\mathbf{x}$, and t. Applying the divergence theorem to the surface integral, one can transfer the relation (35) into the form

$$\int_V [\partial c/\partial t + \nabla\cdot\mathbf{J} - f(c, \mathbf{x}, t)]dv = 0,\ \nabla\cdot\mathbf{J} = \partial J_x/\partial x + \partial J_y/\partial y + \partial J_z/\partial z. \tag{36}$$

Since the volume V is arbitrary, (36) implies the so-called *conservation equation* for c:

$$\partial c/\partial t + \nabla\cdot\mathbf{J} = f(c, \mathbf{x}, t). \tag{37}$$

In a cellular context, c is the cell density n, with logistic population growth $f = rn(N-n)$, where r is the initial proliferative rate and N is the maximum cell density in the absence of any other effects. In addition, the flux $\mathbf{J}$ can be at the expense of not only the classical, Fickian diffusion, but also of more general diffusion, including long range contribution as well as convection, haptotaxis, galvanotaxis, and chemotaxis (see the last equation in (62) below). Still more complicated models can be found in [31and 129]. Introducing, for example, the pressure field p($\mathbf{x}$, t), we obtain the so-called Navier-Stokes equations; in particular, the Euler's equations (see C. Doering, J. Gibbon [40]).

The kinetic equation with regard to interactions of components and diffusion has the form

$$\partial x_i/\partial t = Fi(x_1, x_2, \dots, x_n) + \partial/\partial r[\Sigma\, D_{ij}(x_1, x_2, \dots, x_n, r)\partial x_j/\partial r\ (j = 1, \dots, n)],$$

$$i = 1, 2, \dots, n, \tag{38}$$

where kinetic variables $x_i = x_i(t, r)$ depend on both the time t and coordinate r, D_{ij} are coefficients of diffusion, and functions F_i determine the total rates of variations x_i under their interactions. In biochemistry, variables x_i are concentrations of reacting substances; in biology, they are enzyme-substrate complex, biomass or the numbers of organisms of the given species in the unit of volume, etc. The particular cases of the system (38) are widely studied Fisher–Kolmogorov equations. The system (1) is the point system corresponding to the spatial system (38).

Let us consider in more detail the case $D_{ii} = D_i$, $D_{ij} = 0$, and $\partial D_i/\partial r = 0$. Then the linearized system for (38) in the vicinity of the stationary values $x^\wedge$ of homogenous state of the initial system, i.e., the root of the system equations (6), has the form

$$\partial x'_i/\partial t = \Sigma\, [a_{ij}x'_j + D_i\partial^2 x_j/\partial r^2]\ (j = 1, \ldots, n),\ a_{ij} = (\partial F_i/\partial x_j)^\wedge,$$

$$x'_i = x'_i - x_i^\wedge,\ i = 1, 2, \ldots, n, \tag{39}$$

The solutions of (30) are linear combinations of the waves

$$x'_i = A_i \exp\,(p_k t + \mathbf{i}kr),\ \ i = 1, 2, \ldots, n, \mathbf{i} = (-1)^{1/2}, \tag{40}$$

where k is the wave number, determining the wave length $\lambda_k = 2\pi/k$, and $p_k = \delta_k + i^\wedge\omega_k$ are the solutions of the so-called *equation of variance*

$$\text{Det}\ \{a_{ij} - D_i k^2 - p_k\} = 0. \tag{41}$$

If the point $x^\wedge$ is unstable, then there is at least one k for which $\delta_k > 0$. Unsteadiness here is divided into two forms. If the equation (41) for the wavelength λ_k has an even number of roots with $\delta_k > 0$, then *unsteadiness* is called *vibrating*. In the case of odd roots we have *unsteadiness by Turing*, reducing to the so-called stationary *dissipative* or damping *structures* (DS).

For $n = 2$, $D_{ij} = 0$, and $\partial D_i/\partial r = 0$, the system (38) can be rewritten in the form

$$\partial x/\partial t = P(x,y) + D_x\partial^2 x/\partial r^2,\ \partial y/\partial t = Q(x,y) + D_y\partial^2 y/\partial r^2. \tag{42}$$

In this case,

$$p_k = -\,\text{A}/2 \pm ((\text{A}/2)^2 - \text{B})^{1/2}, \tag{43}$$

where

$$\text{A} = k^2(D_x + D_y) - a_{11} - a_{22},$$

$$\text{B} = a_{11}a_{22} - a_{21}a_{12} + k^4 D_x D_y - k^2(a_{11}D_y + a_{22}D_x. \tag{44}$$

The particular case of the system (42) is the *Van-der-Pol self-vibrating system*:

$$\partial x/\partial t = y + D_x\,\partial^2 x/\partial r^2,\ \partial y/\partial t = -\,\omega^2 x + 2\delta_0 y - 2\delta_2 x^2 y + D_y\,\partial^2 y/\partial r^2. \tag{45}$$

In this case,

$$p_k = -\,\text{A}'/2 \pm ((\text{A}'/2)^2 - \text{B}')^{1/2}, \tag{46}$$

where

$$A' = 2\delta_0 - k^2(D_x + D_y),\ B' = \omega^2 + k^4 D_x D_y - 2\delta_0 k^2 D_x. \tag{47}$$

Another class of MM for partial systems arises in quantum mechanics and its applications in biology (J. Adam [3]; A. Davydov [38]).

For examples, we have *Schrödinger time-independent* equation in the form

$$\nabla^2\psi(\mathbf{x}) + [E - V(\mathbf{x})]\psi(\mathbf{x}) = 0, \tag{48}$$

where $|\psi(\mathbf{x})|^2$ is a measure of the probability that the particle of energy $E > 0$ will be found at location $\mathbf{x}$, given that the particular potential $V(\mathbf{x})$ is present, or the particular case of *Schrödinger nonlinear equation* in the form

$$\{i\hat{}\hbar\partial/\partial t + E - V + J\partial^2/\partial\xi^2 + G|\psi(\xi, t)|^2\}\psi(\xi, t) = 0,$$

$$V = D + 2J,\ J = d^2R^{-3}(3\cos^2\theta - 1), \tag{49}$$

where $\hbar = 1.054\cdot 10^{-27}$ erg; E, V, and $|\psi(\xi, t)|^2$ have the same sense as above; D is the strain potential of the molecular chain with distance R between adjacent molecules interactions, which are determined by the potential $2J$ due to the electric dipole moment $\boldsymbol{d}$ directed under the angle θ with respect to the chain line; G is the parameter of non-linearity determining the part Δz of the chain that is energizing circuit; and $\Delta z = \pi RJ/4G$.

In the case (39), $\mathbf{x} = x$, $\nabla^2 = d^2/dx^2$, and $V(x) = V > E > 0$, $0 < x < a$, $V(x) = 0$, otherwise, denoting $k = (E)^{1/2}$ and $K = (V{-}E)^{1/2}$, the solution has the following form

$$\psi(x) = Ae^{ikx} + Be^{-ikx}\ (x < 0),\ Ce^{-Kx} + De^{Kx}\ (0 < x < a),$$

$$AS(E)e^{ik(x-a)}\ (x > a), \tag{50}$$

where the constants A, B, C, and D can be easily found from the condition of continuity at $x = 0$ and $x = a$; the transmission amplitude

$$S(E) = 2\mathbf{i}kK/\{2\mathbf{i}kK\text{ch}(Ka) + (k^2{-}K^2)\text{sh}^2(Ka)]\},\ \mathbf{i} = (-1)^{1/2}. \tag{51}$$

The quantity $AS(E) \neq 0$, meaning that the wave or quantum particle has a non-zero probability of being found to the right of the step. This phenomenon is called *tunneling*.

A related quantity $|S(E)|^2$ can be regarded as the energy transmission factor:

$$|S(E)|^2 = 4E(V{-}E)/[4e(V{-}E) + V^2\text{sh}^2(Ka)]. \tag{52}$$

In the case of (49), the exact solution with the condition $\int_{-\infty}^{\infty} |\psi(\xi, t)|d\xi = 1$ has the following form

$$\psi(\xi, t) = (\mu/2)^{1/2}\exp\{ i[\hbar v/(2J)(\xi-\xi_0)-Et/\hbar]\}/\mathrm{ch}[\mu(\xi-\xi_0-vt)],\ \mu = G/(4J), \tag{53}$$

where $vR = V$ is the rate of perturbation propagation along the chain $V < V^\wedge$ and $V^\wedge$ is the rate of sound propagation along the chain. In addition,

$$|\psi(\xi, t)|^2 = \mu/\{2\mathrm{ch}^2[\mu[\xi-\xi_0-vt)]\} \tag{54}$$

is the probability of perturbation of separate molecules along the chain. The value ξ_0 is the argument of the maximal probability and deformation at the instant t = 0.

The perturbations described by the functions (53) and (54) are called *solitons*. They are capable to carry over perturbation energy very effectively. In particular, they arose in the processes of ATP hydrolyses [38].

The important tool of mathematical modeling is a connection between the system of ODE with a random term to the end of it and with the respective system of PDE.

A typical form of the equations describing biochemical reactions is

$$dC(t)/dt = Ar(C), \tag{55}$$

where C, t, A, and r represent the concentrations, time, stoichiometric matrix, and the rate law, respectively. Molecular fluctuations can be incorporated explicitly, for example, by including a white noise term to the end of (55):

$$dC(t)/dt = Ar(C) + x(t). \tag{56}$$

Then the evolution of the probability density function $p(C,t)$ is described by the Fokker-Planck or *Kolmogorov*'s forward equation

$$\partial p(C,t)/\partial t = -\nabla\cdot[Ar(C)p(C,t)] + 1/2\textstyle\sum_{i,j} \partial^2\sigma_{ij}p(C,t)/(\partial C_i\partial C_j), \tag{57}$$

where the matrix σ_{ij} is the covariance of the noise process $x(t)$.

In the case of cells, the quantity $p(C,t)\partial C$ is the probability of finding a cell with a concentration of a certain chemical between C and $C + \partial C$ at a time t.

MM (57) is usually analyzed using Monte-Carlo methods, i.e., solving numerically the equation (56) many times and then using statistics to estimate the probability density function.

All these relations (36)--(57) are widespread under investigation of various biological events, in particular, events in cells. In particular, works by C. Rao, D. Wolf, A. Arkin [158] apply (55)--(57) for modeling of control, exploitation and tolerance of intracellular noise.

3.1. MM OF TUMOR ANGIOGENESIS

MM of tumor angiogenesis are described in detail by M. Chaplain [33]. *Tumor angioginesis factor* (TAF) having concentration $c(\mathbf{x}, t)$ is secreted by the solid tumor and diffuses into the surrounding tissue. Upon reaching endothelial cells in, for example, the

Limbal vessels, the TAF stimulate the release of enzymes by the endothelial cells, which degrade their basement membrane. The initial response of the endothelial cells is to migrate towards the source of TAF. Capillary sprouts are formed and cells begin to proliferate at a later stage.

MM for TAF and endothelial cells having concentration $n(\mathbf{x}, t)$ are

$$\partial c/\partial t = D_c\nabla^2 c - Qcn/[(K + c)n_0] - dc,$$

$$\partial n/\partial t = D_n\nabla^2 n - \chi\nabla(n\nabla c) + rn(1-n/n_0),$$

$$G(c) - kn, G(c) = 0\ (c \le c^*), (c-c^*)/c^*\ (c > c^*),$$

$$\nabla = \partial/\partial x\boldsymbol{i} + \partial/\partial y\boldsymbol{j} + \partial/\partial z\boldsymbol{k}, \nabla^2 = \nabla\cdot\nabla = \partial^2/\partial x^2 + \partial^2/\partial y^2 + \partial^2/\partial z^2, \tag{58}$$

where the first, second, and third term into the first equation mean accordingly diffusion, production, and decay of moiety and the first two, next, and last term in the second equation mean accordingly cell migration, mitotic generation, and cell loss.

In addition,

$$c(\mathbf{x}, 0) = c_0(\mathbf{x}), c(0, t) = c^*, c(L, t) = 0,$$

$$n(\mathbf{x}, 0) = n_0\ (x = L), 0\ (x < L), n(0, t) = 0, n(L, t) = n_0, \tag{59}$$

where $c_0(\mathbf{x})$ is a prescribed function chosen to describe qualitatively the profile of TAF in the external tissue when it reaches the limbal vessels, c^* is a constant value c on the boundary of the tumor, L is a distance to the limbus, and n_0 is an initial endothelial cell constant density at the limbus.

Given the complex nature of tumors' micro-ecology, authors of [146] have developed a series of models, which address the interactive nature of the sub-populations making it up.

Assume, for example, a tumor is composed of two different kinds of cells, X and Y, that represent accordingly the proliferating tumor cells and the hypoxic/anoxic tumor population. Then the micro-ecology in which it is growing can be modeled by

$$dX/dt = r_1X(1 - X/K_1 - c_1Y) - mX,$$

$$dY/dt = r_2Y(1 - K_2 - c_2X) + mX, \tag{60}$$

where r's are the Malthusian growth parameters of the individual population, the K's are the carrying capacities for the individual populations, the c's are the inter-specific competition rates (and could represent negative growth factor signaling), and m is the transition rate from the X-type to the Y-type.

Anyone also can find in M. Mackey [123] a more sophisticated approach and certain results of simulation.

3.2. MECHANICAL MODELS FOR MESENCHYMAL MORPHOGENESIS

According to J. Murray [143], the time-scale of embryonic motions during development is long (hours) and the spatial scale is small (less than a millimeter or two). We can therefore ignore inertial effect in the mechanical equation for the cell-extracellular matrix (ECM) interaction.

The mechanical cell-matrix equation is then

$$\Delta\cdot\sigma + \rho\mathbf{F} = 0, \tag{61}$$

where **F** is the external force acting on the matrix and σ is the stress tensor (applied force is balanced by elastic force).

Here are the various models of σ, **F**, and the matrix material $\rho = \rho(\mathbf{r},t)$:

$$\sigma = \sigma_{\mathrm{ECM}} + \sigma_{\mathrm{cell}},\ \sigma_{\mathrm{ECM}} = \sigma_{\mathrm{viscous}} + \sigma_{\mathrm{elastic}},$$

$$\sigma_{\mathrm{viscous}} = [\mu_1\boldsymbol{\varepsilon}_t + \mu_2\vartheta_t\mathbf{I}],\ \sigma_{\mathrm{elastic}} = E/(1+\nu)[\boldsymbol{\varepsilon} + \nu/(1-2\nu)\vartheta\mathbf{I}],$$

$$\boldsymbol{\varepsilon} = \tfrac{1}{2}(\nabla\mathbf{u} + \nabla\mathbf{u}^T),$$

$$\sigma_{\mathrm{cell}} = \tau n(1-\lambda n^2)^{-1}(\rho + \gamma\nabla^2\rho)\mathbf{I},$$

$$\mathbf{F} = -s\mathbf{u},$$

$$\partial\rho/\partial t + \nabla\cdot(\rho\mathbf{u}_t) = S(n,\rho,\mathbf{u}),$$

$$\partial n/\partial t = -\nabla\cdot[n\partial\mathbf{u}/\partial t] + \nabla\cdot[D_1\nabla n - D_2\nabla(\nabla^2 n)] - \nabla\cdot n[a_1\nabla\rho - a_2\nabla^3\rho] + rn(N-n), \tag{62}$$

where the subscript t denotes partial differentiation, **I** is the unit tensor, μ_1 and μ_2 are shear and bulk viscosities of ECM, **ε** is the strain tensor, ϑ is the dilation, $\vartheta = \nabla\cdot\mathbf{u}$, E and ν are the Young's modulus and Poisson ratio respectively, τ is a measure of the traction force generated by a cell, λ is a measure of how the force is reduced because of neighboring cells of number n, γ is the measure of the non-local long range cell-ECM interaction, s is an elastic parameter characterizing the substrate attachment, and $S(n,\rho,\mathbf{u})$ is the rate of secretion of the matrix by the cells. D_1, D_2, a_1, a_2, r, and N are positive parameters.

The last equality in (62) is the cell conservation equation, where the first term corresponds to the convention, the second to diffusion, the third to haptotaxis, and the fourth to the example of mitosis.

The additional galvanotactic and chemotactic contributions to the flux can be written respectively as

$$\mathbf{J}_G = gn\nabla\varphi,\ \mathbf{J}_C = \chi n\nabla c, \tag{63}$$

where φ is the electrical potential and c is the concentration of a chemotactic chemical. The parameters g and χ are positive.

J. Murray [143] contains detailed analysis of the complicated systems (62) and (63) with applications to the various processes of pattern formation and to certain medical problems.

4. Possibility for Description of Biological Events on Molecular Level

If external conditions are constant, then after a certain time, called the *time of relaxation*, there is thermodynamic equilibrium of the system with the external surrounding.

The state of such system in quantum statistical physics is determined by the *statistical density operator* (A. Davydov [38])

$$\rho = \exp\,[(G - H + \mu N^{\wedge} - PV)/\theta],\ \theta = kT, \tag{64}$$

where k is *Boltzmann constant*, $k = 1.4\cdot 10^{-23}$ joules per degree; T is the absolute temperature; H is the *Hamiltonian operator* of energy; $N^{\wedge}$ is the operator of a number of particles; μ is the chemical potential; P is the pressure; V is the volume of the system; G is Gibbous thermodynamic potential or *Gibbous free energy*.

From the condition of normalization $\mathrm{Sp}\rho = 1$, one can find implicit expression for Gibbous free energy

$$G = -\,\theta \ln \mathrm{Sp}[(\mu N^{\wedge} - H - PV)/\theta], \tag{65}$$

where the symbol Sp[A] means the speak of the operator A, i.e., the sum of its diagonal matrix elements by all quantum states of the system.

The magnitude

$$Z = \mathrm{Sp}[(\mu N^{\wedge} - H - PV)/\theta] \tag{66}$$

is called the sum of states of the system. The mean values of energy E, number of particles N, and volume V are determined by the following formulae

$$E = \mathrm{Sp}[\rho H] = -\,\theta^2[\partial/\partial\theta(G/\theta)],\ N = \mathrm{Sp}[\rho \mathrm{N}^{\wedge}] = -\,\partial G/\partial\mu,\ V = \mathrm{Sp}[\rho V] = \partial G/\partial P. \tag{67}$$

Thus, to determine all the mean values it is sufficiently to know the sum of states Z.

Entropy of the system is determined as

$$\mathrm{S} = -\,k\mathrm{SP}[\rho \ln \rho] \tag{68}$$

On the strength of (64), (65), (67), and (68), we have

$$G = E - TS - \mu N + PV. \tag{69}$$

For the quasi-stationary processes (the processes running their course sufficiently slow so that the system considered is always very closed to the state of equilibrium), from (69) it follows that the variation of Gibbous free energy under constant T, μ, and P

$$\Delta G = \Delta E - T\Delta S - \mu\Delta N + P\Delta V, \tag{70}$$

where ΔE is the variation of internal energy of the system, $T\Delta S = Q$ is a quantity of warmth incoming from the external surrounding, $\mu\Delta N$ is the work connecting to variation of number of particles in the system, and $P\Delta V$ is the work under increasing volume of the system.

Besides, since all extensive variables (the variables depending on size of the system) are usually relative to one mole, ΔG, ΔE, ΔS, and ΔV characterize the respective variations of one mole of the system, and ΔN characterizes the variation of number of moles in the system.

If such variations take place in 25° C and in normal atmospheric pressure (P = 1), then they are standard and are designated by the same symbols, but with the subscript or superscript of 0.

In biological systems, ions eZ (Z = ± 1, ± 2, ...) transpose frequently unlike neutral particles.

Under the transfer of electric charge Δq from the system with the electric potential U_{int} to the external surrounding with the potential U_{ext} the system results in the work

$$\Delta W = \Delta q(U_{\text{int}} - U_{\text{ext}}). \tag{71}$$

Since the charge of one mole of ions is equal to eZN_0 ($N_0 = 6\times10^{23}$), transfer of ΔN moles requires

$$\Delta q = ZF\Delta N,\ F = e\mathrm{N}_0 = 96{,}500 \text{ Kcal}, \tag{72}$$

where F is *Faraday constant*.

Therefore, (71) can be rewritten in the form

$$\Delta W = ZF(U_{\text{int}} - U_{\text{ext}})\Delta N. \tag{73}$$

Combining this work with the work $-\mu\Delta N$ for transfer of neutral particles, we have the general work for transfer of ions

$$\Delta W\hat{} = -\mu\hat{}\Delta N,\ \mu\hat{} = \mu + ZF(U_{\text{int}} - U_{\text{ext}}). \tag{74}$$

The magnitude $\mu\hat{}$ is called *electrochemical potential.* Thus, $\mu\hat{}$ characterizes variation of free Gibbous energy per one mole of additional charged substance.

Therefore, the complete variation of free energy of charged particles

$$\Delta G\hat{} = \Delta E - T\Delta S - \mu\hat{}\Delta N + P\Delta V. \tag{75}$$

Let us consider the system of ions in the solution under constant pressure and temperature, divided by semi-permeable partition on two subsystems: 1 and 2. According to (75), under the variation of number of ions in each subsystem, the variation of free Gibbous energy is determined by the expression

$$\Delta G = \Sigma_i\ (\mu\hat{}_{1i}\Delta N_{1i} + \mu\hat{}_{2i}\Delta N_{2i}), \tag{76}$$

where $\mu\hat{}_{1i}$ and $\mu\hat{}_{2i}$ are electrochemical potentials of ions of types i in the subsystems 1 and 2.

The value of free Gibbous energy under equilibrium is minimal. Therefore,

$$\Sigma_i\ (\mu\hat{}_{1i}\Delta N_{1i} + \mu\hat{}_{2i}\Delta N_{2i}) = 0. \tag{77}$$

If there are no chemical reactions in the system, then the numbers of ions of each type are the same, and the equality (77) takes place under

$$\mu\hat{}_{1i} = \mu\hat{}_{2i}. \tag{78}$$

Thus, the condition of equilibrium of two subsystems is the equality of the chemical potentials of ions of each type in both subsystems.

Free Gibbous energy G, internal energy E, and entropy S are the functions of steady states. Usually, their variations do not depend on the way by which the process is running from one steady state to another.

Biological systems not in equilibrium arise under their effect by various external perturbations. Unsteady states later converge to steady ones.

In addition, the total entropy of the system and the external surrounding usually increases, attaining maximal value under equilibrium. At the same time, free Gibbous energy is attaining minimal value.

Thus, thermodynamics and statistical physics, determining the directions of process passing, do not give us the rate of passing or the respective molecular mechanism. The latter can be achieved on the basis of the dynamic theory.

However, due to a very large number of degrees of freedom, dynamical mathematical description in biology seems to be impossible. Fortunately, this pessimism has little ground.

Many important biological processes do not have a large number of selected, usually collective, degrees of freedom. These processes have a long time of relaxation because of weak connections with the other degrees of freedom. Finally, they have a very short time of relaxation.

The variation of the collective degrees of freedom is related essentially to unsteady processes that can be described by the dynamical laws of quantum or classical physics.

One of the problems that molecular biophysics faces is determination of the main properties of such strong- and weak-relaxations and their respective mechanisms.

4.1. MM OF TRAFFIC THROUGH CELL MEMBRANES

4.1.1. *Passive transport*

Let $c(x,t)$ be molar concentration of the same kind of neutral molecules on the two sides (1 and 2) of homogeneous membrane situated along the y-axis. The rate of change of the amount of material in that region is equal to the rate of flow J across the boundary plus any that is created within the boundary. If no material is created, then $\partial c/\partial t = -\partial J/\partial x$. The gradient $\partial c/\partial x$ causes diffusion (passive transport).

Molar density of flow J of molecules (i.e., the molar number molecules passing through the unit of membrane length per 1 sec) is determined by *Fickian law*, $J = -D\partial c/\partial x$, where D is the material diffusivity (cm/sec). Therefore, we have

$$\partial c/\partial t = \partial(D\partial c/\partial x)/\partial x. \tag{79}$$

If D is constant and $c(x,0) = Q\delta(x)$, where $\delta(x)$ is the Dirac delta function, then the solution of (79) is (see [143])

$$c(x,t) = Q/[2(\pi Dt)^{1/2}]\exp[-x^2/(4Dt)],\ t > 0. \tag{80}$$

Note that sometimes the membrane is characterized by the factor of permeability $P = D/l$, where l, instead of D, is the length of the membrane.

In the case of ions of charge eZ_k, where e is the charge of a proton and $Z_k = \pm 1, \pm 2, \ldots$, and where electric field inside the membrane has the potential U, the motion of ions through the membrane is expressed by the density of partial current

$$I_k = -eZ_kD_k\partial c_k/\partial x - eZ_k^2Fv_kc_k\partial U/\partial x, \tag{81}$$

where $F = eN_A = 96{,}500$ Kcal/mole (Faraday constant) and v_k is mobility of ions through the membrane.

Based on the relation by Einstein $D_k = v_kRT$, where T is the absolute temperature measured in Kelvin and where R is the *universal gaseous constant*, (81) can be rewritten in the form

$$I_k = -eZ_kv_kc_k[RT/c_k\partial c_k/\partial x + Z_kF\partial U/\partial x]. \tag{82}$$

The expression in the brackets of (82) is called *electrochemical gradient*.

According to (82), the mobility of ions of k-type does not depend on other ions (principle of independence of flows). This is usually true for small concentrations of ions. However, this principle is not true at least for potassium (P) ions in the case of the cell membrane [38].

Let n be the total number of ions of different type. Then the equations (82) for $k = 1, 2, \ldots, n$, are the system of n equations relative $n+1$ functions $c_k(x)$ and $U(x)$. Adding to this system the condition of electric neutrality inside the membrane

$$\Sigma\, Z_kc_k(x)\ (k = 1, \ldots, n) = 0, \tag{83}$$

and introducing the boundary conditions, i.e., the values $c_k(0)$, $U(0)$, $c_k(1)$, and $U(1)$, we obtain the determined system for the functions $c_k(x)$ and $U(x)$.

4.1.2. *Active transport*

Along with passive transport, living cells also have active transport, in which molecules and ions are pumped from low concentrations to high concentrations against their gradient.

To transfer one mole of substance from the side of the membrane with low concentration c_1 to the side with high concentration c_2 it is necessary to use the work that is equal to the change of Gibbous free energy

$$\Delta G = RT \ln (c_2/c_1) \approx 0.863 \lg (c_2/c_1) \text{ [kcal/mole]}. \tag{84}$$

This work has to be accompanied by another process creating and releasing the respective energy.

To transfer ions, it is necessary to have the respective difference of electrical potentials:

$$\Delta U = -RT/F \ln [(\Sigma_k P_k c^+_{kin} + \Sigma_k P_k c^-_{kout})/(\Sigma_k P_k c^+_{kout} + \Sigma_k P_k c^-_{kin})], \tag{85}$$

where c^+_{kj} and c^-_{kj} are the molar concentrations of positive and negative ions inside (j is in) and outside of the membrane (j is out). The values of ΔU for living cells in their rest state can be situated in the range from 20 to 200 mV.

For the nerve axon of the giant squid, the values of the molar concentrations of three base ions P^+, Na^+, and Cl^- are equal respectively to 340, 49, and 114 (inside) and 10.4, 463, and 592 (outside). In addition, in the rest state of the nerve axon, permeability factors can be found as the relation $P_P : P_{Na} : P_{Cl} = 1 : 0.04 : 0.45$. Using (85), one can find

$$\Delta U = -RT/F \ln [(340 + 0.45 \cdot 592)/(10.4 + 0.45 \cdot 114)] \approx -59.7 \text{ mV}, \tag{86}$$

which is in good agreement with the experimental value of – 60 mV.

Transport systems of membranes creating gradients of concentrations of transferring substances are usually called *pumps*.

There are Na^{+}/K^{+}ATPase pumps, Ca^{2+} pumps, proton pumps, etc. Na^{+}/K^{+}ATPases provide a small ratio concentration of Na^+ to K^+ inside cells in spite of the fact that such a ratio outside cells is usually large. Na^{+}/K^{+}ATPases and Ca^{2+} pumps work at the expense of consumption of energy. ATP molecules are hydrolyzed to ADP and to the inorganic phosphate (P_i). These pumps can work backwards, using gradients of ion concentrations for the synthesis of ATP from the molecules ADP and P_i.

It should be noted that we considered here only the simplest well-known MM of traffic substances through membranes. Significantly complete and complicated MM of transport of matter through biological membranes, including transport by vesicle formation, are given in [1; E. Fromter [50]; 134; H. Othmer, et al. [146]].

4.2. ON THE PROBLEM OF STRUCTURE AND PROTEIN FOLDING

The *problem of structure* (C. Delisi [39]) is to find the minimum free energy structure. An accurate prediction depends on an accurate assignment of free energies to various types of structures (Fig. 4). Q in Fig. 4 is the progress variable, representing the fraction of native contacts present at a given stage of folding and E is the average free energy.

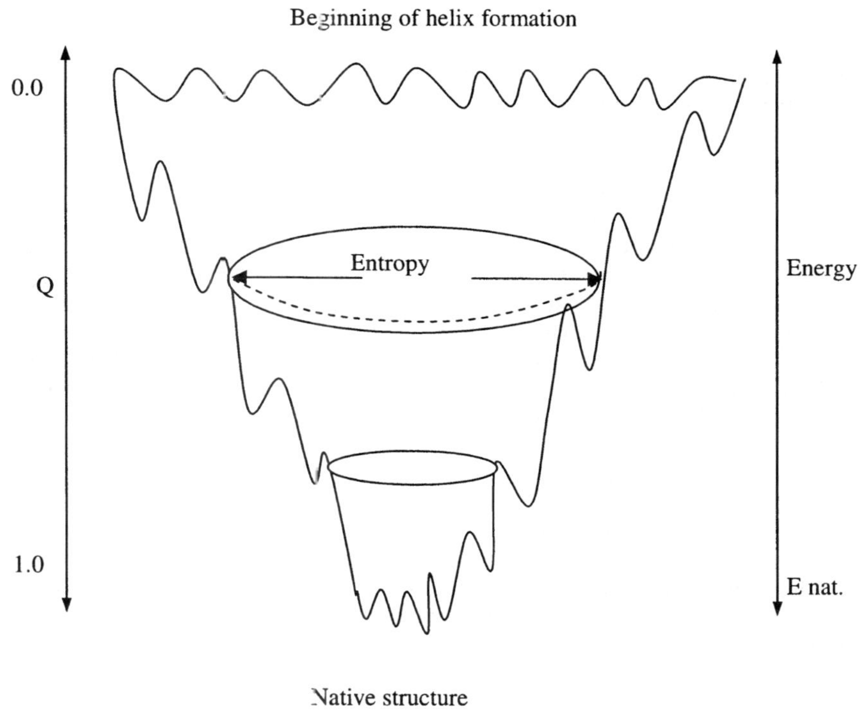

Figure 4. A three-dimensional representation of a folding process

The problem was decomposed by noting that the free energy of any structure can be, to a good first approximation, thought of as the sum of free energies of more basic structures, which include helical runs of AU and GC, internal loops, bulges, and hairpin turns.

Suppose a particular structure has a total of n_{GC} and n_{AU} base pairs, n_i bulges of length i, n_j internal loops of length j, and n_k hairpins of length k (where i, j and k run from 1 through some finite limit, which depends on the chain's sequence and length). Let G_{AU}, G_{GC}, $G_B(l)$, $G_I(l)$, and $G_{HP}(l)$ ($l = 1,2,\ldots$) represent the various base pair and loop free energies. Then G_T, the total free energy of the secondary structure, is

$$G_T = n_{GC}G_{GC} + n_{AU}G_{AU} + \sum_i n_iG_B(i) + \sum_j n_jG_I(j) + \sum_k n_kG_{HP}(k). \tag{87}$$

The free energies for the various types of base pairing as well as for very short loops of various types were obtained experimentally. The free energy for loop closure is proportional to the natural algorithm of the equilibrium constant for bond formation, and this constant

$$K = [S(R^*)A(\theta^*,\varphi^*,\psi^*)]_N\delta r\delta u\delta vK_0, \tag{88}$$

where for a randomly coiling chain $[S(R)A(\theta,\varphi,\psi)]_N$ is the complete (spatial and angular) probability density function for the location of the second bond separated from the first by N units of the chain and where R^* and $(\theta^*,\varphi^*,\psi^*)$ are spatial coordinate and angular orientation of the second bond within δr and $\delta u\delta v$ respectively of the first, K_0 is the geometric requirements on the bases. If the probability density is calculated as a function of the length of the intervening sequence, the angular function quickly becomes uniform and the distance distribution becomes nearly Gaussian:

$$S_N(R) = [2/(2\pi R_N^2)]^{3/2}\exp[-3R^2/(2R_N^2)]. \tag{89}$$

Among the grand challenges is the *problem of protein folding* [39]. Computational approaches start from a fundamental assumption that a system at equilibrium will be at its minimum free energy. Two main obstacles lie in the well-known way of progress. The first requires a useful potential function; the second requires rapidly executable algorithms that can find a structure close to the global free energy minimum. One of possible ways to find such a structure discussed in s. 3.2, ch. 16. For certain additional results in this direction see also L. Adam et al. [4]; [129]; T. Schlick [165].

4.3. SIMPLE KINETIC MM IN PROTEIN FOLDING

4.3.1. *Theories of the reaction rates*

The Van't Hopf-Arrhenius equation

$$k = A\exp(-E_a/k_BT), \tag{90}$$

where k is reaction rate, E_a is activation energy, T is absolute temperature, k_B is the Boltzmann constant, and A is a constant subject to further identification.

Conventional transition-state theory gives

$$k = k_B T/h \exp(-\Delta G^o/k_B T), \tag{91}$$

where ΔG^o is Gibbs free energy and h is the Plank constant. Comparing (90) and (91) and using $\Delta G^o = \Delta H^o - RT$, we find that the pre-exponential factor A in (90) corresponds to $k_B T/h \exp[(\Delta S^o + R)/R]$ and $E_a = \Delta H^o - RT$. The Brownian motion driven by thermal forces of the particle is described by the *Langevin equation*

$$md^2x/dt^2 + \gamma dx/dt + dV/dx = F_{ext}(t), \tag{92}$$

where m is the particle mass, x is a coordinate, V is the potential energy surface of the particle, γ is friction (viscosity) coefficient, and F_{ext} is the external thermal excitation from the bath (solvent). It was shown in Kramer's theory, by solving the diffusion equation that the rate of escape in the high friction limit is

$$k = 1/\gamma \exp(-\Delta G^o/k_B T), \tag{93}$$

4.3.2. *Application to the protein folding without barriers*

The escape rates k_u and k_f, for two-state folding are given (O. Bieri, T. Kiefhaber [25] by

$$k_u = \nu_u \exp(-\Delta G^o/k_B T),\ k_f = \nu_f \exp(-\Delta G^o/k_B T),$$

$$\nu_u = C_u/(\eta_{solvent} + \sigma_N),\ \nu_f = C_f/(\eta_{solvent} + \sigma_U), \tag{94}$$

where C_u and C_f are constants depending on the potential energy surface, $\eta_{solvent}$ is the solvent viscosity, and σ is internal viscosity arising from internal friction.

Since the equilibrium of a reaction is always determined by the escape rates for folding and unfolding, the free energy for folding

$$\Delta G^o = -k_B T \ln k_f/k_u = \Delta G^o{}_f - \Delta G^o{}_u + k_B T \ln \nu_u/\nu_f. \tag{95}$$

Folding starts from a solvated polypeptide chain in water with little internal friction compared to solvent friction. We can neglect the internal friction for the refolding reaction. For the unfolding reaction, starting from the native protein, the internal viscosity σ_N will be significant due to the restricted motion of the polypeptide chain.

This and similar considerations show that in the case of myoglobin the difference in the pre-exponential factors for unfolding and refolding will contribute maximally 1kcal/mol to the equilibrium stability of the native protein.

In general, ν_u and ν_f depend on the solvent conditions such as denaturant concentration, temperature, and pressure.

4.4. PHYSICAL MODEL OF DNA

DNA, organized into two pairs (double helix, see Fig. 5) of 24 distinct types of chromosomes, is a nucleic acid polymer, made up of small molecules called nucleotides that can be distinguished by the four bases: adenine (A), cytosine (C), guanine (G), and thymine (T).

The nucleotides A and T as well as C and G form hydrogen bonds and so A, T and C, G are called the complementary base pairs.

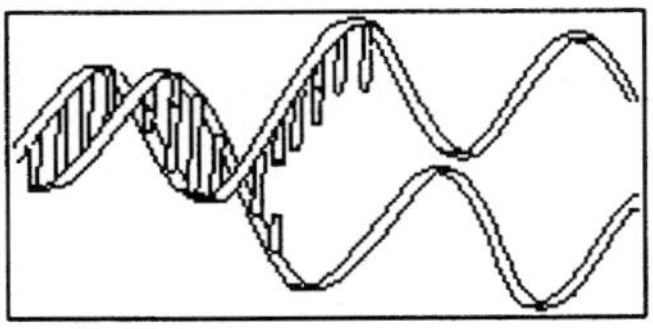

Figure 5. Sketch of a double helix

Linear (for small motions) and nonlinear (for arbitrary motions) physics of DNA are described in a work by L. Yakushevich [211]. It turns out that one of the typical solutions of the nonlinear models is the so-called *solitons*, which require the least energy for their motions.

The rod-like ideal model takes into account three types of internal motions: stretching, twisting and bending.

So, the Hamiltonian of the system

$$H = H_s + H_t + H_b + H_{s\text{-}t} + H_{s\text{-}b} + H_{t\text{-}b}, \tag{96}$$

where the first three terms describe stretching, twisting, and bending, respectively, and the rest of the terms describe interactions among these three types of motions.

The expected general form of the non-linear equations corresponding to Hamiltonian H is

$$\varphi u_{tt} = Y u_{zz} + \text{nonlinear term} + U_{s\text{-}t} + U_{s\text{-}b},$$

$$\mathbf{i}\varphi_{tt} = C\varphi_{zz} + \text{nonlinear term} + \Phi_{t\text{-}s} + \Phi_{t\text{-}b},$$

$$S\rho y_{tt} = -YIy_{zz} + \text{nonlinear term} + Y_{b\text{-}s} + Y_{b\text{-}t}, \tag{97}$$

where φ and u are angular and longitudinal displacements of the nth disk; Y is Young's modulus; C is the torsion rigidity of the rod; S is the area of the rod section; ρ is a density; I is the moment of inertia of the disk; and $U_{s\text{-}t}$, $U_{s\text{-}b}$, $\Phi_{t\text{-}s}$, $\Phi_{t\text{-}b}$, $Y_{b\text{-}s}$, $Y_{b\text{-}t}$ are the terms describing interactions between internal motions in DNA.

The problem of derivation of all the terms in (97) has not yet been solved. A work by L. Yakushevich [211] contains a survey of many existing models and of their original results. The ideal models did not take into account such factors as the environment, non-homogeneity, and the helicity. Effects of all those factors are considered in the book.

In particular, when accounting for the helix, it implies the consideration of the well-known models: the *sine-Gordon equation*

$$\varphi_{zz} - \varphi_{TT} = \sin \varphi, \tag{98}$$

having soliton-like solutions; kinks and anti-kinks:

$$\varphi(Z,T) = 4\arctan \{\exp \pm[(1 - v^2)^{-1/2}(Z - vT - Z_0)]\}, \tag{99}$$

where v is the soliton's velocity and Z_0 is a constant.

Appendix of this book contains the quantitative values of structural and dynamical parameters of DNA such as the radius of DNA helix, the distance between base pairs, the mass of a base pair, the moment of inertia of a base pair, the velocity of sounds in DNA, the velocity of torsion waves, the force constants for torsion, the rolls, the tilts, the rises, the bends, and the hydrogen bond stretching motions.

Chapter 2

MM OF ENZYME REACTIONS

1. Method of Graphs

A series of simplifying assumptions is used to derive the enzyme reaction rates (G. Ivanitsky, V. Krinsky, E. Selkov [105];D. Roberts [161]; E. Voit [197]).

In particular, as a rule, it is accepted that enzyme reactions run their course under the conditions of ideal intermixing, pH- and thermal-stationary state, and that the reactions go to the *quasi-stationary state* very fast. The word "quasi" means that only a part of variables reaches the stationary state, while the other variables continue to change slowly.

This all allows us essentially to simplify analysis of bio-chemical systems. Instead of solving systems of nonlinear differential equations that describe changes of intermediate substrates, we can solve only the system of algebraic equations by connecting quasi-stationary concentrations of those substrates.

The main reason why enzyme reactions go to the quasi-stationary state is that the concentration of an enzyme is by far less than the concentration of substrates interacting with the enzyme. As a rule, those algebraic equations are linear. Therefore, to determine the quasi-stationary concentrations of intermediate substrates and the enzyme reaction rates, we can use the methods of linear algebra and the theory of graphs.

An *enzyme reaction graph* is a set of nodes corresponding to quasi-stationary concentrations of all enzyme complexes X_i (i = 0, 1, ..., n) and a set of branches (that connect nodes) characterized by certain values of k_{ij} equal to the rate constants of a transformation $X_i \rightarrow X_j$. It is possible that k_{ij} will be a function of concentrations of substrates of the transformation. The concentration of each substrate in its quasi-stationary state is constant.

For example, the enzyme reaction $S_1 \Leftrightarrow S_2$, elapsing through intermediate formation of two enzyme-substrate complexes $X_1 = ES_1$ and $X_2 = ES_2$ ($X_0 = E$) can be represented by the graph with three nodes and six branches, the values of which are

$$k_{01} = k_{+1}s_1,\ k_{02} = k_{-2}s_2,\ k_{10} = k_{-1},\ k_{20} = k_{+2},\ k_{12} = k_{+},\ k_{21} = k_{-},$$

where s_1 and s_2 are the concentrations of S_1 and S_2.

A graph tree directed into ith node is a non-closed collection of branches directed from all nodes of the graph to the ith node. The value of the tree is the scalar product of the values of its branches. Base determinant of the ith node is a sum of the values of all trees directed to the ith node (i.e., a base). Determinant of the graph is the sum of all base determinants.

For example,

$$\Delta_0 = k_{-1}k_{-} + k_{-1}k_{+2} + k_{+}k_{+2},$$

$$\Delta_1 = k_{+1}s_1k_{-} + k_{+1}s_1k_{+2} + k_{-2}s_2k_{-},$$

$$\Delta_2 = k_{+1}s_1k_{+} + k_{-1}s_2k_{-2} + k_{-2}s_2k_{+} \quad (1)$$

are determinants of nodes E, ES_1, ES_2 respectively, and

$$\Delta = \Delta_0 + \Delta_1 + \Delta_2 \quad (2)$$

is a determinant of the graph under consideration.

The initial quasi-stationary rate of an enzyme reaction is expressed with the help of the determinants mentioned above by the formula

$$v = e_0\Sigma_i\, k_i\Delta_i/\Delta, \quad (3)$$

where k_i is the rate constant of a product formation for the node X_i, Δ_i is the base determinant of the node X_i, and e_0 is the complete concentration of the enzyme.

In the case of reversible formation of the product, the following agreement about signs is used: $k_i > 0$, if node X_i rejects the product, and $k_i < 0$, if X_i binds the product. For our example,

$$v = e_0(k_{+2}\Delta_2 - k_{-2}s_2\Delta_0)/\Delta. \quad (4)$$

The first term in the parenthesis is positive for disintegration of ES_2, which releases S_2 and the second term is negative for S_2, which is bound with E.

Quasi-stationary concentrations of intermediate complexes are found by the formula

$$x_i = [X_i] = e_0\Delta_i/\Delta, \quad (5)$$

For example,

$$[E] = e_0\Delta_0/\Delta,\ [S_1E] = e_0\Delta_1/\Delta,\ [S_2E] = e_0\Delta_2/\Delta, \quad (6)$$

It should be noted that the graphs of enzyme reactions differ from usual topological graphs by the following ways:

1. For each node, the law of mass conservation is fulfilled, i.e., the sum of mass concurrent in the node is equal to 0, and

2. The concentrations of all nodes are also connected with the law of conservation, i.e., the sum of all node concentrations is equal to the complete concentration of enzyme e_0.

2. MM of Open Reactions with Substratum Inhibition

Let S_1, S_2, E, v, and v_1, v_2 be the substratum, product, enzyme inhibited by substratum, the rate of the reaction, and rates of inflow S_1 and S_2, respectively.

On the basis of the method of graphs (s. 1), we can obtain (see details in G. Ivanitsky, V. Krinsky, E. Selkov [105])

$$v = [V_+ s_1/K_1 - V_- s_2/K_2]/[1 + (s_1/K_1 + s_2/K_2)(1+(s_1/K_i)r)],$$

$$v_1 = v_{1s} - k_1 s_1,\ v_2 = v_{2s} - k_2 s_2, \tag{7}$$

where V_+ and V_- are maximal rates of direct and opposite reactions, K_1 and K_2 are *Michaelis constants*, K_i is *inhibition constant*, $s_1 = [S_1]$, $s_2 = [S_2]$, and v_{1s} and v_{2s} are the rates of formation of S_1 and S_2. Then MM has the form

$$ds_1/dt = v_1 - v,\ ds_2/dt = v_2 + v. \tag{8}$$

The respective dimensionless MM has the form

$$\varepsilon d\sigma_1/ds = \nu_{1s} - \beta_1\sigma_1 - \nu = \varepsilon P(\sigma_1, \sigma_2),\ d\sigma_2/ds = \nu_{2s} - \beta_2\sigma_2 + \nu =$$

$$Q(\sigma_1, \sigma_2),\ \nu = (\sigma_1 - \alpha\sigma_2)/(\xi + (\sigma_1+\sigma_2)(\xi + \sigma_1\gamma), \tag{9}$$

where

$$\sigma_1 = s_1/K_1,\ \sigma_2 = s_2/K_2,\ s = V_+(K_1/K_i)^{-r}/K_2 t,\ \alpha = V_-/V_+,\ \varepsilon = K_1/K_2,\ \xi = (K_i/K_1)^r,$$

$$\nu_{1s} = v_{1s}/V_+\xi,\ \beta_1 = k_1K_1/V_+\xi,\ \nu_{2s} = v_{2s}/V_+\xi,\ \beta_2 = k_2K_2/V_+\xi. \tag{10}$$

For MM (7)--(10), we consider the quasi-stationary output characteristic $\nu\hat{}(\sigma_2)$ determined from the equation

$$\nu_{1s} - \beta_1\sigma_1 - \nu = 0. \tag{11}$$

From the equations (8) and (11) it follows that

$$\sigma_2 = (\sigma\hat{}_1 - \nu\hat{}(\xi + \sigma\hat{}_1(\xi + \sigma\hat{}_1{}^r)))/(\alpha + \nu\hat{}(\xi + \sigma\hat{}_1{}^r)), \tag{12}$$

where $\sigma\hat{}_1 = (\nu_{1s} - \nu\hat{})/\beta_1$ and $\nu\hat{}$ are quasi-stationary values.

The collection of quasi-stationary output characteristics of $\nu\hat{}(\sigma_2)$ is constructed in [105] for the following values of the parameters: $\nu_{1s} = 0.55$, $\beta_1 = 0.165$, $\xi = 0.1$, r = 2, α = 0, 0.5, 10.

Under large values of σ_2 the quasi-stationary rate of reaction $\nu\hat{}$ becomes negative; that is, the reaction runs in the opposite direction from S2 → S1.

If the value of α is sufficiently small, then for some σ_2 dependence $\nu\hat{}(\sigma_2)$ is ambiguous, which is connected with extremes of the function $\sigma_2(\nu\hat{})$. The values of $\nu\hat{}$ are positive for extreme points.

The equation of neutrality curve of the model (9) has the form

$$-1/\varepsilon(\beta 1 + \partial\nu/\partial\sigma_1) + \partial\nu/\partial\sigma_2 - \beta_2 = 0, \quad (13)$$

the mono-stationary curve -

$$\beta_1\beta_2 - \beta_1\partial\nu/\partial\sigma_2 + \beta_2\partial\nu/\partial\sigma_1 = 0, \quad (14)$$

the multiplicity curve -

$$(-1/\varepsilon(\beta_1 - \partial\nu/\partial\sigma_1) + \beta_2 - \partial\nu/\partial\sigma_2)^2 - 4/\varepsilon\partial\nu/\partial\sigma_1\partial\nu/\partial\sigma_2 = 0. \quad (15)$$

The solution of the equations (9) jointly with the system

$$P(\sigma_1, \sigma_2) = 0, Q(\sigma_1, \sigma_2) = 0 \quad (16)$$

allows us to obtain the parametric form of all three curves in planes (β_1, ν_{1s}) and (β_2, ν_{2s}) with the free parameter $\sigma\hat{}_1$. The result shows that it is possible for the reaction to have one, two, or three stationary states.

In certain domains of parametric portraits, the condition of existence of stable limited cycle takes place. This condition is sufficient because of the instability of the model on infinity. It is also sufficient due to the fact that the axes σ_1 and σ_2 phase trajectories go inside the positive quadrant on the phase plane.

3. MM of Two-Substratum Reactions with Inhibition

Let S_1 and S_2, $S'_1$and S'_2, E, v, v_1, and v_2 be correspondingly two substratum, products, enzyme, the rate of reaction, the rate of S_1 inflow, and the rate of S'_2 into S_2 regeneration. According to the s. 2., the initial rate has the form

$$v = Vs_1s_2/(K_1 + s_1^3/K_i^2)(K_2 + s_2), \ s_1 = [S1], \ s_2 = [S2], \quad (17)$$

where V is maximal rate, and K_1, K_2, K_i are constants of dissociation of enzyme-substratum complexes; summarized flow of S_2 and S'_2 is constant,

$$s_2 + s'_2 = c_0; \quad (18)$$

and the rates v_1 and v_2 have the form

$$v_1 = v_{1s}k_1s_1, \ v_2 = k_2s'_2. \quad (19)$$

Then the reaction can be described as follows:

$$ds_1/dt = v_{1s} - k_1 s_1 - v,\ ds'_2/dt = k_2(c_0 - s_2) - v \tag{20}$$

or in dimensionless values as

$$\varepsilon d\sigma_1/ds = \nu_{1s} - \beta_1\sigma_1 - v,\ d\sigma_2/ds = 1 - \sigma_2 - \nu, \tag{21}$$

where

$$\nu = \alpha\sigma\ \sigma_2/((1+\delta\sigma_2)(\xi + \sigma_1(1+\sigma_1^2))),$$

$$\sigma_1 = s_1/K_i,\ \sigma_2 = s_2/c_0,\ \nu = v/(k_2 c_0),\ s = k_2 t,$$

$$\alpha = V/(k_2 K_2),\ \beta_1 = k_1 K_i/(k_2 c_0),\ \delta = c_0/K_2,$$

$$\varepsilon = K_i/c_0,\ \nu_{1s} = v_{1s}/(k_2 c_0),\ \xi = K_1/K_i. \tag{22}$$

From the equation

$$\nu_{1s} - \beta_1\sigma_1 - \nu = 0 \tag{23}$$

and from the formula for ν, one can obtain the following expression for quasi-stationary input characteristic $\nu\text{^}(\sigma_2)$:

$$\sigma_2 = \nu\text{^}(\xi + \sigma\text{^}_1(1+\sigma\text{^}_1{}^2))/(\alpha\sigma_1 - \nu\text{^}\delta(\xi + \sigma\text{^}_1(1+\sigma\text{^}_1{}^2))),$$

$$\sigma_1 = (\nu_{1s} - \nu\text{^})/\beta_1. \tag{24}$$

Hysteresis, or *N*-shape, of the curve $\nu\text{^}(\sigma_2)$ remains when $\delta \to 0$. So, for the simplicity of the analysis we can assume that $\delta\sigma_2 \ll 1$. Then MM can be described in the form

$$\varepsilon d\sigma_1/ds = \nu_{1s} - \beta_1\sigma_1 - \nu,\ \ d\sigma_2/ds = 1 - \sigma_2 - \nu, \tag{25}$$

where $\nu = \alpha\sigma_1\sigma_2/(\xi + \sigma_1 + \sigma_1^2 + 1)$.

Setting the right sides of the equations in (25) equal to 0, we find σ_2 from the second equation and insert it into the first one. As a result we have

$$f(\sigma_1) = \beta_1\sigma_1^4 - \nu_{1s}\sigma_1^3 + \beta_1(1+\alpha)\sigma_1^2 + (\alpha + \beta_1\xi - \nu_{1s}(1+\alpha)\sigma_1 - \nu_{1s}\xi = 0. \tag{26}$$

The equation (26) does not have more than 3 positive roots because the number of sign changes in sequence of its coefficients is equal to 3. So, the cases of 3, 2 or 1 positive stationary states are possible for the system (25). The condition of double multiplicity of roots can be described as

$$f(\sigma_1) = 0,\ df(\sigma_1)/d\sigma_1 = 0. \tag{27}$$

Under $\alpha = 1$ we obtain the mono-stationary curve on the plane (β_1, ν_{1s}):

$$\nu_{1s} = \sigma\hat{}_1^{\,2}(3\sigma\hat{}_1^{\,2}+2)/(\sigma\hat{}_1^{\,3}+2\sigma\hat{}_1+\xi)2 \tag{28}$$

and also

$$\beta_1 = (2\sigma\hat{}_1^{\,3}-\xi)/(\sigma\hat{}_1^{\,3}+2\sigma\hat{}_1+\xi)2, \tag{29}$$

where the free parameter is a stationary value of $\sigma\hat{}_1$. The neutrality curve is given by the formulae

$$\nu_{1s} = \sigma\hat{}_1(3\sigma\hat{}_1^{\,3}+\sigma\hat{}_1-\varepsilon(\sigma\hat{}_1^{\,3}+2\sigma\hat{}_1+\xi)2)/(\sigma\hat{}_1^{\,3}+2\sigma\hat{}_1+\xi)(\sigma\hat{}_1^{\,3}+\sigma\hat{}_1+\xi)$$

$$\beta_1 = (2\sigma\hat{}_1-\xi-\varepsilon(\sigma\hat{}_1^{\,3}+2\sigma\hat{}_1+\xi)2)/(\sigma\hat{}_1^{\,3}+2\sigma\hat{}_1+\xi)(\sigma\hat{}_1^{\,3}+\sigma\hat{}_1+\xi). \tag{30}$$

The parametric pictures for the system (25) and (9) appear topology equivalent. Besides, the unlimited point for the system (25) is also unstable, and by the axes σ_1 and σ_2 its phase trajectories go inside the positive quadrant. So, the phase trajectories also go to the stable limiting cycle.

The phase picture for the system (25) in more detail [105] shows diverse dynamic behavior even in the case of this comparatively simple reaction.

4. On Multi-Enzyme Systems

4.1. GENERAL NOTIONS

The term "*multi-enzyme systems*" is usually used to describe interconnected reactions, fulfilling a certain bio-chemical function. The important property of these systems is their relative autonomy, which allows us to model them without any connection to other cell metabolisms. Of course, such an isolated consideration makes sense until there is no need to investigate their interactions with each other.

The interconnected reactions have a very complicated functional organization based on structural accordance of micro- and macromolecules. This structural accordance is the basis for specificity of action of enzymes and various iso- and allosteric regulator factors. The structural accordance is also the basis for specificity of interactions of proteins and nucleic acids among themselves and with each other. This organization is called organization on the basis of specificity. It can be functioning even under the condition of ideal intermixing in the cell medium.

Prolonged evolution created well distinguishable hierarchy of interactions in which one can distinguish three structures. The primary structure of cell metabolism is a frame of metabolism made up by interactions between adjacent reactions through common intermediate substrates.

A well-known map of metabolic ways by Nicholson represents namely the primary structure. Also, the structure of non-enzyme glycolysis of carbohydrates can serve as an example of the primary structure.

The secondary or stoichiometric structure is made up from the primary structure and from interactions among reactions through common co-factors. This structure is a collection of all stoichiometric interactions. A sketchbook of metabolic ways by Dagly and Nicholson represents the secondary structure of various multi-enzyme systems. Also, the structure of enzyme glycolysis of carbonates can serve as an example of the secondary structure. There is a similarity between the secondary structure of macromolecules and the multi-enzyme systems--repeated interactions of reactions through common co-factors can cause partial spiraling of multi-enzyme systems. The set of enzyme reactions connected with a common co-factor makes up the so-called *core of reactions*. There is a close connection among separate parts of the primary structure due to this core. MM of the core of reactions can be obtained by combining MM considered above.

The tertiary structure is made up, in addition, of interactions that control activity of enzymes. Regulators or control bonds of the tertiary structure have improved properties of the secondary structure. The example of the tertiary structure of glycolysis can be found in [105].

On the strength of analysis of the secondary structure, it is possible theoretically to find those controlled interactions, which can improve the structure properties. If they really exist, then they play the important role in functional organization of cells. This method and the general approach for creation of MM of multi-enzyme systems will be illustrated through analysis of energetic metabolism below.

4.2. SIMPLE MM OF ENERGETIC METABOLISM

Energetic metabolism in cells takes place at the expense of oxidation of nourishment and breathing. Under oxidation, electrons of organic substances are transferred to oxygen. The respective energy of oxidation is used initially by mitochondria for the synthesis of ATP from the molecules of ADP and phosphoric acid.

The respective chemical *reaction* ADP + $H_3PO_4 \rightarrow$ ATP + H_20 is called *phosphorylation*. Involving special enzymes, a reversible reaction that takes place is called *hydrolysis* of ATP. Hydrolysis of ATP releases about 0.54 eV (12.5 Kcal/mole) of energy that is used by cells for active transport through their membranes of various substances and ions as well as for many other important processes including synthesis of amino-acids, proteins, fulfillment of different motions, etc.

The aggregate of energetic metabolism reactions can be described [105] by the kinetic model shown on the following scheme Fig. 6):

$$\begin{array}{l} \qquad\qquad\qquad\qquad\qquad\qquad\qquad\qquad\qquad\qquad\qquad\qquad\qquad\qquad\quad v_5 \\ \rightarrow S \rightarrow \mathrm{ATP} \Rightarrow \mathrm{ADP} \rightarrow I_1 \rightarrow \times g_2 \rightarrow I_2 \rightarrow \times g_3 \rightarrow \mathrm{ADP} \Leftrightarrow \mathrm{ATP} \rightarrow P \\ \;\; v_0 \qquad\quad v_1 \qquad\quad g_1 v_1 \quad\; v_2 \qquad g_2 v_2 \downarrow \;\; v_3 \qquad\quad g_3 v_3 \qquad\qquad\quad v_6 v_4 \end{array}$$

Figure 6. Scheme of energetic metabolism reactions

In this scheme, S is a substratum, I_1 and I_2 are inter-mediators, P is a product of oxidation, v_0 is the rate of source, v_1 is the rate of initiator stage in which g_1 molecules of ATP are consumed for activation of one molecule of S, v_2 is the rate of decomposition of I_1 into g_2 molecules of I_2, v_3 is the rate of generator stage in which oxidation of one molecule of I_2 until P is accompanied by phosphorylation of g_3 molecules of ADP, v_4 is the rate of leakage, v_5 is the rate of ATP hydrolysis by load, and v_6 is the rate of ATP creation by alternative source.

We assume that the rates of reactions on the scheme can be described by the simple expressions:

$$v_0 = v_{0s} - k_0[\mathrm{S}],\ v_1 = k_1[\mathrm{ATP}][\mathrm{S}]/(K_1+[\mathrm{S}]),\ v_2 = k_2[\mathrm{I}_1],\ v_3 = k_3[\mathrm{I}_2][\mathrm{ADP}],$$

$$v_4 = k_4[\mathrm{I}_2],\ v_5 = V_5[\mathrm{ATP}]/(K_5+[\mathrm{ATP}]),\ v_6 = k_6[\mathrm{ADP}], \tag{31}$$

We also assume that there exists the linear integral:

$$[\mathrm{ATP}] + [\mathrm{ADT}] = A_0 = \text{constant}. \tag{32}$$

Regarding that reactions run their course in the medium of ideal intermixing and under the conditions of pH- and thermal-static, we have the following dimensionless MM:

$$d\sigma/ds = \nu_0 - \nu_1,\ d\alpha_3/ds = g_3/g_1\nu_3 - \nu_1 - \nu_5 + \nu_6,$$

$$\varepsilon_1 d\iota_1/ds = \nu_1 - \nu_2,\ \varepsilon_2 d\iota_2/ds = g_2\nu_2 - \nu_3 - \nu_4, \tag{33}$$

where

$$\nu_0 = \nu_{0s} - \beta\sigma,\ \nu_1 = \alpha_3\sigma/(1+\sigma),\ \nu_2 = \iota_2,\ \nu_3 = \iota_2\alpha_2,$$

$$\nu_4 = \beta_4\iota_2,\ \nu_5 = \alpha_3/(\kappa_5 + \alpha_3),\ \nu_6 = \beta_6\alpha_2. \tag{34}$$

In MM (33)-(34), dimensionless variables have the sense

$$\sigma = [\mathrm{S}]/K_1,\ \alpha_3 = [\mathrm{ATP}]/A_0,\ \alpha_2 = [\mathrm{ADT}]/A_0 = 1 - \alpha_3,$$

$$\iota_1 = k_2[\mathrm{I}_1]/k_1A_0,\ \iota_2 = k_3[\mathrm{I}_2]/k_1,\ s = k_1A_0t/K_1,$$

$$\varepsilon = A_0/K_1g_1,\ \varepsilon_1 = k_3A_0/k_2K_1,\ \varepsilon_2 = k_1/k_3K_1,$$

$$\nu_{0s} = v_{0s}/k_1A_0,\ \beta = k_0K_1/k_1A_0,\ \beta_4 = k_4/k_3A_0,$$

$$\beta_5 = V_5/g_1k_1A_0,\ \beta_6 = k_6/g_1k_1,\ \kappa_5 = K_5/A_0. \tag{35}$$

If the values of ε_1 and ε_2 are sufficiently small, then with the help of passage to the limit under $\varepsilon_1, \varepsilon_2 \to 0$ MM (33) is reduced to the system of two differential equations:

$$d\sigma/ds = \nu_0 - \nu_1 = P(\sigma, \alpha_3),$$

$$\varepsilon d\alpha_3/ds = \nu_{out} - \nu_5 + \nu_6 = \varepsilon Q(\sigma, \alpha_3), \tag{36}$$

in which $\nu_{out} = (\eta\alpha_2-\beta_4)/(\alpha_2+\beta_4)\alpha_3\sigma/(1+\sigma)$ is the output characteristic and $\eta = g_2g_3/g_1 - 1$ is stoihiometric efficiency of the source of energy consisting of the reactions from the first to the fourth stages of scheme 1.

The equation for quasi-staticnary states $Q(\sigma, \alpha_3) = 0$ has one to three solutions depending on the values of σ. It means that the quasi-stationary curve on the plane (σ, ν_1) has the hysteresis-type shape.

A very important quantitative characteristic of an energetic metabolism is the so-called *loading characteristic*; that is, dependence of the stationary concentration of ATP on the load consisting of the total activity of all biochemical processes that use ATP. According to the model (36), the loading characteristic is the dependence $\alpha\hat{}_3(\beta_5)$ obtained from the condition of stationary state $P(\sigma, \alpha_3) = Q(\sigma, \alpha_3) = 0$.

Under small activity of the leakage (β_4) the loading characteristic has an important property. This property allows us to change the load in the wide range $0 < \beta_5 < \beta_{5cr}$, where $\beta_{5cr} \approx \nu_{0s}(\eta-\beta_5)/(1+\beta_5)$ is a critical value of the load. The stationary concentration changes of ATP are very small ($\alpha\hat{}_3 \approx 1$).

Another important peculiarity of loading characteristic is its hysteresis-type shape. Under $\beta_5 > \beta_{5cr}$, these results jump to stationary concentrations of ATP to very low levels determined only by the activity of the alternative source (β_6). Many investigators (see [1; 51, 129, 146] have confirmed these two important properties. Thus, the plateau and *hysteresis* of the loading characteristic are fundamental properties of energetic metabolism.

The tertiary structure of energetic metabolism has to contain controls that weaken or strengthen the plateau of loading characteristic as well as creates or removes its hysteresis. Indeed, inhibition of generator stage of ATP and activation of the same stage of ADP as well as inhibition of the initiator stage of ATP by ADP can serve as examples of such controls. Phenomenological description of the first kind of control may be

$$U_1 = 1/(1+c\alpha_3^n) \text{ and } U = \alpha_2^r,\ c > 0,\ n,\ r > 1, \tag{37}$$

where c is a dimensionless factor of inhibition, n and r are exponents of inhibition. The rate of generator stage ν_3 has to be multiplied by U_1. Phenomenological description of the second kind of control may be

$$U_2 = 1/(1+c\alpha_2^n),\ c > 0,\ n > 1. \tag{38}$$

Controls of hysteresis of loading characteristic can provide generation of auto-oscillations, which is an important and maybe a unique form for prolonged existence of energetic metabolism. The period of such auto-oscillations is about twenty-four hours. This period allows us to think that the cell's clock is none other than the energetic metabolism [105].

5. Effects of *pH* and *T* on the Reaction Rates

A description of the effects of *pH* and temperature *T* on the enzyme-catalyzed reactions and computer programs for enzyme kinetics can be found in D. Roberts [161].

The effects of *pH* can be understood by using the example of ionization of the enzyme-substrate complex. The idea can be represented by the following scheme (Fig. 7):

$$
\begin{array}{c}
ES^- \\
K_{1ES} \uparrow H^+ \\
K_S \downarrow \quad k_2 \\
S + EH \Leftrightarrow EHS \rightarrow EH + P \\
K_{2ES} \uparrow \\
\downarrow H^+ \\
EH_2S^+
\end{array}
$$

Figure 7. Scheme of ionization of the enzyme-substrate complex

Now, by analogy with the s. 1--4 and according to the scheme above, we have

$$e_0 = eh + es^- + (ehs) + (eh_2s^+),\ v = dp/dt = k_2(ehs) = k_2/K_S(eh)s$$

$$= k_2/K_S[e_0 - es^- - (ehs) - (eh_2s^+)]s, \quad (39)$$

from where

$$v = V's/(K_m + s),\ V' = V/K_+,\ K_m = K_S/K_+,$$

$$K_+ = (1 + h^+/K_{2ES} + K_{1ES}/h^+),\ V = k_2e_0,\ K_S = (eh)s/ehs,$$

$$K_{1ES} = (es^-)h^+/ehs,\ K_{2ES} = (ehs)h^+/eh_2s^+, \quad (40)$$

so both V' and K_m are *pH*-dependent.

It is easy to see that

$$h^+_{max} = (K_{1ES}K_{2ES})^{1/2},\ v_{max} = Vs/(K_S + [1 + 2(K_{1ES}/K_{2ES})^{1/2}]s),$$

$$h^+_{min} = 0, \infty,\ v_{min} = 0. \quad (41)$$

At low *pH*, $pH = -\lg(h^+)$, where $h^+ \gg K_{1ES} \gg K_{2ES}$, we have

$$V' = VK_{1ES}/h^{+},\ K_m = K_S K_{1ES}/h+, \tag{42}$$

and at high *pH*, where $K_{1ES} \gg K_{2ES} \gg h+$, we have

$$V' = Vh+/K_{2ES},\ K_m = K_S h+/K_{2ES}. \tag{43}$$

It should be noted that similar relations can be obtained in the cases of ionization of a free enzyme and both of a free enzyme and an E^S complex. h^+ above has to be replaced by H_3O^+. We will use these relations later in the Part III of this book.

To determine the effects of absolute temperature *T*, let us consider the reaction A + B ⇔ X ⇔ C + D, where X is the transition state or the activated complex [161]. If this complex is in equilibrium, then the equilibrium constant is $k = ab/x$ and the rate of the reaction is

$$da/dt = db/dt = x\omega = ab/k\omega, \tag{44}$$

where ω is the rate of passage over the energy barrier,

$$\omega = E/\hbar = kT/\hbar, \tag{45}$$

where E is the average energy of the vibration that leads to decomposition, $\hbar$ is *Planck constant*, and k is *Boltzmann constant*. But

$$\Delta G = \Delta H - T\Delta S = RT\ln k = E_a - RT - T\Delta S, \tag{46}$$

where ΔG, ΔH, *T*ΔS are the standard correspondingly free energy changes of formation, enthalpy (heat content), entropy of activation change for the transition state, and E_a is the energy activation change.

It follows from the equations (44)-(46), that the equilibrium constant *k* of the reaction (43) is

$$k = ab/x = \exp\,[E_a/(RT) - \Delta S/R - 1]. \tag{47}$$

Combining the formulae (40) and (47), we have

$$v = V's/(K_m + s),\ V' = V/K^{\wedge}_{+},\ V = k_2 e_0,\ K_m = K^{\wedge}_S/K^{\wedge}_{+},$$

$$K^{\wedge}_S = (eh)s/ehs = \exp\,[E_a/(RT) - \Delta S/R - 1],$$

$$K^{\wedge}_{+} = (1 + h^{+}/K^{\wedge}_{2ES} + K^{\wedge}_{1ES}/h^{+}),$$

$$K^{\wedge}_{1ES} = (es^{-})h^{+}/ehs = \exp\,[E_{a,1}/(RT) - \Delta S_1/R - 1],$$

$$K^{\wedge}_{2ES} = (ehs)h^{+}/eh_2s^{+} = \exp\,[E_{a,2}/(RT) - \Delta S_2/R - 1], \tag{48}$$

where E_a, $E_{a,1}$, and $E_{a,2}$ correspond to energy activation changes of the reactions, with the equilibrium constants $K^\wedge_S$, $K^\wedge_{1ES}$, and $K^\wedge_{2ES}$, respectively. Instead of (43), we have a more complicated situation now.

Using (48), let us consider the system

$$\partial v/\partial h^+ = v^2/(Vs)(K^\wedge_{1ES}/h^{+2} - 1/K^\wedge_{2ES}) = 0,$$

$$\partial v/\partial T = v^2/(VsRT^2)[K^\wedge_S E_a + s(K^\wedge_{1ES}E_{a,1}/h^+ - 1/K^\wedge_{2ES}E_{a,2}h^+)]. \quad (49)$$

It is easy to see that $T^*_{max} = \infty$ and the root of the first equation and the maximum of v are

$$h^+_{max} = (K^\wedge_{1ES}K^\wedge_{2ES})^{1/2} = \exp[-(\Delta S_1+\Delta S_2)/(2R)-1],$$

$$v_{max} = Vs/(K^\wedge_S + [1+2(K^\wedge_{1ES}/K^\wedge_{2ES})^{1/2}]s),$$

$$K^\wedge_S = \exp[-\Delta S/R-1]. \quad (50)$$

In addition, we obviously have

$$T_{min} = 0;\ h^+_{min} = 0, \infty;\ v_{min} = 0. \quad (51)$$

Thus, the system (49) always has unique solutions in which the rate v achieves its maximum and minimum.

6. MM of Cell-to-Cell Interactions

Most works in this direction [1; 51; J. Jacquez [106]; V. Kuznetsov, 117; 129] use the approach similar to that for interactions of chemicals in accordance with MM of enzyme reactions.

It should be noted that the book [106] is devoted in general to the compartmental analysis in biology and medicine that is also the main direction of this book. However, it seems difficult to find the common problems and especially the common results of these two books.

It should be also noted that the cell-cell interaction model (named the "harpoon recognition model" [117]) is related to interaction natural killer cells (NK cells) with cancer cells. Nonspecific NK recognition is associated with selective "*harpooning*" of the target cells by lipid vesicles of the cells that express multiple disorders in structure and organization of the extra-membrane macromolecular framework and of the sub-membrane cyto-skeleton assembly.

The classical MM of two cells interaction can be seen in S. Smale [173]. This model is based on Turing's equation of the form

$$dx_k/dt = R(x_k) + \sum \mu_{ik}(x_i - x_k) \ (i = 1, \ldots, N),\ k = 1, \ldots, N, \tag{52}$$

where x_k mean all the concentrations for kth cell, $R(x_k)$ determines the chemical kinetics of kth cell, the sum spreads to the set of the cells adherent to kth cell, and μ_{ik} are matrices.

The main result is two cells, each having the equilibrium of its enzyme concentrations after their interaction, can gain new structure-stable state in which enzyme concentrations of each cell oscillate arbitrarily long. It is in contrast to A. Sherman [170], where coupling of two pancreatic β-cells with fast spiking voltages can result in inhibition of those spikes and in gaining of new desired properties.

For cell-cell interactions in neural networks based on electrical and chemical reciprocity see [134; G. Rowe, 163; J. Tyson, K. Chen, B. Novak, 189; W. Yamada, C. Koch, P. Adams, 212].

The interactions of cells in the process of *morphogenesis* are considered in detail in [J. Murray, 143; G. Rowe, 163] (see also ch. 1 and 7). The main idea is that there exists the chemical-mechanical pre-pattern of morphogenesis so that the cell can read out its position in the coordinate of chemical concentration and differentiate, undergo appropriate cell shape change, or migrate accordingly. MM of cell-cell interactions in the process of evolutionary genetics are considered in detail in Yu. Svirezhev, V. Passekov, [177].

We do not dwell on these subjects since they extend beyond the main goals of this book. However, we consider the process of organism development later in Part II with the help of MM of development.

Note that cell-cell interactions among different sub-populations of the immune network as well as of their interactions with other cells, the cancer cells in particular are considered in detail in separate chapters of parts II and III of this book.

Chapter 3

GENERAL KINETIC CELLULAR THEORY

The cell-to-cell interactions, cell population, and tumor-host relationships are identified in N. Bellomo, L. Preziosi [20] by means of the following assumptions.

1. Assumption 1

Cells that actually take part to this process can be subdivided into three main populations: tumor cells, characterized by an anomalous proliferative activity and the difficulty to receiving inhibitory and apoptotic signals; environment cells characterized by promoting (feeding) influence over tumor cells; immune cells potentially able to either strongly hamper or favor tumor growth.

Each population can be further subdivided. In general, we will deal with a number p of interacting populations. Soluble factors and cytokines and cell membrane signals regulate the activity of all these cell populations.

2. Assumption 2

The functional state of each cell of a given population is described by the real number u with values in the interval [−1, 1]. *Positive values of u denote: aggressive proliferation for tumor cells, adequate energy supply by the environment, and strong anti-tumor activity by the immune cells.*

Negative values correspond to opposite activities: dormancy of tumor cells, tumor ischemic necrosis due to poor blood supply by the environment, tumor stimulating activity by the immune cells.

Activation, or conversely inhibition, of the immune cells is regulated by cytokine and stimulatory signals, or by the lack of co-stimulation signals.

3. Assumption 3

The statistical distribution within each population is defined by the density functions $N = N_i$, $i = 1, \ldots, p$; $N_i = N_i(t,u)$, *where* $N_i(t,u)du$ *defines the number of cells per unit volume at time t, with their state in the interval* $[u, u + du]$ *and where the number of inter-acting populations is generically indicated by p, although simulation will be developed for* $p = 3$.

Admittedly, the definition of cell populations involved in this game for life and death and the definition of their functions are a gross simplification. Even if it is difficult to sketch the critical features that shape the process, one can consider that, as a result of the accumulation of several gene mutations, tumor cells proliferate without the usual restraints and respond no longer to the signals from the environment. Each tumor is characterized by a particular set of gene mutations that confers on it a more or less rapid expansion and independence from environmental signals that regulate normal proliferation or induce cells to die apoptotically. The accumulation of gene mutations and the intense proliferate activity also make the tumor different from the normal cells of the same tissue. The membrane expression of the altered proteins coded by the mutated genes and other traits (collectively defined *tumor associate antigens*, TAA) allow tumor recognition by the immune cells.

The effector arm of the immune system hampers tumor growth in four main ways:

a) Direct killing by lymphocytes;

b) Release of cytokines by T lymphocytes that triggers and pilots the anti-tumor activity of several leukocytes;

c) Binding the antibodies to tumor cells, thus causing their death in the presence of serum factors. These antibodies also adhere to the membrane of leukocytes and guide their anti-tumor activity;

d) Induction of tumor ischemic necrosis by alteration of tumor vessels.

Bearing in mind that we have decided to operate in a simplified framework, we can now return to the kinetic model and consider, in the case of scalar u, the problem of computing the result of the tumor-host relationship by knowing the densities N_i. In particular, the number of cells of the *i-th* population per unit volume at time t is

$$n_i(t) = \int_{-1}^{1} N_i(t,u)du. \tag{1}$$

Another important quantity is the mean value of u over the *i-th* population

$$E_i(u)(t) = \int_{-1}^{1} uN_i(t,u)du / \int_{-1}^{1} N_i(t,u)du. \tag{2}$$

The solution of mathematical problems related to the evolution equations for the densities N could give quantitative description of their evolution in time. By means of moments, as the ones defined by Equations (1-2), the desired information on the behavior of the macroscopic observables of the system can finally be obtained.

Before dealing with the derivation of the evolution equation of the densities N_i, it is useful to identify the cellular regimes that characterize the behavior of the system. A classification can be proposed in order to distinguish the cellular regime from the condensed cell regime. One is the free (dispersed) cells regime, a situation where the cells move freely and homogeneously in space interacting by means of binary encounters. Another is the condensed (packed) cell regime, a physical situation where the tumor cells are already packed in a quasi-spherical volume.

The modeling of the cell-to-cell interactions can be developed according to the following hypotheses.

4. Assumption 4

Cell interactions can be divided into mass conservative encounters and interactions with cell proliferation or destruction..

Cell interactions in the case of mass conservative encounters will be described by two physical quantities: the encounter rate η and the transition probability density ψ. Similarly, proliferative encounters will be described by two quantities: the proliferation rate p and the proliferation probability density φ. Destructive encounters will be described by the destruction rate d.

5. Assumption 5

Conservative encounters between the cell of the population (with state v) and the cell of the j-population (with state w), the quantitatively described by the transition rate

$$T_{ij}(v,w;\,u) = \eta_{ij}(v,w)\psi_{ij}(v,w;\,u), \tag{3}$$

where:

$T_{ij}(v,w;\,u)$ *denotes the number of encounters, per unit volume and time, between cell pairs of the (ij)-populations, respectively with states v and w with transition* of the *i-cells into the state u*;

$\eta_{ij}(v,w)$ is *the encounter rate, which denotes the* number of encounters *per unit volume and time between* cell *pairs* of *the (ij)-populations with states v and w, respectively;*

$\psi_{ij}(v,w;\,u)$ *is the transition probability density, which is the density of transition of the i-cell to the state u given its initial state v and the* state *of the encountering* cells *belonging to the j-th population.*

6. Assumption 6

Proliferative encounters occur between cell pairs of the same or different populations, and cause generation of new cells in one or both populations. These interactions are quantitatively described by the proliferation rate

$$P_{ij}(v,w;\,u) = p_{ij}(v,w)\varphi_{ij}(v,w;\,u), \tag{4}$$

where:

$P_{ij}(v,w;\,u)$ *denotes the number* of *the i-cells per unit volume and time, which proliferate in the state u due* to *the encounter between cell pairs* (*with* states *v and w, respectively*) of *the (ij)-populations*;

$p_{ij}(v,w)$ is *the proliferation rate, which denotes the number* of *i-cells that* are *produced per unit volume and time due to the encounter of cell pairs* (*with states v and w, respectively*) *of the* (*ij*)*-populations;*

$\varphi_{ij}(v,w;\, u)$ is *the proliferation probability density, which is the density of the proliferation of the i-cell in the state u due to the encounter of two* cells *belonging, respectively, to the i-th population with state v and to the j-th population with state w.*

7. Assumption 7

Destructive encounters occur between cell pairs of different populations, and cause death of one or both cells.

These interactions are quantitatively described by the destruction rate $d_{ij}(v,\mathrm{w})$, *that is the number of i-cells with state v that are suppressed in the interaction with the j-cells with state w.*

Note that the terms φ and ψ have the structure of a probability density

$$\int_{-1}^{1} \varphi_{ij}(v,w;\, u)du = \int_{-1}^{1} \psi_{ij}(v,w;\, u)du = 1,\ \ i, j = 1, \ldots, p. \tag{5}$$

The evolution equations of the densities N_i, for $i = 1, \ldots, p$, can be derived by suitable balance equations, which equate the total derivative of N_i to the gain term G_i minus the loss term L_i, where G_i and L_i are generated by the conservative interactions between cell pairs and their related change of state, plus the proliferation term S_i, minus the destruction term D_i.

An additional term can be introduced in order to simulate the production g_i of cells of the *i-th* population due to the bone marrow or/and to any external source (e.g. injection)

$$\partial N_i/\partial t(t,u) = J_i[N](t,u) = G_i[N](t,u) - L_i[N](t,u) + S_i\,[N](t,u) - D_i[N](t,u) + g_i(t,u), \tag{6}$$

where all the terms G_i, L_i, S_i, and D_i are assumed to depend on N in order to take into account the relationship between the different cell-types. In particular,

G_i is obtained by summing all transitions, ruled by η and ψ, of the *i-cells* from the state v into u,

$$G_i = \Sigma \int_{-1}^{1} \int_{-1}^{1} \eta_{ij}(v,w)\,\psi_{ij}\,(v,w;\, u)N_i(t,v)N_j(t,w)dvdw \quad (j = 1, \ldots, p), \tag{7}$$

where the transitions occur after the encounter of the i-cell with all j-cells, which requires the sum over j. Furthermore, integration over v refers to the original state distribution before the interaction, while integration over w refers to all possible states of the interacting j-cell.

L_i is obtained by summing all loss, ruled by η, from the state u due to the encounter with the cell with state v,

$$L_i = N_i(t,u)\, \Sigma \int_{-1}^{1} \eta_{ij}(v,w)\psi_{ij}(u;\, v)N_j(t,v)dv \; (j = 1, \ldots, p), \tag{8}$$

where the loss occurs after the encounter of the *i-cell* with all *j-cells* having all possible states, which requires the sum over j and the integration over v, respectively.

S_i is obtained by summing all proliferation rates, ruled by p and φ, into u,

$$S_i = \Sigma \int_{-1}^{1}\int_{-1}^{1} p_{ij}(v,w)\varphi_{ij}(v,w;\, u)N_i(t,v)N_j(t,w)dvdw$$

$$(j = 1, \ldots, p), \tag{9}$$

where proliferation occur after the encounter of the i-cell with all j-cells, which requires the sum over j and the integration over v and w, respectively.

D_i is obtained by summing all destruction, ruled by d, of the cells with state u due to the encounter with cells with state v,

$$D_i = N_i(t,u)\Sigma \int_{-1}^{1} d_{ij}(u,v)N_j(t,v)dv \; (j = 1, \ldots, p), \tag{10}$$

where the destruction can occur after the encounter of the i-cell with all j-cells, which requires the sum over j and the integration over v, respectively.

g_i is a source term due to inner (for instance, from bone marrow) or outer inlets.

Remark. *The computation implies* a *statistical independence of the distributions* N_i *and* N_j. *This hypothesis is generally used in phenomenological kinetic theory, but is not genuinely true. It can be accepted as an approximation of physical reality for bounded times.*

Then the following evolution equation is obtained

$$\partial N_i/\partial t(t,u) = \Sigma \int_{-1}^{1} \int_{-1}^{1} [A_{ij}(v,w;\, u)N_i(t,v)N_j(t,\, w) -$$

$$1/2N_i(t,u)B_{ij}(u,v)N_i(t,v)N_j(t,v)]dvdw \; (j = 1, \ldots, p) + g_i(t,u), \tag{11}$$

where

$$A_{ij}(v,w;\, u) = \eta_{ij}(v,w)\psi_{ij}(v,w;\, u) + p_{ij}(v,w)\varphi_{ij}(v,w;\, u),$$

$$B_{ij}(u,v) = \eta_{ij}(u,v) + d_{ij}(u,v). \tag{12}$$

We point out that this model can be regarded as a superposition of a kinetic and a population dynamics model. In particular, one can have either totally conservative models: $p_{ij}(v,w) = d_{ij}(u,v) = 0$ or totally competitive models: $\eta_{ij}(u,v) = 0$.

An immediate simplification is obtained assuming, as is reasonable, that newborn cells retain the same activation state as the generating cells. That is,

$$\varphi_{ij}(v,w;\, u) = \delta(u-v) \qquad (13)$$

for all pairs i, j.

Further simplifications are introduced in [20] for three interacting populations: tumor cells (i=1), cells of the immune system (i=2), and environmental cells (i=3).

In addition, it is assumed that N_3 is constant in time and is in a given distribution over the state u.

Thus, we obtain a system of two population dynamics with respect to N_1 and N_2, where the characterization of the terms ψ can be done, for example, by delta functions over the output of the variable v

$$\psi(v,w;\, u) = \delta(u - m(v,w)), \qquad (14)$$

where m is the mean value or the most probable value.

Referring to conservative encounters, the following two assumptions are stated.

8. Assumption 8

The temperature of the environment remains constant during observation time, so that the encounter rates η_{ij} are also constant. The interactions that generate change of state (nontrivial interactions) are significant only for interactions between cells of tumor and immune system

$$\eta_{12} = \eta_{21} = \alpha, \qquad (15)$$

where α is constant in time. The assumption that the other encounters do not generate change of state yields

$$\psi_{11} = \psi_{13} = \psi_{22} = \psi_{23} = \delta(v-u), \qquad (16)$$

which means that the contributions to the gain and lost terms are equal and cancel out.

9. Assumption 9

The transition probability densities are probability densities with two parameters: variance σ and most probable output value defined by

$$m_{12}(v,w) = v\ (w > 0, v \le 0),\ v - \beta^{+}_{12}wv\ (w > 0, v < 0),$$

$$v - \beta^{-}_{12}w(1-v)\ (w \le 0), \qquad 17)$$

and also

$$m_{21}(v,w) = v - \beta_{21}\, w(1+ v)\ (w > 0),\ v\ (w \le 0). \qquad (18)$$

where β^{+}_{12} *characterizes the ability of the active immune system to dis-activate the aggressive tumor cells,* β^{-}_{12} *characterizes the negative action of the suppressed immune system to activate the aggressive tumor cells,* β_{21} *characterizes the ability of the tumor cells to inhibit the immune system (their all value* $\ni$ [0, 1]).

Assumptions with a similar mathematical structure can also be made referring to proliferative and destructive encounters.

10. Assumption 10

Proliferative and destructive encounters are ruled by the following phenomenological assumptions

Both production and destruction are localized in the state v independently of the value w (see (150)).

Proliferation of tumor cells is significant only for interactions with the suppressed immune cells, i.e., w < 0, *and with the environmental cells. Proliferation only occurs for aggressive states u* > 0, *and is directly proportional to the aggressive power of the tumor u and to the negative activation of the suppressed immune cells -w and to the activation of the environmental cells.*

Proliferation of immune cells only occurs in the interaction with active tumor cells and is independent on the activation state of the pair.

Destruction of immune cells only occurs in the interaction with the host environment (or by natural death) and is independent on the activation state of the cells.

Destruction of aggressive tumor cells only occurs in the interaction with active immune cells and is directly proportional to the activation state of the immune cells.

11. Assumption 11

The bone marrow constantly produces active immune cells uniformly distributed over the (positive) activation state

$$g_2(t,u) = g_2\ (u > 0),\ 0\ (u < 0). \qquad (19)$$

With regard to these assumptions, we have the following MM:

$$\partial N_1/\partial t(t,u) = \alpha\int_{-1}^{1}\int_{-1}^{1} [\psi_{12}(u,m_{12}(v,w))N_1(t,v)N_2(t,w) - 1/2N_1(t,u)N_2(t,v)]dvdw -$$

$$N_1(t,u)H(u)[\gamma_{12}u\int_{-1}^{0} vN_2(t,v)dv + \delta_{12}\int_{0}^{1} vN_2(t,v)dv - \gamma_{13}u\int_{-1}^{0} vN_3(v)dv,$$

$$\partial N_2/\partial t(t,u) = \alpha\int_{-1}^{1}\int_{-1}^{1} [\psi_{21}(u,m_{21}(v,w))N_1(t,w)N_2(t,v) - 1/2N_2(t,u)N_1(t,v)]dvdw +$$

$$N_2(t,u)[\gamma_{21}\int_{-1}^{0} N_1(t,v)dv - \delta_{23}\int_{0}^{1} N_3(v)dv + g_2H(u), \quad (20)$$

where the terms ψ are given by (14) and (16), $H(u) = 1$ $(u > 0)$, 0 $(u < 0)$ (the Heaviside function), γ_{12} characterizes the proliferative behavior of the tumor cells in their encounter with the suppressed immune system, γ_{13} characterizes the proliferative behavior of the tumor cells in their encounter with the host environment, γ_{21} characterizes the proliferative behavior of the immune cells in their encounter with tumor cells, δ_{12} characterizes the suppression of the tumor cells in their encounter with the active immune system, δ_{23} characterizes the suppression of the immune cells in their encounter with the host environment or with their natural extinction. Such a specialization of the parameters is useful for their identification.

The model (20) was numerically investigated by N. Bellomo, L. Preziosi [20].

12. On Generalized Kinetic Models in Applied Sciences

The cellular kinetic theory is based on the Bolzman equation for a system of interacting particles (see N. Bellomo, S. Stocker [22]). The authors discuss applications of that theory to the immune system modeling. We continue this discussion in Ch. 8.

The generalized kinetic models (N. Bellomo, M. Pulvirenti [21]) are the cases when i) entering objects may not be particles; ii) the physical state of each object may not be simply identified by position and velocity; iii) two or more object interactions should be considered.

The overall state of the system is described by the density distribution $f = f(t,u)$, $t \in [0, T]$, $u \in D_u \subseteq R^n$, that assigns, by means of the expectation value

$$n_{1,2}(t) = \int_D f(t,u)du, \quad (21)$$

the number of individuals that at time t can be found in a state lying in the range $D = [u_1, u_2]$.

The dynamics of the system is ruled by an evolution equation for the distribution function f obtained by equating the time derivative of f to the difference between a gain and loss term in the state u.

Chapter 4

SOME OTHER MM

1. Delay Differential Equations (DDE)

These MM have gradually become more and more popular (see Y. Kuang [116]).

The ground for their consideration is the fact that a more realistic MM must include some of the past history of the system. As it appears, even small delays can have large effects. Let us demonstrate this by some important examples.

It is easy to see that the trivial solution $x(t) = 0$ of the equation $x'(t) + 2x'(t) = -x(t)$ is stable. However, the trivial solution of the equation

$$x'(t) + 2x(t-\tau) = -x(t) \tag{1}$$

is unstable for any positive delay τ.

A more general result in [115] states that if $|b_n| > 1$, then the trivial solution of the equation

$$\Sigma\ [a_k d^k/dt^k x(t) + b_k d^k t/dt^k x(t-\tau)]\ (k = 1, \ldots, n) = 0 \tag{2}$$

is unstable.

For the *delayed logistic equation*

$$x'(t) = rx(t)[1-x(t-\tau)/K] \tag{3}$$

the critical value $\tau\hat{} = \pi/(2r)$ destabilizes the local stability of the steady state $x(t) = K$. For large r, $\tau\hat{}$ can be viewed as being small.

Thus, in a series of important cases we cannot neglect small delays, and in each individual equation we should decide whether we could neglect delay or not. Certain applications of DDE to the cell biology can be seen in J. Jacquez [106]; M. Mackey [123].

2. Mathematics in Population Biology

Population biology traditionally deals with the problems of demography, ecology, and epidemiology. However, certain tools of mathematical modeling in population biology can also be used for problems concerning the cell and CAO. We consider here, in short, connections of those tools with MM of development.

The first similar models for ecological systems were proposed by F. Sharpe and A. Lotka [169]. These models have the form

$$B(t) = \int_t^T m(u)l(u)n(u{-}t)du + \int_0^t m(u)l(u)B(t{-}u)du,\ T \geq t \geq 0, \tag{4}$$

where $B(t)$ is the birthrate, the intensity of the number of individuals being born in the unit of the time at the instant t; $m(u)$ is the specific age birthrate, the average number of descendants being born by an individual of age u in the unit of the time; $l(u)$ is the fatigue function, a share of individuals attaining the age u; $n(u)$ is the initial (under $t = 0$) distribution of individuals by the age u; T is the upper bound of the age.

The equation (4) allows us to find the dynamics of the birthrate B under given m, l, and n. After substituting $s \rightarrow t\text{-}u$ we obtain the particular case of MM (1) in Introduction, namely

$$B(t) = \int_t^{t-T} m(t{-}s)l(t{-}s)B(s)ds + \Delta B. \tag{5}$$

Differential MM with partial derivatives are alternatives to *integral MM*. For the linear integral-type model (5) the respective analogies (see R. Poluectov, Yu. Pykh, I. Shvytov [152]) have the form

the evolutionary equation

$$\partial x/\partial t + \partial x/\partial u = -d(u)x(t,u) \tag{6}$$

and the birthrate equation

$$x(t,0) = \int_0^T m(u)x(t,u)du, \tag{7}$$

where $x(t,u)$ is the number of individuals of the age u at the instant t, $x(0,u) = n(u)$; $m(u)$ is the same as in (5) and $d(u)$ is the specific age death rate.

It is easy to see that the equation (6) is equivalent to

$$x(t,u) = x(t{-}u,0)\exp\left[-\int_0^{t-u} d(v)dv\right]. \tag{8}$$

Let us substitute (8) into the equation (7) and introduce the desired function $B(t) = x(t,0)$. Then

$$B(t) = \int_t^{t-T} m(t\text{-}s)\exp\left[-\int_0^{v-s} d(v)dv\right]B(s)ds + \Delta B. \tag{9}$$

From (9) it follows that the fatigue function is connected to the death rate by

$$l(u) = \exp\left[-\int_0^{v-u} d(v)dv\right]. \tag{10}$$

The transition from *differential MM* to *integral ones* in the case of a non-stationary external environment, where the given functions $m(t,u)$, $d(t,u)$, and $l(t,u)$ depend on t, is quite similar.

In more general cases when the scantiness of resources, intro-specific competitions, etc. are also considered, the transition from nonlinear differential MM to integral ones can be found in S. Michelson, J. Leith [135] and in H. Othmer et al. [146]. One of the advantages of *integral MM* compared to *differential ones* is the reduction of dimensionality of the desired functions. If one needs to find, for example, the density $x(t,u)$ one can do it by the formula (8), where $x(t\text{-}u,0) = B(t\text{-}u)$. Another advantage is the explicit, well-posed representation of the desired solution. For the proof of well posedness in the case of differential MM, we still have to transform them to integral MM.

Let us dwell on comparison of mathematics in population biology (H. Tiemme [181]) with mathematics of development (V. Ivanov [92] (see also Introduction and the subsequent Ch.). Let the stage be life, and $u(t,a)$ be the density of individuals at time t that have age a, $B(t)$ be the birthrate of population, $P(a)$ be the probability of still being in the stage a, c be the maximum age, and 0 be the time at which we start to consider the development of the stage. Then we have [181]

$$u(t,a) = B(t-a)P(a),\ t > a \geq 0;\ u(t,a) = 0,\ a > c, \tag{11}$$

from where

$$\partial[u(t,a)/P(a)]/\partial t + \partial[u(t,a)/P(a)]\partial a = 0,\ t \neq a. \tag{12}$$

On the other hand, we have the relation (see MM (1) in Introduction)

$$R(t) = \int_{a(t)}^{t} m(s)ds, \tag{13}$$

where in the case of population, $R(t)$ is the quantity of individuals at time t; $t-a(t)$ is the life span at time t; and $m(s)$ is the rate of origin at time s of the number of new individuals in the population under consideration, $m(s) \geq 0$. It is clear that the birthrate

$$B(t) = dR(t)/dt = m(t) - m(a(t))da(t)/dt,\ \mathrm{c} = t - a(t). \tag{14}$$

Therefore, we can rewrite MM (11) and (12) in population biology in terms of MM of development with the help of (13) and (14). In addition, we can obtain another interpretation of all equations and results of their investigation. Note that the maximum age in our MM is dependent of t.

Adding other equations of MM (1) in Introduction and considering optimization problem (3)-type in the Introduction, we can investigate the problems of survival or productivity for various populations (see also J. Murray [143]; L. Segel [166]).

We consider in this book the problems of survival in the case of HIV population and cancer cells. At the same time, we can enrich mathematics of development at the expense of introducing the probability functions of $P(t)$-type and introducing the case of the models, when the equations are forming the so-called semi-group (G. Webb [207]). However, it would exceed the limits of this book.

Note in conclusion that ES are not monotone dynamical systems and note that MM of development are different from MM by H. Smith [174] and from the so-called evolutionary equations by X. Zhao [216]. From the position of functional analysis, integral-functional MM of development are rather general and strongly nonlinear. The only common restriction on the corresponding mappings or operators is their non-negativity.

3. Some Problems and MM in Computational Biology and Bioinformatics

The physical models of DNA have been described in Ch. 1, s. 4.4. From the information point of view, DNA is a word over four-letter alphabet: A, C, G, and T.

Another nucleic acid in the cell, RNA is a word over another four-letter alphabet where U (uracil) replaces T. These molecules have a distinguishable direction: one end (the left) is labeled 5′ and the other 3′. Proteins are also polymers and here the word is over an alphabet of 20 *amino acids* or *amino acid residues*. The name of these amino acids and their abbreviations are given in Fig. 8.

Alanine	A	Ala	Leucine	L	Leu
Arginine	R	Arg	Lysine	K	Lys
Asparagine	N	Asn	Methionine	M	Met
Aspartic acid	D	Asp	Phenylalanine	F	Phe
Cysteine	C	Cys	Proline	P	Pro
Glutamic acid	E	Glu	Serine	S	Ser
Glutamine	Q	Gln	Threonine	T	Thr
Glycine	G	Gly	Tryptophan	W	Trp
Histidine	H	His	Tyrosine	Y	Tyr
Isoleucine	I	Ile	Valine	V	Val

Figure 8. Amino acids and their abbreviations

Amino acids are encoded by blocks of three nucleotides known as *codons*. There are 64 possible codons. Each codon encodes only one amino acid. The end of a protein being built is signaled by any of the three stop *codons*.

Some proteins give organisms their physical structure. Protein catalysts are called enzymes, and their names can be recognized by the suffix – *ase*. Proteins do not assume a predictable shape analogous to DNA's double helix. Instead, protein shapes are determined by complicated interactions among their residues in the chain. A protein's shape and electrical charge distribution determine its function. Predicting the shape of a given amino acid sequence in vivo is the protein folding problem, which is one of the most important and difficult problems of computational molecular biology.

In DNA blueprint, a code for a single protein is called a *gene*. DNA remains in the nucleus, whereas proteins are made by ribosomes in the cytoplasm outside of the nucleus. The intermediate molecule carrying information out of the nucleus is RNA. The process of copying DNA into RNA is called *transcription*. The RNA blueprints are read and used to assemble proteins from amino acids in a process known as *translation*.

The information flow in biology is summarized by the "central dogma' (M. Waterman [204)]: once information has passed into protein it cannot be retrieved. This dogma was reconsidered later in connection with retroviruses.

The *genetic structure* of *DNA* consists of exons separated by non-coding introns. Transcription of DNA to RNA is followed by intron excision and exon splicing to form an uninterrupted stretch of coding RNA. The information resides in the local, three-dimensional structure of the nucleic acids. The sequence of DNA encodes a great deal of information, not just the code connecting it to protein sequence, but also to the information required for regulating all aspects of protein synthesis.

There is the *Human Genome Project* (HGP), i.e. the national (USA) program to locate and analyze all human genes. The human genome consists of approximately 10^5 to 10^6 genes distributed over 22 pair of autosomes and 2 sex chromosomes (about 3×10^9 nucleotides). The goal to sequence the entire genome early in the 21st century was recently achieved.

Large quantities of DNA were obtained by cloning, i.e. by inserting fragments of the DNA into a virus that would then replicate in an appropriate host. The assemblage problem of those fragments must be solved algorithmically [204].

Let us consider some of the typical problems of computational molecular biology (R. Dwyer [41]; C. Fall et al. [45]). One of them is the so-called *double digest problem*, or DDP. A restriction enzyme cuts linear DNA molecule of length L bps or base pairs at all occurrences of a short specific pattern, and the lengths of the resulting fragments are recorded. In DDP, we have the list of fragment lengths when each of two enzymes is used alone, say

$$\mathbf{a} = \{a_i\!: 1 \le i \le n\},\ \mathbf{b} = \{b_i\!: 1 \le i \le m\} \tag{15}$$

from the first, second digest, as well as a list of double digest fragment lengths, when two restriction enzymes are used in combination and the DNA is cut at all occurrences of both restriction patterns, say

$$\mathbf{c} = \{\ c_i\!: 1 \le i \le l\}. \tag{16}$$

We assume that the sets **a**, **b**, and **c** are ordered, that is, for example, $a_i \le a_j$ for $i < j$. Of course,

$$\Sigma\ a_i = \Sigma\ b_i = \Sigma\ c_i = L, \tag{17}$$

as fragment lengths are measured in numbers of bases with no errors.

Given the above data, the problem is to find the order for set **a** and **b** such that the double digest implied in these orderings is, in a sense below, **c**. By ordering **a** and **b** according to a permutation σ of (1, 2,..., n) and μ of (1, 2,..., m), respectively, we obtain the set of locations of cut sites

$$S = \{s: s = \sum_{1\leq j\leq r} a_{\sigma(j)} \text{ or}$$

$$s = \sum_{1\leq j\leq t} b_{\mu(j)};\ 0 \leq r \leq n,\ 0 \leq t \leq m\}. \tag{18}$$

Since we want to record only the location of cut sites, the set S is not allowed to be repeated. Now label the elements of S such that

$$S = \{s_j : 0 \leq j \leq l\} \text{ with } s_i < s_j \text{ for } i < j. \tag{19}$$

The double digest implied by the configuration (σ, μ) can now be defined by

$$C(\sigma, \mu) = \{c_j(\sigma, \mu): c_j(\sigma, \mu) = s_j - s_{j-1} \text{ for some } 1 \leq j \leq l\}. \tag{20}$$

The DPP is then to find a configuration (σ, μ) such that $\boldsymbol{c}$ = C(σ, μ), where $\boldsymbol{c}$ is determined through experiment. The simple combinatorial analysis gives $n!m!$ map configurations. It is proved (M. Waterman [204]) that the DDP belongs to the so-called *NP-Complete* problems (see M. Garey, D. Johnson [53]).

It means that there is most likely no simple algorithm for the general case of DDP (see Ch. 16). For an effective solution of different problems of mapping DNA, we have to improve the existing approach or find a new approach to the respective experiments.

Many other problems of computational biology are transcription, determination of RNA second structure, finding protein motifs, construction of the respective database PROSITE, and so on.

Transcription in Perl (R. Dwyer [41]) is a program that reads strands of DNA from input and writes out the corresponding RNA and protein sequences.

Determination of RNA second structure with certain constraints is a program that adopts a strategy of finding the best folding of shorter sub-strings first and building on these to find the best folding of longer sub-strings. It is typical for the so-called method of dynamic programming (M. Waterman, T. Smith [205]) (see also Ch. 16).

A motif is a highly conserved segment in the protein sequence with a recognized biochemical function. PROSITE is a database of motifs to which a new sequence can be compared. The respective programs in Perl, using the suffix trees, are rather effective. The suffix tree is a data structure for quick identification of sub-strings of a fixed target string.

In conclusion, we mention some problems of exploration and analysis of DNA microarrays (or array, for short) and protein array data described in detail by D. Amaratunga and J. Cabrera [9].

One of the key areas of *Genomics* (that studies the structure and functions of genes) is *functional genomics*, which is the study of the functions of genes. It is important to realize that genes (and proteins) operate collectively in pathways. The interplay between genes, proteins, and external factors makes functional genomics a complex subject.

The great influx of genomics information has spawned a new discipline called *bioinformatics*, whose one of divisions is the creation and the analysis of DNA and proteins arrays. Briefly, DNA array experiment proceeds as follows. Divide a glass slide into a series of imaginary square cells to form a rectangular grid. Onto each square cell, stick a tiny amount of liquid that contains a gene of known sequence. Prepare a solution that contains a mixture of mRNA with unknown sequence. Add to this solution a substance that fluoresces by light. Pore the solution onto the slide. If the mRNA molecules have a match, the solution will stick to the slide. Otherwise, the solution will not stick and can be washed away. Use a laser scanner to detect and measure the fluorescent signal emitted.

There are various types of array experiments:

i) Tissue-specific gene expression,
ii) Developmental genetics,
iii) Genetic diseases,
iv) Complex diseases,
v) Pharmacological agents,
vi) Plant breeding, etc.

Each of those experiments consists of the following basic steps: preparing the array and the labeled sample, hybridizing the labeled sample to the array and washing the array, scanning the array, processing and interpreting the scanned image. The scanned image must be converted into spot intensities for further, mainly statistical, analysis.

Actually, all powerful apparatus of mathematical statistics was applied in the book by D. Amaratunga and J. Cabrera [9] for analysis of DNA and protein array.

4. On the Theory of Catastrophes

The *theory of catastrophes* (see, for example, R. Thom [182]) is closed to the theory of bifurcation (H. Othmer et al. [146]).

In the theory of catastrophes, the phenomenon under study is assumed to be governed by a potential function, i.e., $V = V(x_1, x_2, \ldots, x_n; a_1, a_2, \ldots, a_m)$, where x_i are known as the state or internal variables and where parameters a_j are known as the control or external variables. The stable states of the system are regarded as the minimum of this potential and can be usually found from the equations

$$\partial V/\partial x_i = 0, \; i = 1, 2, \ldots, n. \tag{21}$$

Changing the control parameters smoothly may alter the positions, relative heights, or total number of local minimums in a discontinuous way. These discontinuous changes in this state are called *catastrophes*. Bearing in mind their further applications, we consider in more detail the simplest catastrophes. Let us consider MM with only one variable of the form

$$dx/dt = P(x) = -dV(x)/dx. \tag{22}$$

To investigate MM in the vicinity of extremes of $V(x)$, the function $P(x)$ is expanded in a fragment of power series in the vicinity of its stationary values. Minimal numbers of the terms of the fragment n and parameters k that should be considered are determined by the degree of 0 of the stationary state. The number k is equal to $n-1$ and is called the co-dimension of a catastrophe. The simplest cases are

i) Isolated and stable stationary point. Then the minimal model

$$P(x) = -x,\ V(x) = x^2/2\ (n = 1, k = 0). \tag{23}$$

Here and later x is a deviation from the stationary value. The form $dx/dt = -x$ is general in the sense that any linear one-dimensional model can be reduced to this form.

ii) Fusion of two critical points. Then the minimal model is

$$P(x) = u + x^2,\ \ V(x) = -ux - x^3/3\ (n = 2, k = 1). \tag{24}$$

The form $P(x)$ in (24) is also general--any quadratic polynomial can be reduced to this form. Under $u < 0$, there are two stationary states, one of which is stable and the other which is unstable. Under $u = 0$, these states are combined. And under $u > 0$, they disappear. On the contrary, when the parameter is changed in reverse order, two stationary states appear. The null-cline (attractor) $dx/dt = 0$ is a parabola, i.e., two-value function x of u. So, this situation is called the *fold catastrophe* (see Fig. 9)

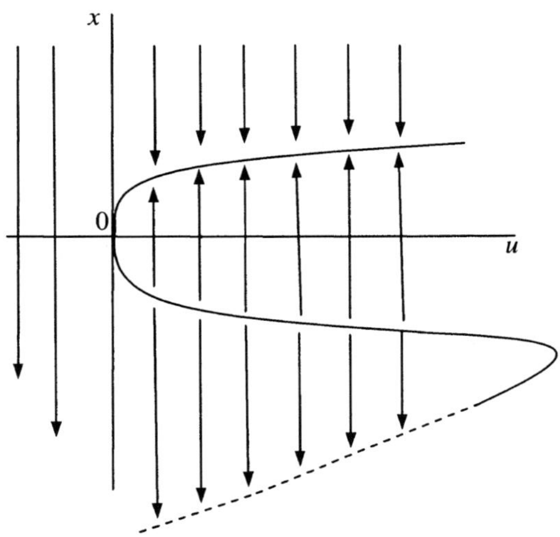

Figure 9. Folding on the plane x, u.

iii) Fusion of three critical points. Then the minimal model

$$P(x) = u_1x + u_2 + x^3,\ V(x) = -u_1x^2/2 - u_2x - x^4/4\ (n = 3, k = 2) \tag{25}$$

The ISO-cline surface or attractor $dx/dt = 0$ has the form of assembly in the space of (x, u_1 , u_2), of which vertex is the origin of coordinate, where three critical points are combined. So, this situation is called the *cusp or assembly catastrophe* (see Fig. 10).

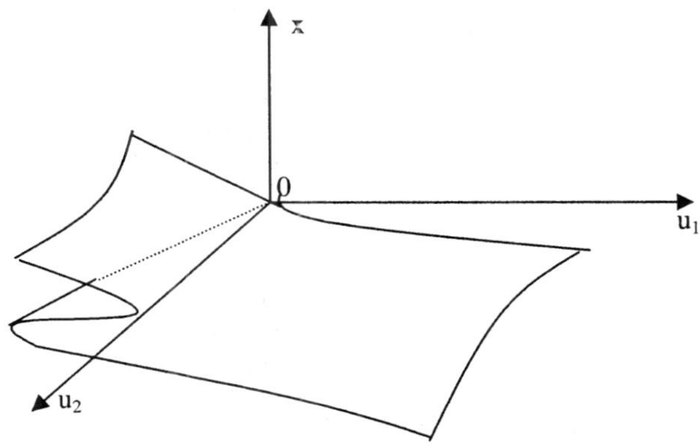

Figure 10. Assembling in the space of x, u_1, u_2.

The so-called *strange attractor* may arise in the case of this catastrophe at the expense of falls of depicted points from edges of the cusp or assembly.

iv) Fusion of four or five critical points--then respective catastrophe is called a *swallow-tail* or *butterfly*.

It should be emphasized that the form of the functions of *P*-types remains the same under variation of their parameters. In other words, the character of the catastrophe is structure-stable. It means that qualitative conclusions based on the structure-stable model remain valid even if parameters of the model are determined inaccurately and may be changed by chance. For applications in biology, this property is especially important.

Let us consider as yet a differentiable function of two real variables and its expansion up to, for example, the third order:

$$f = a_1x + b_1y; f = a_1x + b_1y + a_2x^2 + 2b_2xy + c_2y^2;$$

$$f = a_1x + b_1y + a_2x^2 + 2b_2xy + c_2y^2 + a_3x^3 + 3b_3x^2y + 3c_3xy^2 + d_3y^3. \quad (26)$$

The successive spaces of coefficients $J^1(a_1, b_1)$, $J^2(a_1, b_1, a_2, b_2, c_2)$, $J^3(a_1, ..., c_3, d_3)$,... project canonically: $J^1 \leftarrow J^2 \leftarrow J^3 \leftarrow \cdots$ by the operators p_2, p_3, ... respectively.

To the first order (in J^1) the topological classification of functions f is easy--either both coefficients a_1, b_1 are not zero or $a_1 = b_1 = 0$. In the first case, one of the derivatives $\partial f/\partial x$, $\partial f/\partial y$ is nonzero at the origin. By using the classical theorem on implicit functions, f can be changed locally into a linear function by changing it to curvilinear coordinates in the *Oxy*-plane, so that the form of the function and its destiny are irreversibly fixed no matter what the higher coefficients are. In the second case, when $a_1 = b_1 = 0$, nothing can be said about the topological nature of the function. We say that the origin

$a_1 = b_1 = 0$ represents, in J^1, the bifurcation set K^1, the analogue of the germinal line in embryology. To the second order the points of J^2 projecting by p_2 outside K^1 have their destiny already decided.

The points projecting to $O = K^1 \subset J^1$ have the following classification:

1. If the discriminant of $a_2x^2 + 2b_2xy + c_2y^2$ is nonzero, then (by a classical theorem of Morse) the addition of higher-order terms does not alter the topological type of the function, which remains that of a quadratic form.

2. If $b_2^2 - a_2c_2 = 0$, the topological type is not yet decided. Thus the surface with equation $b_2^2 - a_2c_2 = 0$ defines the bifurcation set K^2 in $J^2 \cap p_2^{-1}$.

The rank of the quadratic form is then used to classify K^2: only the zero form can be considered as totipotent and a form of the rank one, for example, $f = x^2$, is not topologically determined; adjoining a term of third order not divisible by x (e.g., $x^2 + y^3$), gives a form whose topological type cannot be altered by adding terms of higher order; adjoining a term of third order divisible by x gives a form whose topological type is once again not determined (for example, $f = x^2 + 2xy^2$ and $g = x^2 + 2xy^2 + y^4$ are different types, for $g = (x + y^2)^2 = 0$ defines a double curve, whereas $f = 0$ does not). Thus, the general nature of these phenomena is as follows. Corresponding to each order k is an algebraic bifurcation subset K^k, and under the canonical map p_k: $J^k \to J^{k-1}$, we have $p_k(K^k) \subset K^{k-1}$. The sequence of bifurcation sets gives the formal analogue of the germinal line in embryology.

5. On the Respective Software

Dynamical systems with applications using MATLAB are considered in E. Allman, J. Rhodes [8].; and S. Lynch [122]. General optimization methods and software with application, in particular, to biology and medicine can be found in series Optimization Methods & Software, Ed. Yu. Evtushenko, [145]. Software PLUS created by A. Ferreira for modeling of biochemical reactions can be found in E. Voit [197].

Problems of numerical methods and software for modeling of neurons can be found in M. Mascagni, A. Sherman [130]. The techniques of MAPLE, which is both a programming language for mathematical operations and a tool for the visualization of mathematical relations, with applications to biology can be found in M. Kofler [114] and H. Thieme [181]. The techniques of C programming see in W. Press et al. [154].

For many basic problems of computational biology, working programs to solve them in the Perl programming language are contained in R. Dwyer [41]. Software packages for the analysis of microarray data include R, SPLUS, SAS, JMP, DataDesk, Matlab, MINITAB, and STATISTICA see D. Amaratunga, J. Cabrera [9].

See also Appendix, Ch. 16 concerning the software for MM of development and their applications.

Part II:

MM OF DEVELOPMENT

According to the examples of ES and the cell and CAO functioning structure (see Introduction), mathematical models (MM) of development are introduced. These MM are exact by definition and different from each other depending on interpretation and the extent of detailing or aggregation. Some additional MM are introduced concerning the natural balance relations for the cell and CAO. The comparison with some other MM is noted. The problem of estimating input data and the problem of MM completeness are analyzed. Statements of typical problems of modeling are given.

Examples of specific CAO and their MM, including hyper-cycle and DNA packing, dehydration and ATP synthesis, regulatory processes in cells with the respective biological background, pattern formation, and protein folding are considered to show new possibilities and results arising from the novel techniques.

In short, biology and MM of the cell are introduced and analyzed. Application of these models to modeling of the cell cycle and morphogenesis are considered with biological background based on the respective results of Nobel laureates of 2001 and 2002.

MM of the immune network are introduced and investigated in details. As a result, it is shown that MM of the immune network, based on known experimental data turn out to be particular cases of the general MM of development. MM of some other CAO, including stem cells, blood systems cells, are also considered.

Chapter 5

BASE MM

1. Base Minimal MM

The *base minimal MM of CAO* has the form

$$m(t) = \int_0^t \alpha(t,s)\lambda(t,s)y(s)m(s)ds + \Delta m(t),\ 0 \le y, \lambda \le 1,$$

$$c(t) = \int_0^t \beta(t,s)\mu(t,s)z(s)m(s)ds + \Delta c(t),\ 0 \le z, \mu \le 1,$$

$$R(t) = \int_0^t [\lambda(t,s)y(s) + \mu(t,s)z(s)]m(s)ds,\ y(s) + z(s) = 1,$$

$$M(t) = \int_0^t m(s)ds,\ G(t) = M(t) - R(t),$$

$$f^{\wedge}(t) \ge f(t) = m(t) + c(t),\ t \ge t^{\wedge} \ge 0, \qquad (1)$$

where $m(t)$ is the rate of creation of the first kind of *new generalized product* (*resource*) quantity at the time instant t, which provides the fulfillment of the internal functions of CAO, that is, restoration of itself and creation of the second kind product; $y(t)m(t)$ is a share of $m(t)$ for fulfillment of internal functions in the *subsystem* A of restoration and perfection of the system as a whole; $\lambda(t,s)$ is the *intensity* of $y(s)m(s)$ product use at the instant t; $\alpha(t,s)$ is the *efficiency index* for functioning of the subsystem A along the channel $\lambda(t,s)y(s)m(s)$ -- $m(t)$, i.e., the number of units of $m(t)$ created in the unit of time starting from the instant t per one unit of $\lambda(t,s)y(s)m(s)$; $\Delta m(t)$ is an increment of $m(t)$ (may be negative); $c(t)$ is the rate of creation of the second kind of *new generalized product* quantity at the instant t, which provides the realization of the external functions of ES; $z(s)m(s)$, $\mu(t,s)$, $\beta(t,s)$, and $\Delta c(t)$ are similar to ym, λ, α, and Δm, respectively, but for the *subsystem* B of creation of the second kind of product; $R(t)$ is the total quantity of the first kind of product functioning at the instant t; $M(t)$ is the total quantity of the first kind of product to be created during the time $t - 0$; $G(t)$ is the total quantity of the out-of-date or *obsolete product* at the instant t; $f^{\wedge}(t)$ is the rate of the resource inflow from the outside ($m(t)$ and $c(t)$ are measured in the units of $f^{\wedge}(t)$); $t^{\wedge}$ is the starting point for modeling; $[0, t^{\wedge}]$ is the *prehistory* of CAO for which all the functions are given (their value will be noted by the same symbols but with the index "^," e g, $m(t) = m^{\wedge}(t)$, $t \in [0, t^{\wedge}]$). It is obvious that all the relations (1) are faithful representations by definition. The indices α and β depend on m, Δm, c, Δc, λ, μ, y, P, and $f^{\wedge}$ in the general case.

Thus, (1) is the system of nonlinear integro-functional relations, which consists of 7 equalities and 11 inequalities connecting 18 values: m, Δm, c, Δc, α, β, λ, μ, y, z, R, M, G, t, $t^\wedge$, f, $f^\wedge$, 0. All of these values are nonnegative (except maybe Δm and Δc).

We note three special cases of MM (1):

$$\lambda = \lambda(t-s),\ \mu = \mu(t-s); \tag{2}$$

$$\lambda, \mu = 0,\ 0 \le s < a(t);\ = 1,\ a(t) \le s \le t; \tag{3}$$

$$\lambda = 0,\ 0 \le s < a(t);\ = \lambda(t-s),\ a(t) \le s \le t;$$

$$\mu = 0,\ 0 \le s < b(t);\ = \mu(t-s),\ b(t) \le s \le t. \tag{4}$$

Case (2) corresponds to the *stationary process*; case (3) means that the product created before $a(t)$, $a(t) \le t$, is never used at the instant t; case (4) expands the previous cases.

Instead of the relations (1), in particular, we have ($\Delta m, \Delta c = 0$)

$$m(t) = \int_{a(t)}^{t} \alpha(t,s)y(s)m(s)ds,\ R(t) = \int_{a(t)}^{t} m(s)ds\ ; \tag{5}$$

$$c(t) = \int_{a(t)}^{t} \beta(t,s)[1-y(s)]m(s)ds; \tag{6}$$

and, in addition,

$$a'(t) \ge 0,\ t \ge 0. \tag{7}$$

Denoting $R'(t)$ by $r(t)$, we convert (5) to the form

$$m(t) = \int_{a(t)}^{t} \alpha(t,s)y(s)m(s)ds,\ \ a'(t) = [m(t)-r(t)]/m(a(t)). \tag{8}$$

Even in the simplified formulation, MM (1) is reduced to the relations (8), in which, along with the nonlinear integral equation of the unusual form (the lower bound $a(t)$ can be unknown function), we have the *nonlinear functional equation.*

The important case of MM (1), which expands (5)-(8), is

$$m(t) = \int_{a(t)}^{t} \alpha(t,s)y(s)m(s)ds,\ R_1(t) = \int_{a(t)}^{t} y(s)m(s)ds,$$

$$c(t) = \int_{b(t)}^{t} \beta(t,s)[1-y(s)]m(s)ds,\ R_2(t) = \int_{b(t)}^{t} [1-y(s)]m(s)ds,\ R(t) = R_1(t) + R_2(t),$$

$$f(t) = m(t) + c(t),\ 0 \le a(t),\ b(t) \le t,\ t \ge t^\wedge \ge \min\,[a(t^\wedge), b(t^\wedge)] = 0. \tag{9}$$

For MM (5)-(8) the functions α, β, y, and R or α, β, y, and f are usually to be given and the other elements to be determined.

For MM (9) the functions α, β, y, R_1, and R_2 or α, β, y, f, and c are usually to be given and the other elements to be found.

Typical examples to interpret all the values are illustrated in the following:

1. The simplest enzyme-catalyzed reaction

$$\text{E+S } (k_{-1}) \leftrightarrow (k_{-1}) \text{ ES} \rightarrow (k_2) \text{ E+P}, \; e_0 = e+es, \; s_0 = s+es+p, \tag{10}$$

where E, S, ES, and P are enzyme, substrate, enzyme-substrate complex, and product, respectively, k_i are the rate constants for the respective reactions, and e_0 and s_0 the initial values of the concentrations e and s at the instant $t = 0$.

Then m is the concentration of new molecules of ES per unit of time, c is the concentration of new molecules of P per unit of time, A is the subsystem of recreation of ES and B is the subsystem of creation of P, α and β are efficiency of functioning the subsystem A and B, R = [ES] = es, i.e., the total concentration of functioning ES, G is the total concentration of obsolete ES.

It is not difficult to see that for our reaction

$$des/dt = k_1 e \cdot s - (k_{-1} + k_2)es, \; dp/dt = k_2 es. \tag{11}$$

Since $s \gg e$, we have $de/dt = -des/dt < 0$, $ds/dt < 0$, and therefore, setting $f = k_1 e \cdot s$ and $r = k_{-1} + k_2$, we have

$$es(t) = e^{-rt} \int_o^t f(s) e^{rs} ds,$$

$$0 < des/dt = f(t) - re^{-rt} \int_0^t f(s) e^{rs} ds < f(t) e^{-rt} =$$

$$k_1 e \cdot s \exp[-(k_{-1} + k_2)t] < k_1 e_0 \cdot s_0 \exp[-(k_{-1} + k_2)t]. \tag{12}$$

Thus, it is generally accepted that $des/dt = 0$ for $t > x$, $\exp[-(k_{-1} + k_2)x]$ is sufficiently small.

In the case of MM (5)-(6), we have

$$m(t) = k_1 e \cdot s, \; R(\text{t}) = es = e_0 \cdot s/(K_m + s) = \int_{a(t)}^t m(s) ds,$$

$$K_m = (k_{-1} + k_2)/k_1, \; dR/dt = m(t) - m(a) da/dt = 0,$$

$$dG(t)/dt = m(a) da/dt = m(t) = (k_{-1} + k_2)R,$$

$$m(t) = m(a), \; da/dt = 0, \; a = t - t^\wedge, \; t^\wedge = R/m = 1/(k_{-1} + k_2),$$

$$c(\text{t}) = d\text{p}/dt = k_2 R, \; \alpha(t,s) y(s) = k_{-1} + k_2,$$

$$\beta(t,s)[1 - y(s)] = k_2, \; t - t^\wedge, \; t^\wedge > x. \tag{13}$$

Deriving from (10)-(12) in the vicinity of 0, we have

$$m = k_1 e \cdot s,\ R(t) = k_1 e_0 \cdot s_0(1 - e^{-rt})/r,\ dR/dt = m - k_1 e_0 \cdot s_0(1 - e^{-rt}),$$

$$m(a)da/dt = k_1 e_0 \cdot s_0\, da/dt = k_1 e_0 \cdot s_0(1 - e^{-rt}),\ a(t) = t + e^{-rt}/r - 1/r,\ r = k_{-1} + k_2. \qquad (14)$$

The models (13)-(14) are not unique possible. If we put

$$R(t) = \int_o^t \lambda(t,s)m(s)ds,\ dR/dt = m(t) + \int_o^t \lambda'_t(t,s)m(s)ds$$

$$= m(t) - (k_{-1} + k_2)\int_o^t \lambda(t,s)m(s)ds, \qquad (15)$$

then we easily find that intensity

$$\lambda(t,s) = \exp\,[-(k_{-1} + k_2)(t-s)]. \qquad (16)$$

Phenomenological description of R (15)-(16) is as good as (13)-(14), but we cannot conclude what kind of a model corresponds better to real enzyme dynamics without experiments.

In general, MM of enzyme reactions are focused on the so-called quasi-stationary processes exterior to the vicinity of $t = 0$, when we can put $dR/dt = 0$ and reduce matter to the solution of linear algebraic systems and/or simple systems of differential equations with respect to substrates and products. In this regard, real enzyme dynamics remain inaccessible.

Probably, the models of (1)-types in the vicinity of $t = 0$ can help to investigate and to understand the structure and functions of substrate-enzyme complexes.

Generally speaking, under modeling metabolic and catabolic pathway reactions, we usually have enzyme-power-substrate- or (E, P, S)-complexes (analogies to work places in economy), and also e, p, and s mean respectively the concentrations of the enzyme, power, and substrate, and sometimes even (E, P, S; I)-complexes, where I means the information and/or control molecules that are necessary for the reactions.

2. The cell mitochondria are self-reproducing. Then m is the number of new mitochondria per unit of time, c is the concentration of new molecules of ATP and other energy-preserving molecules per unit of time, A is the subsystem of recreations of mitochondria, B is the subsystem creation of mitochondria power, R is the total number of functioning mitochondria in a cell, G is the total number of obsolete and/or preserving in reserve mitochondria, and f is the flux of substrate into mitochondria.

3. The cell can self-reproduce with a known target function. Then m is the number of new cells per unit of time, c is the number of the unit (molecules, ions, organelles, cells, etc.) for the realization of the target function, A is the subsystem of proliferation or differentiation of cells, α is the rate of proliferation or differentiation, B is the subsystem of creation of the product for the realization of the target and β is the efficiency of this realization, R is the total number of proliferating or differentiating cells, G is the total number of the remaining cells, and f is the external flux of the general resources.

2. More General MM

It is not hard to introduce different additions and generalizations of MM (1) (V. Ivanov [92]).

The *n-product MM* of CAO, $n > 2$, can be formally written almost in the same form as (1) under the condition that m and c (and Δm, Δc) are the vector functions: $m = (m_1, \ldots, m_r)$, $c = (c_1, \ldots, c_s)$, $r + s = n$; y and z are the column matrices and $\alpha\lambda$ and $\beta\mu$ are appropriate matrix functions (where the inequalities for the vectors and matrices signify the inequalities of the same name for all of their appropriate components).

However, the relations for f, R, and M should be replaced by

$$f(t) = \sum_{i=1}^{r} m_i(t) + \sum_{i=1}^{s} c_i(t),\ r + s = n,$$

$$R(t) = \sum_{i=1}^{r} R_i(t),\ M(t) = \sum_{i=1}^{r} M_i(t),$$

$$G_i(t) = M_i(t) - R_i(t),\ M_i(t) = \int_o^t m_i(s)ds,\ R_i(t) = \int_o^t [\sum_{j=1}^{r} \lambda_{ij}(t,u)y_j(u)$$

$$+ \sum_{j=1}^{s} \mu_{ij}(t,u)z_j(u)]m_i(u)du. \tag{17}$$

We can consider also the more general case when y and z are nonnegative matrices with the restriction

$$\sum_{j=1}^{r} y_{ij}(t) + \sum_{j=1}^{s} z_{ij}(t) = 1,\ i = 1, \ldots, r. \tag{18}$$

Stochastic MM similar to (1) can be obtained by considering α, β, λ, μ, and y as the functions of a random factor ω.

Discrete MM can be represented in the same form if the integrals in (1) are understood in the sense of Stieltjes.

The *MM*, regarding the *spatial structure*, have the form

$$m(t,X) = \int_{a(t,X)}^{t} \int_0^X \alpha(t,X,s,Y)m(^1)(s,Y)dYds,$$

$$M(t,X) = \int_0^t m(s,X)ds,\ M(t) = \int_V M(t,X)dX,$$

$$c(t,X) = \int_{b(t,X)}^{t} \int_0^X \beta(t,X,s,Y)m(^2)(s,Y)dYds,$$

$$C(t,X) = \int_{t^\wedge}^{t} c(s,X)ds,\ C(t) = \int_V C(t,X)dX,$$

$$R(t,X) = \int\int m(^1)(s,Y)dYds + \int\int m(^2)(s,Y)dYds,$$

$$R(t) = \int_V R(t,X)dX,\ m(^1)(t,X) + m(^2)(t,X) = m(t,X),\ G(t,X) = M(t,X) - R(t,X) > 0,$$

$$G(t) = M(t) - R(t), f(t) = \int_V m(t,X)dX + \int_V c(t,X)dX,\ F(t) = \int_0^t f(s)ds,$$

$$0 \le a, b \le t,\ t \ge t\hat{}\ \ge 0,\ O,\ X \in V \subset R^3, \tag{19}$$

where V is the domain that CAO occupies in R^3: $X = (x_1, x_2, x_3)$; $m(t,X)$ is the new CAO part or internal product (resource) quantity at the instant t in its point X per unit of time (this means that X is simultaneous notation for the part $m(t,X)$ of X-type and by the same token various parts of CAO are distributed in space); $M(t,X)$ is the total CAO internal product in its point X (created during the time t); $M(t)$ is the total product created CAO; $R(t)$ is the total functioning product of CAO; $G(t)$ is the total obsolete product of CAO; $c(t,X)$ is the new resource quantity in the unit of time in the point X that is necessary for fulfillment of the external functions of CAO, $\alpha(t,X,s,Y)$ is the *index of efficiency* of CAO functioning along the channel $m(^1)(s,Y)$ -- $m(t,X)$, etc.

In general, we can consider that $m(^1)$, $m(^2)$, m, a, b, c, R, and f are the vector functions of t,X and α and β are the appropriate matrices.

Then the first equality in (19) means

$$m_i(t,X) = \sum_{j=1}^{n} \int_{a,i(t,X)}^{t} \int_0^X \alpha_{ij}(t,X,s,Y)m(^1)_j(s,Y)dYds,\ i = 1, \dots, n; \tag{20}$$

the sense of the other equalities in (19) is similar to (20); inequalities for vectors mean, as usually, the appropriate inequalities for the respective components of these vectors.

Setting, in particular, $a,i(t,X) = 0$ and

$$\alpha_{ij}(t,X,s,Y)m(^1)_j(s,Y) = [F_{ij}(M_1,\dots,M_n) + D_{ij}(M_1,\dots,M_n)\partial^2 M_i/\partial X^2]m_j(s,X)/M_j(t,X),$$

$$i = 1, \dots, n, \tag{21}$$

where $M_1, \dots, M_n$ depend on t,X, we result in the system of the nonlinear equations with partial derivatives:

$$m_i(t,X) = \partial M_i/\partial t = F_i(M_1,\dots,M_n) + \sum_{j=1}^{n} D_{ij}(M_1,\dots,M_n)\partial^2 M_i/\partial X^2],\ i = 1, \dots, n, \tag{22}$$

which describes many well known, self-oscillating processes (G. Ivanitsky et al. [105]; Y. Romanovsky et al. [162]; A. Zhabotinsiy [215]), various dissipative structures [162], and also very complicated processes of cells' proliferation and differentiation (B. Sendov [168]; see also Ch. 1 and 7).

Experience of investigation and application of more general models of the form (1) and (2) show that for explanation of cells' proliferation and differentiation, there are new possibilities that, in particular, do not connect with the mechanism of stepwise formation of the model elements.

It will be shown in Ch. 7 and 14 that in a certain case of the oscillating functions α and β-types, the equations (19)-type themselves and different optimization problems for the class of the models under consideration have non-unique solutions.

Assuming, that cells' development pursues the solution of determined optimization problems (e.g., cells' duplication maximizes reliability of their functioning) and that cells' differentiation maximizes efficiency of fulfillment of given functions, we conclude that there is the possibility of explaining cells' duplication and differentiation by the non-uniqueness of solutions of the appropriate optimization problems. Starting from the prehistory, for the first cell we can obtain (under decrease of functions of α and β-types stimulated by the internal and/or external factors through the membrane of the cell) two different solutions of model (19), which can be interpreted as the models of development and growth for two new cells (under the increase of α and β values), whose prehistory (and new decrease of α and β) results in the creation of models for four new cells, etc. Thus created cells can probably be used, with the help of the respective spontaneous processes (see Ch. 1) and the program encoded in DNA of the cells, for the construction of different tissues, organs, subsystems of an organism, and an organism as a whole.

For the spatial MM, we can also consider field-kind models of our class.

Let V be the domain in R^3 for isolated CAO and $R^3\backslash V$ be the complement of V to R^3. We accept that the external bio-field with respect to CAO will be decided if the vector function

$$c(t,x) = \{c_1(t,X), c_2(t,X), c_3(t,X)\}, X \in R^3\backslash V, \quad (23)$$

is given.

Here t is the time and (x_1, x_2, x_3) are Cartesian coordinates of the point X; c_1 is the rate of creation along t,X (i.e., along t, x_1, x_2, and x_3) of the (generalized) *new material product* of t,X-type in the point X at the instant t; c_2 is the rate of creation along t,X of the (generalized) *new energy product* of t,X-type in the point X at the instant t; c_3 is the rate of creation along t,X of the *new information product* of t,X-type in the point X at the instant t.

As the base *MM of a bio-field of CAO*, we can naturally assume the following system:

$$m(t,X) = \int_{a(t,X)}^{t} \int_{0}^{X} \alpha(t,X,s,Y)m(^{1})(s,Y)dYds,\ X \in V;$$

$$c(t,X) = \int_{b(t,X)}^{t} \int_{0}^{X} \beta(t,X,s,Y)m(^{2})(s,Y)dYds,\ X \in \mathrm{R}^3;$$

$$R(t,X) = \int\int m(s,Y)dYds, f(t,X) = m(t,X) + c(\mathrm{t,X}),\ t \geq t\hat{}>0,$$

$$m(^{1})+m(^{2}) = m,\ m(t,X) = \{m_1(t,X), m_2(t,X), m_3(t,X)\},\ X \in V, \tag{24}$$

where $m_i(t,X)$ are the rates of creation along t,X of the (generalized) internal products of t,X-type, respectively bio-mass (i = 1), energy (i = 2), and information (i = 3) in the X at the instant t; α and β are matrices of the order three whose components are indices of efficiencies of functioning of a bio-object along the corresponding channels; the definitions of all the other values are similar to the previous ones.

It is natural to call V with (24) as *MM of bio-field of the internal sphere of CAO* or the *MM of internal bio-field*, and $\mathrm{R}^3\backslash\mathrm{V}$ with (23), (24) as the *MM of external bio-field*.

The conditions under which the significant values of $c_2(t,X)$ can arise for the values of X outside V and far removed from the boundary of V are shown in [7]. Of course, the notion "bio-fields" means the usual physical fields for biological objects.

3. Additional MM

We consider, in addition, the *balance equation* in the form

$$f(t) = k(f^*,t)c(t),\ 1 \leq k(f^*,t), \tag{25}$$

where the coefficient $k = k(t) = k(f^*) = k(f^*,t)$ depends on the qualitative structure $f^* = f^*(t)$ of the external resources $f\hat{}(t) = f(t)$. The equation (25) can be obtained from the laws of conservation and the conversion of matter and energy [118, 180] (see also Ch. 1).

Equation (25) means that for CAO, usually c, f, R, α, and β are to be given and y, z, m, and λ, μ (or $a(t)$) to be found. If the equation in (25) is indeterminate, then for CAO, usually f, R, α, β are to be given and y, z, m, c, and λ, μ (or $a(t)$, $b(t)$) to be found.

We can also introduce the *equation* of *homeostasis* of the kind

$$R'(t) = 0 \tag{26}$$

For MM (1), in order to receive the equation of homeostasis, we should differentiate the respective representation of $R(t)$.

For MM (5), it has the form

$$m(t) - m((a(t))a'(t) = 0. \qquad (26')$$

In the case of two CAO interactions according to Fig. 2 (see Introduction), one can obtain *MM of interactions* if in addition to the equations of the (1)--(12)-types for each CAOi, i = 1, 2, the following bonds have to be added:

$$m,i = m.i' + m,i'', c,i = c,i' + c,i'', G,i = G,i' + G,i'', f = f,1 + f,2,$$

$$\Delta m,i = m,i{+}1'' - m,i'', \Delta c,i = c,i{+}1'' - c,i'', f,i = f,i' + G,i{+}1'', i = 1, 2, \qquad (27)$$

where i+1 should be comprehended as addition by module 2. Also, the respective changes should be done in each of MM for each CAO. In the case of the interaction of two CAO according to Fig. 3 (Introduction), we have the same MM, regarding $m,1'' = c,1'' = 0$. More general MM of two CAO interactions can be found in K. Girlin, V. Ivanov [57].

4. On Comparison with Well-Known MM

We have dwelled on the connection of MM in question with the classical MM and some well-known ones in the fields of macro-economy and population biology (see Introduction, s. 5.2. and Ch. 4, s. 2). It is obvious that delay differential equations (DDE) (see Ch. 4, s. 1) are a particular case of MM in question. We dwell on the comparison of our MM with DDE later, in Chapter 14.

Most similar to our class of MM is the integral-type of the models of cell-cell interactions in the immune system (see Ch. 8). We dwell on the comparison between these two types of MM in Ch. 8 after introducing detailed models of the immune network. MM of the immune network with application to the problems of AIDS were reported to the Conference on Mathematical Models in Medical and Health Sciences, Nashville, TN, May 1997. M. Novak, a Shanks lecturer of Oxford University, emphasized that one of the important problems is the creation of the evolutionary theory of a disease progression.

In this regard, we would like to note that the evolutionary theory and its applications, in particular, to modeling of CAO and their development in the norm and pathology are the main subjects of this book.

5. Methods for Estimation of MM Input Data

The estimation of efficiency indices of the α, β-types is the most difficult. Let us consider several MM of α and some methods of the estimate of parameters of these MM.

The construction will be similar for β.

We divide the given ES prehistory [0, t^] into two segments: [0, t_1] and [t_1, t_2], $0 < t_1 < t_2 \le t$^. If α has the form

$$\alpha = \alpha(t,s) = \alpha(s)\exp\,[-c_\alpha(t-s)],\ c_\alpha \ge 0, \tag{28}$$

where c is a parameter considered as given for the prehistory [0, t_1], then $\alpha(s)$ can be found on [0, t_1] from the relations

$$m\text{^}(t) = m(0) + \int_{a(t)}^{t} \alpha(s)y(s)m\text{^}(s)ds = m(t)\exp\,(c_\alpha t),\ R(t) = \int_{a(t)}^{t} m(s)ds, \tag{29}$$

under the condition that m, y, and R are given. Indeed, one can find $a(t)$ from the second equation in (29) which is equivalent to $R'(t) = m(t) - m(a(t))a'(t)$, $a'(0) = 0$.

Let M be the anti-derivative for m. Then $R(t) = M(t) - M(a)$ and

$$a(t) = M^{-1}[M(t) - R(t)],\ a'(0) = [m(0) - R'(0)]/m(0), \tag{30}$$

where M^{-1} is the reciprocal function of M (which always exists since m is nonnegative). After this, one can find α from the first equation in (29):

$$\alpha(t) = [m\text{^}'(t)+\alpha(a(t))y(a(t))m\text{^}(a(t))a'(t)]/[y(t)m((t)],$$

$$\alpha(0) = m\text{^}'(0)/[y(0)P'(0)]. \tag{31}$$

We prolong $\alpha(t)$ on the segment [t_1, t_2] with the help of one of the methods of prediction and find $m(t)$ (and $a(t)$) under the given $y(t)$ from the system of equations (29). This system with respect to $m(t)$ and $a(t)$ has the unique solution under very general conditions (it will be shown in Appendix, Ch. 14).

Let us denote this solution by $m^*(t)$. We can select the parameter c_α with the help of the method of simple scanning from, for example, the condition

$$I_1 = \int [m(t) - m^*(t)]^2 dt\ (t \in [t_1, t_2]) = \min. \tag{32}$$

After this, we prolong $\alpha(t)$ on the given future segment [t^, T].

In a similar way we can select, for example, three parameters, $c1$, $c2$ and $c3$, if α has the form

$$\alpha = \alpha(s)\exp\,[-c1(t-s)-c2(t-s)^2-c3(t-s)^3],\ c1, c2, c3 \ge 0. \tag{33}$$

If α has the form of the so-called *degenerate kernel*:

$$\alpha = \sum_{k=1}^{n} \phi_k(t)\psi_k(s), \tag{34}$$

we can find all functions ϕ_k and ψ_k from the condition (see R. Baglay, K. Smirnov [13]):

$$I_2 = \int [m(t) - \sum_{k=1}^{n} \phi_k(t)\int_{a(t)}^{t} \psi_k(s)y(s)ds]^2dt \ \ (dt \in [0, t_2]) = \min, \qquad (35)$$

where $a(t)$ is given by the formula (30). In the particular case of (34), when $\phi_k(t) = c_k\chi_k(t)$, c_k are parameters, and the functions χ_k and ψ_k are given, we can find all the parameters c_k from (35) by the well-known *method of least squares* (V. Ivanov [89]).

Let $\alpha(t,s) = \alpha(t\text{-}s)$ and let R, m, and y be given in the prehistory. To find $\alpha(t\text{-}s)$, we can solve the Volterra equation of the first kind:

$$m(t) = \int_{t-a(t)}^{0} \alpha(s)y(t-s)m(t-s)ds. \qquad (36)$$

In the general case it is required to find the function α from the condition:

$$\min \left| 2m - \int_{a(t)}^{t} \alpha ymds \right| \ (\alpha \in V) \qquad (37)$$

where V is a reasonable set given, $|\bullet|$ is a sign of the measure of $\bullet$.

After estimating the functions α and β-types on their prehistory, we can obtain their estimates on the future period by the well-known *prediction methods* ([10 46]). Dependence of α, β on internal and external factors can be found by *data processing methods* based, for example, on K. Astrom, P. Eykhoff [12] or R. Bethea et al. [24].

Until now we have a deal with the *problem of identification* for *CAO operating*.

For *CAO design*, all elements of the CAO MM have to be given and the CAO itself has to be designed. This can be done by analogy with the general theory of synthesis or with the design of dynamic systems. That is, the CAO MM have to be approximated by MM with sufficient accuracy such that they can be realized physically (V. Ivanov et al. [93]) and/or biologically (J. Rifkin [160]).

6. On the Problem of MM Completeness

We still need to improve our approach. By a theoretical plan, the most important problem is that of completing the given class of MM. The *problem of MM completeness* is the constriction of an additional reasonable set of relations such that it would be possible to find all the elements of CAO in the future by knowing the prehistory of CAO and predicting only separate elements or parameters of the exogenous nature.

In other words, one of our objectives has to do with the so-called "light box." This completion can be based at least on the following approaches:

i) system *de- and aggregation methods*;
ii) *detailing the structure of CAO*;
iii) *using extreme concepts*.

6.1. METHODS OF CAO DE- AND AGGREGATION

The function of efficiency $\alpha(t,s)$ is usually increasing of s and decreasing of t. Anyone can accept many probable reasons for increasing and decreasing of efficiency for any particular CAO. The simpler the system the more reliable the reasons.

Let

$$m_i(t) = \int_{a(t)}^{t} \sum_{j=1}^{n} \alpha_{ij}(t,s)y_{ij}(s)m_j(s)ds, \ i = 1, \ldots, n;$$

$$m(t) = \sum_{i=1}^{n} m_i(t) = \int_{a(t)}^{t} \alpha(t,s)y(s)m(s)ds. \tag{38}$$

Then

$$\alpha(t,s) = [\sum_{i,j=1}^{n} \alpha_{ij}(t,s)y_{ij}(s)m_j(s)]/[y(s)\sum_{i=1}^{n} m_i(s)]. \tag{39}$$

An example of elementary indices of efficiency is

$$\alpha_{ij}(t,s) = \alpha_{ij}\hat{}\exp[-d_i(t-s)], \ \alpha_{ij}\hat{}, \ d_i > 0, \ i, j = 1, \ldots, n, \tag{40}$$

where $\alpha_{ij}\hat{}$, and d_i are constants. Then

$$m_i'(t) + d_i m_i(t) = \sum_{j=1}^{n} \alpha_{ij}\hat{}\{y_{ij}(t)m_j(t) - y_{ij}(a(t))\exp[-d_i(t-a(t))]a'(t)\}, \ i = 1, \ldots, n. \tag{41}$$

Thus, the matter reduces to the solution of the linear system of ODE with the prehistory. In particular, if $y_{ij}(t) = y_{ij}$, where y_{ij} are constants, then the matter reduces to the solution of the linear stationary system of ODE with its prehistory.

Another example of elementary indices of efficiency is

$$\alpha_{ij}(t,s) = \alpha_{ij}\hat{}/y_{ij}(s)s/t, \ \alpha_{ij}\hat{} > 0, \ i, j = 1, \ldots, n, \ t > 0. \tag{42}$$

Then the matter reduces to the solution of the linear stationary system with the prehistory

$$z_i'(t) = \sum_{j=1}^{n} \alpha_{ij}\hat{}\{z_j(t) - z_j(a(t))a'(t)\}, \ z_j(t) = tm_j(t), \ i = 1, \ldots, n. \tag{43}$$

6.2. DETAILING THE CAO STRUCTURE

6.2.1. *Case of self-development of the indices* α- *and* β-*types*

Another *method* for the construction of MM of the functions α and β is *detailing the structure of CAO*. Decreasing of α and β of t depends on many factors in the theory of deterioration (aging). As reasonable simple hypotheses, we accept

$$\alpha_1(t,s) = \alpha_1(s)\exp\,[-c_1(t-s)],\ \beta_1(t,s) = \beta_1(s)\exp\,[-d_1(t-s)]; \tag{44}$$

or

$$\alpha_2(t,s) = \alpha_2(t-s)\exp\,(-c_2 t),\ \beta_2(t,s) = \beta_2(t-s)\exp\,(-d_2 t), \tag{45}$$

where c_i and d_i are given constants, $c_i, d_i \geq 0$.

Reasoning about the structure of CAO as above (see Introduction) results in

$$\alpha_\mu^{\nu}(t) = \int_{a(t)}^{t} \alpha_\mu(t,s)x(s)m(s)ds,\ \nu = 0,\ 1(\equiv');$$

$$m(t) = \int_{a(t)}^{t} \alpha_\mu(t,s)y(s)m(s)ds,\ M(t) = \int_0^t m(s)ds;$$

$$\beta_\kappa(t) = \int_{a(t)}^{t} \alpha_\mu(t,s)z(s)m(s)ds,\ \mu,\ \kappa = 1,\ 2;\ c(t) = \int_{a(t)}^{t} \beta_\kappa(t,s)[1-x(s)-y(s)-z(s)]m(s)ds;$$

$$R(t) = \int_{a(t)}^{t} m(s)ds,\ G(t) = \int_0^{a(t)} m(s)ds = M(t) - R(t);$$

$$f(t) = \alpha_\mu^{\nu}(t) + m(t) + \beta_\kappa(t) + c(t);\ 0 \leq x,\ y,\ z,\ 1-x-y-z \leq 1;$$

$$0 \leq a(t) \leq t,\ T \geq t \geq t\hat{} \geq 0, \tag{46}$$

where $\alpha_\mu^{\nu}(t)$ is νth derivative of $\alpha_\mu(t)$, $\alpha_\mu^0 = \alpha_\mu$, $\alpha_\mu(t\hat{}) = \alpha_\mu\hat{}$, $x(s)m(s)$--$\alpha_\mu^{\nu}(t)$ and $z(s)m(s)$--$\beta_\kappa(t)$ are the channels of creation of the new technology quantities in the respective subsystems (for CAO they are likely parts of *Genome* – *G* with regard to certain natural laws), and the rest of the notations were previously noted. Clearly $\alpha_\mu(t)$ is always increasing in the case when $\nu = 1$, since $\alpha_\mu'(t) > 0$.

For CAO, *x, y, z*, and *G* (or *m*) are usually given. The distributions *x*, *y*, and *z* can be usually found from the solution of some extreme problems. For CAO, we may probably count on the fact that those extreme problems have been solved by nature during CAO evolution.

Note that we should eliminate the case $\mu = 1$, $\nu = 0$ since it seems senseless. Indeed, we have in this case where *x* and *y* are positive constants and $m > q > 0$, $R = \int_{a(t)}^{t} m(s)ds > xd > 1$,

$$y/x\alpha(t) = m(t) = \int_{a(t)}^{t} xm^2(s)ds > qxd,\ q(xd)^2, \ldots,\ q(xd)^n, \ldots \tag{47}$$

and where $m < q\hat{}$, $R < xd < 1$,

$$m(t) = \int_{a(t)}^{t} xm^2(s)ds < q\hat{}xd,\ q\hat{}(xd)^2, \ldots,\ q\hat{}(xd)^n, \ldots. \tag{48}$$

This means that our equations are inconsistent in this case.

The natural generalization of MM (46) is

$$\alpha_\mu^{\nu}(t) = \int_{a,1(t)}^{t} \alpha_\mu(t,s)x(s)m(s)ds,\ \nu = 0,\ 1;\ m(t) = \int_{a,2(t)}^{t} \alpha_\mu(t,s)y(s)m(s)ds,$$

$$M(t) = \int_0^t m(s)ds;\ \beta_\kappa(t) = \int_{a,3(t)}^{t} \alpha_\mu(t,s)z(s)m(s)ds,\ \mu,\ \kappa = 1,\ 2;$$

$$c(t) = \int_{a,4(t)}^{t} \beta_\kappa(t,s)[1-x(s)-y(s)-z(s)]m(s)ds;\ R_i(t) = \int_{a,i(t)}^{t} m_i(s)ds,$$

$$G_i = \int_0^{a,i(t)} m_i(s)ds,\ i = 1, 2, 3, 4;\ m_1 = xm,\ m_2 = ym,\ m_3 = zm,\ m_4 = (1\text{-}x\text{-}y\text{-}z)m;$$

$$R(t) = R_1(t) + R_2(t) + R_3(t) + R_4(t), G(t) = M(t)-R(t);\ f(t) = \alpha_\mu^{\nu}(t) + m(t) + \beta_\kappa(t) + c(t);$$

$$0 \le x,\ y,\ z, 1-x-y-z \le 1;\ 0 \le a,i(t) \le t,\ i = 1, 2, 3, 4;\ T \ge t \ge t^\wedge \ge 0, \qquad (49)$$

For MM (49) x, y, z, and R_i (or m_i) are given.

We can generalize the models (46) and (49) in a way similar to (1), considering all the elements in (46) and (49) as appropriate function matrices (1 is the unit matrix), interpreting integrals in the sense of Stieltjes, and so on.

6.2.2. *Case of the additional "executive" subsystem* (*E*)

Let us consider the case when the *executive subsystem* (*E*) is introduced in addition to (49). Then

$$\alpha_\mu^{\nu}(t) = \int_{a,1(t)}^{t} \alpha_\mu(t,s)x(s)m(s)ds,\ \nu = 0,\ 1;\ m(t) = \int_{a,2(t)}^{t} \alpha_\mu(t,s)y(s)m(s)ds,$$

$$M(t) = \int_0^t m(s)ds;\ \beta_\kappa(t) = \int_{a,3(t)}^{t} \alpha_\mu(t,s)z(s)m(s)ds,\ \mu,\ \kappa = 1,\ 2;$$

$$\gamma_\kappa(t) = \int_{a,4(t)}^{t} \alpha_\mu(t,s)e(s)m(s)ds,\ \gamma_1(t,s) = \gamma_1(s)\exp\,(e_1(t\text{-}s)),$$

$$\gamma_2(t,s) = \gamma_2(t-s)\exp\,(e_2 t);\ f(t) = \alpha_\mu^{\nu}(t) + m(t) + \beta_\kappa(t) + \gamma_\kappa(t) +$$

$$\mathrm{cost}(f) + c(t) = \int_{a,5(t)}^{t} \gamma_\kappa(t,s)g(s)m(s)ds;$$

$$c(t) = \int_{a,6(t)}^{t} \beta_\kappa(t,s)[1 - x(s) - y(s) - z(s) - e(s) - g(s)]m(s)ds;$$

$$R_i(t) = \int_{a,i(t)}^{t} m_i(s)ds,\ G_i = \int_{a,i(t)}^{0} m_i(s)ds,\ i = 1, \ldots, 6;$$

$$m_1 = xm,\ m_2 = ym,\ m_3 = zm,\ m_4 = em,\ m_5 = gm,$$

$$m_6 = (1\text{-}x\text{-}y\text{-}z\text{-}e\text{-}g)m;\ R(t) = R_1(t)+R_2(t)+R_3(t)+R_4(t)+R_5(t)+R_6(t),$$

$$G(t) = M(t) - R(t);\ 0 \le x, y, z, e, g, 1-x-y-z-e-g \le 1;\ \ 0 \le a{,}i(t) \le t,$$

$$i = 1, \ldots, 6;\ T \ge t \ge t^{\wedge} \ge 0, \tag{50}$$

where cost(f) is the "cost" or the necessary resource for f extraction, which should be naturally less than f itself; $\gamma_\kappa(t,s)$ is the new technology for the subsystem E; and the remaining notations have the same definitions as previously stated.

6.2.3. *Case of the additional "recreation" subsystem* (*R*)
In this case

$$\alpha_\mu{}^{\nu}(t) = \int_{a,1(t)}^{t} \alpha_\mu(t,s)x(s)m(s)ds,\ \nu = 0, 1;\ m(t) = \int_{a,2(t)}^{t} \alpha_\mu(t,s)y(s)m(s)ds,$$

$$M(t) = \int_{0}^{t} m(s)ds;\ \beta_\kappa(t) = \int_{a,3(t)}^{t} \alpha_\mu(t,s)z(s)m(s)ds,\ \mu, \kappa = 1, 2;$$

$$\gamma_\kappa(t) = \int_{a,4(t)}^{t} \alpha_\mu(t,s)e(s)m(s)ds,\ \delta_\kappa(t) = \int_{a,5(t)}^{t} \alpha_\mu(t,s)r(s)m(s)ds,$$

$$\delta_1(t,s) = \delta_1(s)\exp\,(r_1(t-s)),\ \delta_2(t,s) = \delta_2(t-s)\exp\,(r_2 t);\ \ N(t) = \int_{a,6(t)}^{t} \delta_\kappa(t,s)w(s)m(s)ds;$$

$$f(t) = \alpha_\mu{}^{\nu}(t) + m(t) + \beta_\kappa(t) + \gamma_\kappa(t) + \delta_\kappa(t) + R(t) + \text{cost}(f) + c(t) = \int_{a,7(t)}^{t} \gamma_\kappa(t,s)g(s)m(s)ds;$$

$$c(t) = \int_{a,8(t)}^{t} \beta_\kappa(t,s)[1-x(s) - y(s) - z(s) - e(s) - g(s) - r(s) - w(s)]m(s)ds,\ \mu, \kappa = 1, 2;$$

$$R_i(t) = \int_{a,i(t)}^{t} m_i(s)ds,\ G_i = \int_{a,i(t)}^{0} m_i(s)ds,\ i = 1, \ldots, 8;$$

$$m_1 = xm,\ m_2 = ym,\ m_3 = zm,\ m_4 = em,\ m_5 = gm,$$

$$m_6 = rm,\ m_7 = wm,\ m_8 = (1-x-y-z-e-g-r-w)m;$$

$$R(t) = R_1(t) + R_2(t) + R_3(t) + R_4(t) + R_5(t) +$$

$$R_6(t) + R_7(t) + R_8(t),\ G(t) = M(t) - R(t);$$

$$0 \le x, y, z, e, g, r, w, 1-x-y-z-e-g-r-w \le 1;$$

$$0 \le a{,}i(t) \le t,\ \ i = 1, \ldots, 8;\ T \ge t \ge t^{\wedge} \ge 0, \tag{51}$$

where $N(t)$ is the amount of a generalized product created in the subsystem R, $\delta_\kappa(t,s)$ is the new technology for the subsystem R, and the remaining notations have the same definitions as previously stated.

6.2.4. *Further factorization of the indices* α- *and* β-*types*

The idea of *factorization* of the matrix $K(t,s)$ of the pulse transition functions (compare (1) and (4), Introduction) can also be strengthened.

First, we should distinguish the stationary factor $\alpha^*(t-s)$ of the function $\alpha(t,s)$-type and the non-stationary factor $\alpha(t,s)/\alpha^*(t-s)$. The latter, in many important cases, can be represented in the form

$$\alpha(t,s)/\alpha^*(t-s) = u(s)v(t)v^{-1}(s), \tag{52}$$

where $u(s)$ is a control factor corresponding to the non-stationary amplified or relaxed unit, $v(t)v^{-1}(s)$ is another non-stationary control factor signifying the easiest way for changing values of CAO internal element α and m without changing CAO structure.

The last statement follows from the fact that after substituting (52) for (1) or for (49)-(51) and replacing m by vm, we will actually have the same models.

On the whole, we have the following *factorization*

$$K(t,s) = \alpha(t,s)\lambda(t,s)\alpha^*(t-s)u(s)v(t)v^{-1}(s)y(s), \tag{53}$$

where $\alpha(t,s)$ means remained non-stationary factors, including the delay unit, non-stationary sluggishness, etc.

We hope that typically in the case of many CAO we can put $\alpha(t,s) = 1$ in (53).

It should be noted that in the cases of various organs and whole organisms, the factor $u(t)$ is probably connected with a reaction of the endocrine subsystem of an organism and realized with the help of hormones, the factor $v(t)v^{-1}(s)$ is connected with the nervous subsystem and realized by different kind of nerves, and the factor $\alpha^*(t-s)$ is connected with the *subsystem genome* of the organism and realized with the help of the code in DNA.

6.2.5. *On introduction of dimensionless relations*

For simplicity above and below, the *dimensionless equations* can be considered. For this, it is sufficient that any element $x(t)$ of the equations has been rewritten in the form

$$x(t) = x(t[t]/[t])[x]/[x] = x_1(t_1)[x],\ t_1 = t/[t], \tag{54}$$

where $[x]$ and $[t]$ are dimensions of x and t, and x_1 is the respective dimensionless element. After this, all dimensions and the subscript $_1$ can be omitted.

6.3. ON USING EXTREME CONCEPTS

Let us return to the models (5)-(7) and consider some *extreme concepts*. We suppose that $\alpha(t,s)$ has to maximize $m(t)$ under given $R(t)$, constants a, c, and also

$$R''(t) > 0,\ R''(t) \geq 0;\ 0 \leq a(t) \leq at,\ a < 1;\ \alpha(t,s) \leq \alpha(s)\exp[-c(t-s)],$$

$$0 \leq y(t) \leq 1, \tag{55}$$

Then the desired $y(t)$, $a(t)$, $m(t)$, and $\alpha(t,s)$ are

$$y(t) = 1,\ a(t) = at,\ m(t) = \sum R'(a^i t)a^i\ (i = 0, 1, 2, \ldots)$$

$$\alpha(t,s) = \exp(-ct)\sum [R''(a^{i+j}s)a^{2i+j} + cR'(a^{i+j}s)a^{i+j}]\exp(ca^j s)/m(s)\ (i, j = 0, 1, 2, \ldots) \tag{56}$$

We may also suppose that $\alpha(t,s)$ has to maximize $R(t)$ under given $m(t)$ and $a(t)$, constants b and d, and the conditions

$$bt \leq a(t) \leq t,\ 0 \leq b < 1;\ \alpha(t,s) \leq \alpha(s)\exp[-d(t-s)],\ d \geq 0,\ 0 \leq y \leq 1. \tag{57}$$

Then, the desired $y(t)$, $a(t)$, $\alpha(t,s)$, and $R(t)$ are

$$y(t) = 1,\ a(t) = bt,\ \alpha(t,s) = [\exp(-dt)]\sum m'(b^i s)b^i \exp(db^i s)/m(s)\ (i = 0, 1, 2, \ldots),$$

$$R(t) = \int_{bt}^{t} m(s)ds. \tag{58}$$

Suppose yet, that in the MM (5)-(7) α, β have to maximize $R(t)$ under given $f(t)$, $c\hat{}(t)$, and the restrictions

$$bt \leq a(t) \leq t,\ 0 \leq b < 1;\ \beta = \alpha \leq \alpha(s)\exp[-d(t-s)],\ d \geq 0,$$

$$0 \leq y(t) \leq 1,\ c(t) \geq c\hat{}(t) \tag{59}$$

Then the desired $a(t)$, $y(s)$, and $\alpha(t,s)$ are

$$a(t) = bt,\ y(s) = A/(A+B),\ \alpha(t,s) = (A+B)\exp[-d(t-s)], \tag{60}$$

where

$$A = q'(s) + \sum b^i q'(b^i s)/q(s)\ (i = 0, 1, 2, \ldots),$$

$$B = r'(s) + \sum b^i r'(b^i s)/q(s)\ (i = 0, 1, 2, \ldots),$$

$$q(s) = [\exp(ds)][f(s) - c\hat{}(s)],\ r(s) = [\exp(ds)]c\hat{}(s). \tag{61}$$

The proof for all of these relations and certain others can be found in the book by V. Ivanov [92].

7. Statements of the Modeling Problems

We can use the models above for the solutions of the following *modeling problems*:

A. The *CAO identification problems*.

B. The solution of problems of the type "*If*. . ., *then* . . . " and/or "In order to . . . , it is sufficient . . . .".

C. The solution of *CAO optimization problems*.

The main identification problem is that of determination of the functions of α and β-types that actually correspond to genes of cells and/or certain natural laws. Using additional MM, s. 3, and applying the general methods of MM completeness, s. 6, we can reduce the problem of identification to determination of few separate parameters of CAO.

In general, we can use plenty of the well-known methods of identification, but in our case we have to deal with the specific character of the initial experimental data (i.e., non-negativity of all the model functions since they have the sense of new substrate-enzyme complex quantities, new product quantities, new technology quantities, etc).

We would like to emphasize here that not infrequently, due to exact, by definition, MM under consideration, their qualitative investigation does not require learning of quantitative values of the initial data, functions, or parameters. It is sufficient to have only qualitative data about specificity of their behavior.

The examples of modeling problems B. can be:

B.1. If indices α and β increase, then how fast does $m(t)$ increase for certain meanings of the distribution function $y(s)$?

B.2. In order for the CAO functioning bio-mass $R(t)$ to decrease, that is, when $R'(t) < 0$, it is sufficient that $y(s)$ be assigned a certain value.

The optimization problems for CAO can be solved at the expense of distribution functions x, y, z, e, g, r, w -types and the control factors u and v by observing the relations of the models in question, and with regard to the given prehistory of CAO.

Examples of optimization problems are

C.1. The external CAO function maximization:

$$I = I_1 = \int_{t^\wedge}^{T} c(t)dt = \int_{t^\wedge}^{T} \{\int_{a(t)}^{t} \beta(t,s)[1-y(s)]ds\}dt = \max, \qquad (62)$$

C.2. The average functioning CAO resource minimization:

$$I_2 = \int_{t^\wedge}^{T} R(t)dt = \int_{t^\wedge}^{T} \{\int_{a(t)}^{t} m(s)ds\}dt = \min \qquad (63)$$

with the restrictions on $c(t)$, $c(t) \geq \hat{c}(t)$, where $\hat{c}(t)$ is given.

C.3. The problem of CAO high-speed operation:

$$I_3 = T - t^\wedge = \min \tag{64}$$

with the conditions $\int_{t^\wedge}^{T} c(t)dt \in C^*$ and $R(t) \in R^*$, the sets C^* and R^* are given.

C.4. CAO reliability maximization:

$$I_4 = \min[m^+(t)-m(t), m(t)-m^-(t), c^+(t)-c(t), c(t)-c^-(t)] = \max \tag{65}$$

with the conditions $m^-(t) \le m(t) \le m^+(t)$ and $c^-(t) \le c(t) \le c^+(t)$, the functions $m^{\pm}(t)$ and $c^{\pm}(t)$ are given.

C.5. Out-of-date CAO product minimization:

$$I_5 = \int_{t^\wedge}^{T} G(t)dt = \int_{t^\wedge}^{T} [\int_0^{a(t)} m(s)ds]dt = \min. \tag{66}$$

C.6. The external CAO function minimization:

$$I_6 = \int_{t^\wedge}^{T} c(t)dt = \min. \tag{67}$$

C.7. The active CAO "life" maximization:

$$I_7 = T - t^\wedge = \max \tag{68}$$

with the restriction $c(t) \ge c^0(t)$, where $c^0(t)$ is given.

In the case when CAO is a cell as a whole, (62) is maximization of the amount of a cell external product during the period T - $t^\wedge$; (64) is minimization of the time for the norm cell restoration; (65) is one of the problems maintaining of the best condition for the cell functioning; and (66) is the problem of minimization of the cell average waste product.

For some CAO, the problems (67) and (68) can make more sense than (62) and (64). If CAO are populations of viruses in human organism (e.g. the population of HIV), $c(t)$ is the aggressive factor, and the problem (67) makes sense. If CAO are human beings, attractiveness of (68) is obvious.

Considering the *problems of design or synthesis of CAO* and bearing in mind the opportunity of the biotech revolution (J. Rifkin [160]), we can obtain CAO optimization at the expense of the functions α and β-types themselves (i.e., at the expense of the gene technology).

In the next Ch., and especially in Ch. 10 and 11, we consider examples of the solutions of some of these problems.

Theoretical ground for all the modeling problems above can be found in the APPENDIX, Ch. 14, 15, and 16.

Chapter 6

EXAMPLES OF CAO AND THEIR MM

1. On Hyper-Cycle, Origin of Life and DNA Code

We should devise a model in which a lot of information can be maintained in small volume over many generations. There are two main approaches. In the first, in order to give an information system a chance to develop, we must provide a semi-isolated compartment. In the second, some sort of a chemical reaction scheme must be found that is self-perpetuating.

Differential equations studied in the hyper-cycle model (see Ch. 1, s. 2.2) apply to any cyclical chemical reaction system, biological or otherwise. The hyper-cycle is suggested by the cyclic property of the DNA-protein synthesis reactions in modern organisms. DNA contains the information for constructing proteins, which are in turn necessary to construct more DNA, and so on.

We suppose that we have n different species of 'information molecules' I_i (such as RNA or DNA sequences), each of which contains the instructions for manufacturing one protein E_i. This protein is assumed to be a 'replicase' whose function is to recognize and translate the next information molecule I_{i+1} into its replicase E_{i+1}. In our notation, $\alpha = \{[I_i]'\}$, $m = \{[E_i]'\}$, and c is the rate of creation of the protein external product. Then according to (1) and (46), Ch. 5, the *MM of hyper-cycle*

$$\alpha(t) = \int_{a(t)}^{t} \alpha(t-s)x(s)m(s)ds + \Delta\alpha(t),$$

$$m(t) = \int_{a(t)}^{t} \alpha(t-s)y(s)m(s)ds + \Delta m(t),$$

$$M(t) = \int_{0}^{t} m(s)ds; \ \beta(t) = \int_{a(t)}^{t} \alpha(t-s)z(s)m(s)ds,$$

$$c(t) = \int_{a(t)}^{t} \beta(t,s)[1-x(s)-y(s)-z(s)]m(s)ds + \Delta c(t),$$

$$R(t) = \int_{a(t)}^{t} m(s)ds. \ G(t) = \int_{0}^{a(t)} m(s)ds = M(t) - R(t);$$

$$f(t) = \alpha(t) + m(t) + \beta(t) + c(t);$$

$$0 \le x, y, z, 1-x-y-z \le 1; 0 \le a(t) \le t, T \ge t \ge t^{\wedge} \ge 0, \quad (1)$$

where $x(s)m(s)$--$\alpha(t)$ and $z(s)m(s)$--$\beta(t)$ are the channels of creation of the new information (technology) quantity in the respective subsystem (for CAO, it is likely *Genome* - *G*), and the remainder of notations were previously noted.

The main feature and properties of this model explain probably many unsolved problems of the origin of life and the genetic code. Indeed, if $\Delta m(t) \neq 0$, then we may interpret the system (1)-type as originating CAO (S. Girlin [56]) that can be complicated indefinitely depending on the external conditions. If $\Delta m(t) = 0$ and $a(t) = 0$, then the system has only the trivial null solution. According to well-known principles of thermodynamics [180], $a(t)$ must be greater than 0. So, to 'survive' under the temporary condition $\Delta m(t) = 0$, the system (1)-type must be keeping and perfecting its prehistory on $[0, t^\wedge]$ (must be originating its genetic code).

The general mathematical properties of the system (1)-type will be considered in detail in the Appendix, Ch. 14. Now we consider an example of that system:

$$\alpha(t) = x\int_{a(t)}^{t} \alpha(t-s)m(s)ds + \int_0^t \alpha(s)\varphi(s)/h(s)ds,\ m(t) = y\int_{a(t)}^{t} \alpha(t-s)m(s)ds +$$

$$\int_0^t m(s)\varphi(s)/h(s)ds,\ R(t) = \int_{a(t)}^{t} m(s)ds,\ h(t) = \alpha(t) + m(t),\ 0 \leq a(t) \leq t, \tag{2}$$

where x and y are positive constants, $\varphi(t)$ is some overall external flux, and $\varphi(0) > 0$. It is easy to see that

$$\alpha(t) = x/ym(t),\ m'(t) = -xm(t-a(t))m(a(t))a'(t) +$$

$$x\int_{a(t)}^{t} m'(t-s)m(s)ds + y\varphi(t)/(x+y). \tag{2'}$$

One can verify that

$$m(t) = y\varphi(0)/(x+y)t,\ a(t) = t - (1-e^{-dt})/d \sim t - 1/d, \tag{3}$$

where d is a certain adapted constant, under the condition that

$$\varphi(t) = \varphi(0) + xy/(x+y)\varphi^2(0)[t - a(t)]\{a(t))a'(t) - [t + a(t)]/2\} \sim$$

$$\varphi(0) - xy/(x+y)\varphi^2(0)/(2d^2). \tag{4}$$

In the system (2) for $t \geq t^\wedge > 0$, we have the prehistory $[0, t^\wedge]$ on which all the elements of the system are given.

If the equation (2') had $\varphi(t) = 0$ for $t \geq t^\wedge$, this equation still would remain non-homogenous due to the given prehistory.

2. Dehydration Synthesis and Hydrolysis

Cells obtain most of their large molecules by joining smaller organic molecules into chains called *polymers*. The units that serve as the building blocks of polymers are called *monomers*. The key to the great diversity of proteins and DNA molecules is arrangement--variation in the sequence in which the monomers are strung together.

Cells link monomers together to form polymers by a process called *dehydration synthesis*. All unlinked monomers have a hydronium ion (H_3O^+) and also a hydroxyl group (OH^-). For each monomer added to a chain, a water molecule is removed. One monomer loses a hydronium ion and the other monomer loses a hydroxyl group. As this occurs, a new covalent bond forms, linking the two monomers.

Thus, with regard to the models and notation considered, we now have that m is the rate of creation of the (E, P, S; I)-complex where E is some polymerase, P is some ATP, S is two monomers, and I is some information about arrangement of two monomers; c is the rate of creation of the resulting polymer. For simplicity, we do not include the concentration of the enzyme e as one of the products.

The respective MM has the form

$$m(t) = \int_0^t \alpha(t,s)\lambda(t,s)y(s)m(s)ds + \Delta m(t),\ 0 \le y(s),\ \lambda(t,s) \le 1,$$

$$c(t) = \int_{t^\wedge}^t \beta(t,s)\mu(t,s)z(s)m(s)ds + \Delta c(t),\ 0 \le z(s),\ \mu(t,s) \le 1,$$

$$R(t) = \int_0^t [\lambda(t,s)y(s) + \mu(t,s)z(s)]m(s)ds,\ y(s) + z(s) = 1,\ t \ge t^\wedge \ge 0,$$

$$M(t) = \int_0^t m(s)ds,\ G(t) = M(t) - R(t),\ f(t) = m(t) + c(t), \tag{5}$$

or, in particular, the form

$$m(t) = \int_{a(t)}^t \alpha(t,s)y(s)m(s)ds,\ R(t) = \int_{a(t)}^t m(s)ds,\ G(t) = M(t) - R(t);$$

$$c(t) = \int_{a(t)}^t \beta(t,s)[1-y(s)]m(s)ds,\ T \ge t \ge t^\wedge \ge 0, \tag{6}$$

where α and β we consider as given. Assuming $t^\wedge$ such that for $t \ge t^\wedge$, $R(t) = R$, $y(s) = y$ and $\alpha(t,s) = \alpha$ are constants, we find

$$m(t) = y\alpha R = m(a)da/dt = m^\wedge = ke \cdot s, \tag{7}$$

where k is the rate constant for the enzyme-substrate complex. From (7) it follows

$$R = m^\wedge/(y\alpha),\ a(\mathrm{t}) = t - 1/(\ y\alpha). \tag{8}$$

Assuming also that $\beta(t,s) = \beta(t)$, we find

$$c(t) = \beta(t)(1-y)m^\wedge/(y\alpha),\ T \geq t \geq t^\wedge \geq 0. \quad (9)$$

Given in advance $c(t)$, the relation (9) determines the respective index $\beta(t)$.
Note that the resulting polymer has the form

$$C(t) = C^\wedge + \int_{t^\wedge}^{t} c(s)ds,\ T \geq t \geq t^\wedge \geq 0. \quad (10)$$

In the case of hydrolysis, which is almost the inverse to dehydration synthesis, we have MM of the same types as (5), (6), where m is the rate of creation of the (e, p, s; I)-complex, e is some depolymerase, p is some ATP, s is the current polymer and the next monomer, I is some information about arrangement of the initial polymer, and c^{-1} is the rate of creation of the resulting elementary monomers,

$$c^{-1}(t) = c(T + t^\wedge - t),\ T \geq t \geq t^\wedge \geq 0. \quad (11)$$

One can see from (9) that the respective index of efficiency

$$\beta^{-1}(t) = \beta(T + t^\wedge - t),\ T \geq t \geq t^\wedge \geq 0. \quad (12)$$

3. ATP Synthesis

We remind that the set of enzyme reactions connected with a common cofactor makes up the so-called *reaction core*. There is a close connection among separate parts of the primary structure due to this core. MM of the core of reactions can be obtained by combining MM considered above. This method and the general traditional approach for creation of MM of multi-enzyme systems have been illustrated in Ch. 2 on the example of the simple model of energetic metabolism. In this example dynamics of all the enzyme reactions were actually missed by the same reasons as above (see example 1. in s.1, Ch. 5).

Now we will investigate, on the basis of the novel approach, *MM of ATP synthesis*, regarding *glycolysis*, *Krebs cycle*, *electron system transport*, and *chemiosmosis*.

In the case of glycolysis (see Fig. 11), m is the rate of recreation in cytoplasmic fluid of the new (e, p, s)-complex, where the power p = [ATP], the enzymes e = [E], E is degidrogenase, the substrate $s = s_1 + s_2$, s_1 = [glucose], s_2 = [NAD^+], c_1 is the rate of creation of the new acetyl group from pyruvic acid, c_2 is the rate of creation of the new NADH + H^+, $c_3 = de/dt$, R is the overall of functioning complexes, and G is the overall of obsolete product or waste consisting of CO_2 and some heat.

It is known that for each molecule of glucose that enters glycolysis, two molecules of NADH + H^+, one acetyl group, one CO_2, and some heat are the total product.

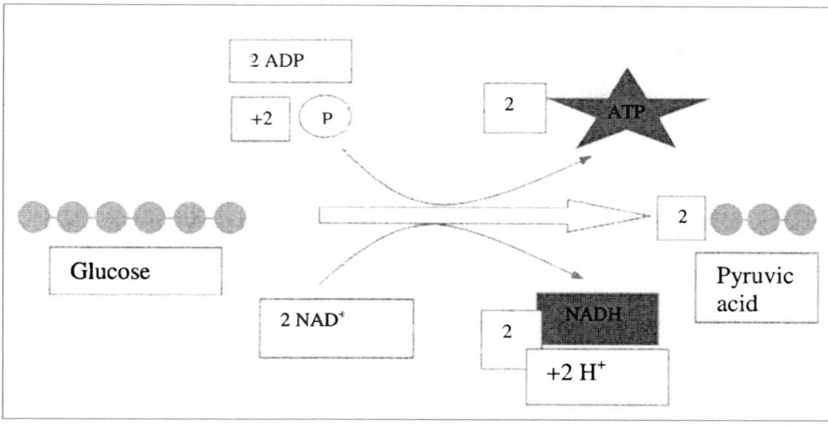

Figure 11. An overview of glycolysis

Thus, on the basis of (1, Ch. 5), we have

$$m(t) = \int_0^t \alpha(t,s)\lambda(t,s)y(s)m(s)ds + \Delta m(t),\ 0 \le y(s),\ \lambda(t,s) \le 1,$$

$$c_i(t) = \int_{t^\wedge}^t \beta_i(t,s)\mu_i(t,s)z_i(s)m(s)ds + \Delta c_i(t),\ 0 \le z(s),\ \mu_i(t,s) \le 1,$$

$$i = 1, 2, 3,\ R(t) = \int_0^t [\lambda(t,s)y(s) + \mu_1(t,s)z_1(s) + \mu_2(t,s)z_2(s) +$$

$$\mu_3(t,s)z_3(s)]m(s)ds,\ M(t) = \int_0^t m(s)ds,\ G(t) = M(t) - R(t),$$

$$f(t) = m(t) + c_1(t) + c_2(t) + c_3(t),\ y(s) + z_1(s) + z_2(s) + z_3(t) = 1,\ t \ge t^\wedge \ge 0, \tag{13}$$

where the remaining magnitudes are similar to those in (1, ch. 5).

In particular, we have

$$m(t) = \int_{a(t)}^t \alpha(t,s)y(s)m(s)ds,\ R(t) = \int_{a(t)}^t m(s)ds,$$

$$c_i(t) = \int_{a(t)}^t \beta_i(t,s)z_i(s)m(s)ds,\ i = 1, 2, 3,\ t \ge t^\wedge \ge 0, \tag{14}$$

where $\alpha(t,s)$ and $\beta_i(t,s)$ are given (by the Human Genome or certain natural law). Assuming $t^\wedge$ such that for $t \ge t^\wedge$, $R(t) = R$, $y(s) = y$, and $\alpha(t,s) = \alpha$ are constants, we find the same formulae as (7) and (8).

In the case of the Krebs cycle (see Fig. 12), 'm is the rate of recreations in the mitochondrion of the substrate-enzyme complex, where the substrate is some combination of the acetyl groups (CH_3CO), four-carbon molecules called oxaloacetic acid, ADP, P, FAD, NAD^+, and the enzyme CoA, and the specific 'enzyme' combines seven known enzymes; 'c_1 is the rate of creation of the main product consisting of energy-rich molecules: ATP, GTP (guanosinetriphosphates), and some others; 'c_2 is the rate of creation of molecules NADH + H^+ and $FADH_2$; 'c_3 and 'c_4 are the rate of releasing enzymes and oxaloacetates, respectively; 'c_5 = 'G is the rate of breakdown of acetyl groups to CO_2. The power that drives the reaction is the chemical energy that was stored in the acetyl group.

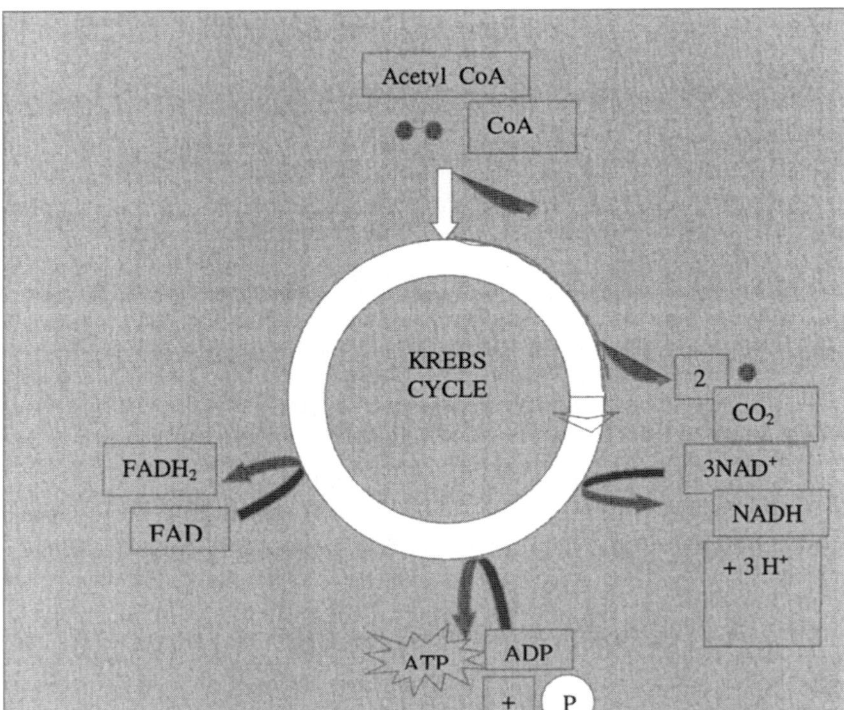

Figure 12. An overview of the Krebs cycle

Thus, on the basis of (1) we have

$$'m(t) = \int_0^t {}'\alpha(t,s)'\lambda(t,s)'y(s)'m(s)ds + {}'m(t),\ 0 \le {}'y(s), '\lambda(t,s) \le 1,$$

$$'c_i(t) = \int_{t^\wedge}^t {}'\beta_i(t,s)'\mu_i(t,s)'z_i(s)'m(s)ds + \Delta' c_i(t),\ 0 \le {}'z_i(s), '\mu_i(t,s) \le 1,$$

$$i = 1, 2, 3, 4, 5,\ 'R(t) = \int_0^t [{}'\lambda(t,s)'y(s) + \textstyle\sum_{i=1}^5 {}'\mu_i(t,s)'z_i(s)]'m(s)ds,$$

$$'y(s) + {}'z_1(s) + {}'z_2(s) + {}'z_3(s) + {}'z_4(s) + {}'z_5(t) = 1,\ t \ge {}'t^\wedge \ge 0, \tag{15}$$

where the remaining magnitudes are similar to those in (1, Ch. 5).

In particular, we have

$$'m(t) = \int_{'a(t)}^t {}'\alpha(t,s)'y(s)'m(s)ds,\ 'R(t) = \int_{'a(t)}^t {}'m(s)ds,$$

$$'c_i(t) = \int_{'a(t)}^t {}'\beta_i(t,s)'z_i(s)'m(s)ds,\ i = 1, 2, 3, 4, 5,\ t \ge {}'t^\wedge \ge 0, \tag{16}$$

where $'\alpha(t,s)$ and $'\beta_i(t,s)$ are given.

Assuming $'t^\wedge$ such that for $t \ge {}'t^\wedge$, $'R(t) = {}'R$, $'y(s) = {}'y$ and $'\alpha(t,s) = {}'\alpha$ are constants, we find the formulae of the same types as (7) and (8). After this, we can find the remaining magnitudes $'c_i(t)$, $i = 1, 2, 3, 4, 5$, on the strength of (16).

In the case of electron transport chain, $m = {}''m$ corresponds to protein complex, to the electron carrier in the inner membrane of the mitochondrion, and to NADH, $FADH_2$, and O_2; $c = {}''c$ corresponds to protein complex and electron carrier ($''c_1$), H^+ ($''c_2$) and H_2O ($''c_3$). The power is released from the electrons in the redox reactions.

Thus, on the basis of the model (1, Ch. 5)-type and (13)-(16), we have

$$''m(t) = {}''k\ ''s \cdot {}''e + m^\wedge \int_{t-1/(y\alpha)}^t \beta_2(t,s)z_2(s)ds + {}'m^\wedge \int_{t-1/('y'\alpha)}^t {}'\beta_2(t,s)'z_2(s)ds,$$

$$''k\ ''s \cdot {}''e = \int_{''a(t)}^t {}''\alpha(t,s)''y(s)''m(s)ds,\ m^\wedge = ks \cdot e,\ 'm^\wedge = {}'k's \cdot {}'e,\ ''R(t) = \int_{''a(t)}^t {}''m(s)ds,$$

$$''c_i(t) = \int_{''a(t)}^t {}''\beta_i(t,s)''z_i(s)''m(s)ds,\ i = 1, 2, 3,\ t \ge {}''t^\wedge \ge 0. \tag{17}$$

Here $''\alpha(t,s)$ and $''\beta_i(t,s)$ are also given.

Assuming $''t^\wedge$ such that for $t \ge \max[t^\wedge, 't^\wedge, ''t^\wedge]$, $''R(t) = {}''R$ is constant, $''y$, $''\alpha(t,s) = {}''\alpha$, z_2, $\beta_2(t,s) = \beta_2$, $'z_2$, and $'\beta_2(t,s) = {}'\beta_2$ are also constants, we have

$$''m(t) = {}''k\ ''s \cdot {}''e + \beta_2 z_2 R + {}'\beta_2 {}'z_2 {}'R = {}''m(''a)d''a/dt = {}''m^\wedge, \tag{18}$$

and

$$''R = {}''k\ ''s \cdot {}''e/(''\alpha''y),\ ''a(t) = t - {}''R/''m^\wedge. \tag{19}$$

In the case of chemiosmosis, m = ‘’’m is the rate of recreations in the cell membrane of the substrate-protein complexes (clusters), where the substrate is some combination of ADP and P and the protein is ATP synthase that resides in the same membrane as the molecules of the electron transport chain; c = ‘’’c is the rate of creation of ATP. The power is created by the concentration gradient dH of H^+ molecules.

MM are similar to above:

$$‘’’m(t) = \int_{'''a(t)}^{t} ‘’’\alpha(t,s)\ ‘’’y(s)\ ‘’’m(s)ds + “m\wedge\int_{t-''t\wedge/''m\wedge}^{t} “\beta_2(t,s)“z_2(s)ds,$$

$$‘’’R(t) = \int_{'''a(t)}^{t} ‘’’m(s)ds,\ ‘’’c(t) = \int_{'''a(t)}^{t} ‘’’\beta(t,s)[1-‘’’y(s)]‘’’m(s)\mathrm{d}s,\ t \geq ‘’’t\wedge \geq 0, \quad (20)$$

where ‘’’$\alpha(t,s)$ and ‘’’$\beta(t,s)$ are also given (in the Human Genome).

Assuming ‘’’$t\wedge$ such that for t > max [$t\wedge$, ’$t\wedge$, “$t\wedge$, ‘’’$t\wedge$] ‘’’$R(t)$ = ‘’’R is constant, ‘’’$y(s)$ = ‘’’y, ‘’’$\alpha(t,s)$ = ‘’’α, “z_2, and “$\beta_2(t,s)$ = “β_2 are also constants, and ‘’’$\beta(t,s)$ is proportional to [C^* - ‘’’$C(t)$], ‘’’$C(t) = C\wedge + \int_{'''t\wedge}^{t}$ ‘’’$c(s)ds$, we find

$$‘’’m(t) = ‘’’\alpha\ ‘’’y\ ‘’’R + “\beta_2\ “z_2“R = ‘’’k\ ‘’’s\cdot‘’’e + “\beta_2“z_2\ “R =$$

$$‘’’m(‘’’a)d‘’’a/dt = ‘’’m\wedge,\ ‘’’C'(t) = ‘’’\beta(1-‘’’y)‘’’R[C^*-‘’’C(t)], \quad (21)$$

from where

$$‘’’R = ‘’’k\ ‘’’s\cdot‘’’e/(‘’’\alpha‘’’y),\ ‘’’a(t) = t - ‘’’R/‘’’m\wedge,$$

$$‘’’C(t) = C^*+(C\wedge-C^*)\exp[-‘’’\beta(1-‘’’y)‘’’R(t-‘’’t\wedge)]. \quad (22)$$

It should be noted that [86] (see also P. Boyer [28]) contains more complete information about the structure and functioning of *ATP synthase*, which is called a *molecular machine* with “binding change mechanism.” This notation allows us to specify more accurately the behavior of some considered functions of α and β-types. In particular, we have

$$C^* = k_1 d\mathrm{H} + k_2,\ ‘’’\beta = k_3 d\mathrm{H},\ k_1 \geq 0,\ k_3 > 0, \quad (23)$$

where k_1, k_2, and k_3 are certain coefficients subjected to identification.

The more dH, the more ATP can be produced and the faster the maximal level of ATP C^* can be established. Thus, by means of chemiosmosis, a cell couples the exergonic reactions of electron transport to the endergonic synthesis of ATP. 36 ATP per glucose molecule is generated by cellular respiration (34 ATP by chemiosmosis). As a final result, a cell can harvest up to 40% of the glucose’s potential energy.

Most food molecules can be utilized by cellular respiration through hydrolysis, in principle, in the same way as above.

4. Biosynthesis

Food also provides the raw materials a cell uses for biosynthesis; that is, to make its own molecules for repair and growth. Cells obtain some raw materials directly from food. Amino acids, for example, can be incorporated without being changed into proteins.

However, the body also needs certain molecules that are not present in the same form in food. Our cells make these molecules using some of the compounds of glycolysis and the Krebs cycle. The respective pathways are almost the reverse of the above, and these pathways consume ATP energy. Instead of glycolysis we have glucose synthesis, and instead of hydrolysis we have dehydration synthesis, resulting in polysaccharides and fats.

Let us consider glucose synthesis in more detail. A certain part of the total product of glycolysis + some ATP (much more than created by glycolysis) may be considered as the initial substrate for glucose synthesis that, in combination with the enzyme dehydrogenase + H_2O, gives the complex m^{-1}.

The product c_1^{-1} of the synthesis is the rate of creation of glucose, c_2^{-1} is the rate of creation of NAD^+, and c_3^{-1} is the rate of creation of dehydrogenase + H_2O.

On the basis of the MM under consideration, we have the following *MM of biosynthesis*

$$m^{-1}(t) = \int_{a^\wedge(t)}^{t} \alpha^{-1}(t,s)y^{-1}(s)m^{-1}(s)ds,\ R(t) = \int_{a^\wedge(t)}^{t} m^{-1}(s)ds,$$

$$c_i^{-1}(t) = \int_{a^\wedge(t)}^{t} \beta_i^{-1}(t,s)z_i^{-1}(s)m^{-1}(s)ds,\ i = 1, 2, 3,\ t \geq t^\wedge \geq 0. \qquad (24)$$

We may assume by analogy with the results just above that

$$m^{-1}(t) = m(T + t^\wedge - t) = \int_{a(T+t^\wedge-t)}^{T+t^\wedge-t} \alpha(T+t^\wedge-t,s)y(s)m(s)ds =$$

$$\int_{t}^{T+t^\wedge-a(T+t^\wedge-t)} \alpha(T+t^\wedge-t,T+t^\wedge-s)y(T+t^\wedge-s)m(T+t^\wedge-s)ds =$$

$$\int_{t}^{T+t^\wedge-a(T+t^\wedge-t)} \alpha(T+t^\wedge-t,T+t^\wedge-s)y^{-1}(s)m^{-1}(s)ds = \int_{t-[T+t^\wedge-t-a(T+t^\wedge-t)]}^{t} \alpha(T+t^\wedge-t,\ t+a(T+t^\wedge$$

$$-t)-s)y(t+a(T+t^\wedge-t)-s)m(t+a(T+t^\wedge-t)-s)/[y(T+t^\wedge-s)m(T+t^\wedge-s)]\ y^{-1}(s)m^{-1}(s)ds, \qquad (25)$$

and also

$$\alpha^{-1}(t,s) = \alpha(T+t^\wedge-t,\ t+a(T+t^\wedge-t)-s)\ y(t+a(T+t^\wedge-t)-s)m(t+a(T+t^\wedge-$$

$$t)-s)/[y(T+t^\wedge-s)m(T+t^\wedge-s)],\ a^\wedge(t) = t-[T+t^\wedge-t-a(T+t^\wedge-t)]. \qquad (26)$$

In particular, if y is constant and $a(t) = t - a$, where a is also constant, then

$$\alpha^{\cdot 1}(t,s) = \alpha(T+t\text{^}-t, T+t\text{^}-a-s)m(T+t\text{^}-a-s)/m(T+t\text{^}-s),\ a\text{^}(t) = t-a. \tag{27}$$

Consumption of ATP, for example, in the process of biosynthesis and creation of ATP as above conceives the notion of *energetic metabolism.* A very important quantitative characteristic of the energetic metabolism is the so-called *loading characteristic*--dependence of the concentration of ATP on the load consisting of the total activity of all biochemical processes using ATP.

According to the model (22) and (23), the loading characteristic is determined by the constant

$$D = -\text{‘’’}\beta(1-\text{‘’’}y)\text{‘’’}R,\ \text{‘’’}\beta = k_3 d\mathrm{H}, \tag{28}$$

The bigger this constant, the sooner the concentration of ATP rises from C^ to its maximal possible value C*.

The tertiary structure of energetic metabolism has to contain controls that weaken or strengthen the plateau of the loading characteristic.

Let us consider the dependence of the constant D on the control functions of u-type in the relation (49, Ch. 5). Assuming

$$\text{‘’’}\alpha = \alpha^*,\ \text{‘’’}t\text{^} < t < t^*,\ \text{‘’’}\alpha = \alpha^* u,\ t^* < t, \tag{29}$$

where α^* and u are constants, on the basis of (20) -- (22) we have

$$\alpha^*\text{‘’’}y\text{‘’’}m\text{^}\int_{\text{‘’’}a(t)}^{t^*} ds + \alpha^* u\text{‘’’}y\text{‘’’}m\text{^}\int_{t^*}^{t} ds = \text{‘’’}k\text{‘’’}s\cdot\text{‘’’}e,$$

$$t^* < t < t',\ \text{‘’’}a(t') = t^*,\ t' = t^* + \text{‘’’}k\text{‘’’}s\cdot\text{‘’’}e/(\alpha^* u\text{‘’’}y\text{‘’’}m\text{^}), \tag{30}$$

from where

$$\text{‘’’}R = \text{‘’’}m\text{^}(t-\text{‘’’}a(t)) = \text{‘’’}m\text{^}[t-t^* + \text{‘’’}k\text{‘’’}s\cdot\text{‘’’}e/(\alpha^*\text{‘’’}y\text{‘’’}m\text{^}) - u(t-t^*)],\ t^* < t < t',$$

$$\text{‘’’}R = \text{‘’’}k\text{‘’’}s\cdot\text{‘’’}e/(\alpha^* u\text{‘’’}y\text{‘’’}m\text{^}),\ t \ge t'. \tag{31}$$

Therefore,

$$\text{‘’’}a(t) = t - \text{‘’’}k\text{‘’’}s\cdot\text{‘’’}e/(\alpha^*\text{‘’’}y\text{‘’’}m\text{^}),\ t\text{^} < t < t^*, \text{‘’’}a(t) = u(t-t^*) + t^* -$$

$$\text{‘’’}k\text{‘’’}s\cdot\text{‘’’}e/(\alpha^*\text{‘’’}y\ \text{‘’’}m\text{^}),\ t^* < t < t',\ \text{‘’’}a(t) = t - \text{‘’’}k\text{‘’’}s\cdot\text{‘’’}e/(\alpha^* u\text{‘’’}y\text{‘’’}m\text{^}),\ t' < t, \tag{32}$$

and also

$$D = -\text{‘’’}\beta(1 - \text{‘’’}y)\ \text{‘’’}k\text{‘’’}s\cdot\text{‘’’}e/(\alpha^*\text{‘’’}y),\ \text{‘’’}t\text{^} < t < t^*,$$

$$D = -\,``'\beta(1 - ``'y)``'m\hat{}[t - t^* + ``'k``'s\cdot``'e/(\alpha^*``'y``'m\hat{}) - u(t - t^*)],\ t^* < t < t'. \quad (33)$$

Hence

$$D = -``'\beta(1 - ``'y)``'k``'s\cdot``'e/(\alpha^* u``'y),\ t' \leq t. \quad (34)$$

Thus, under $u < 1$ we have a strengthening of the plateau of the loading characteristic at the expense of increasing time $t - a(t)$ of the substrate-enzyme complex functioning.

5. On Pattern Formation

We recollect, in short, here that the *pattern formation* is the cellular process from the cell origin to self-creation of the whole organism. This process is often accompanied by differentiation, in which a daughter cell differs from the initial cell. In many cases, however, daughter cells are identical to mother cells. Usually, the cell cycle is divided into the following phases: G_1, S (DNA synthesis), G_2, and M (mitosis). The state of division is called proliferation.

There also exist the resting state R_1, in which a cell can be situated for long periods of time, and the resting state R_2, in which cell proliferation can be blocked in the time of phase G_2. Switching from the resting state to proliferation and vice versa can be under the influence of different factors and may occur spontaneously. Mechanism of this switch is the main object of modeling. The problem is to find the models for the origin of cell signals that influence transition from one phase of the cell cycle to another. In Ch. 1, we considered one of such models. Later, in Ch. 7, we show that these kinds of models can also be obtained with the help of MM under consideration.

Here we show that the novel approach allows us to receive the novel mathematical illustration of the pattern formation.

The *minimal MM* of the *pattern formation* based on our class models has the form

$$f(t) = m(t) + c(t) = \int_{a(t)}^{t} \alpha(t,s)y(s)m(s)ds + \int_{b(t)}^{t} \beta(t,s)[1-y(s)]m(s)ds,$$

$$R(t) = R_1(t) + R_2(t) = \int_{a(t)}^{t} y(s)m(s)ds + \int_{b(t)}^{t} [1-y(s)]m(s)ds,\ t\hat{} < t, \quad (35)$$

where the unknowns are a and b, $0 < a$, $b < t$; m is the rate of recreation of enzyme-power-substrate-information complexes (the complexes may contain the SMF and MPF (S- and M-phase promoting factor)); c is the rate of creation of the cell external product, αy and $\beta(1-y)$ are given (by the Human Genome and certain natural laws).

Let $\alpha = \beta = \alpha(t) > 0$. Then the system (35) has no solution under $f/\alpha \neq R$ and has an infinite number of solutions under $f/\alpha = R$, $R(t) < \int_0^t m(s)ds$.

We find a case when the system (35) has two coincident solutions or a *two-multiple solution*, which is related to a case of a cell production growth and diversity. Let

$\alpha = \alpha(t)$, $(\beta - \alpha)(1-y)m = 2u(t)s + 2v(t)$, $u(t) > 0$, $s \in [0, t\hat{}]$, $t\hat{} > 0$; $\beta = \alpha(t)$, $t \geq s > t\hat{}$;

$0 < (f - \alpha R)/u < t\hat{}^2$, $t\hat{} < t < t^*$, $f(t^*) - \alpha(t^*)R(t^*) = 0$; $0 \leq (f - \alpha R)/u \leq (t - t\hat{})^2$, $t^* \leq t$;

$$\int_0^t [1-y(s)]m(s)ds < R(t) < \int_0^t y(s)m(s)ds. \tag{36}$$

Then

$$f(t) = \int_{b(t)}^t (\beta-\alpha)(1-y)mds + \alpha R = ut\hat{}^2 + 2vt\hat{} - ub^2 - 2vb + \alpha R, \tag{37}$$

from which

$$b(t) = \{-v \pm [v^2 - u(f - \alpha R - ut\hat{}^2 - 2vt\hat{})]^{1/2}\}/u. \tag{38}$$

Therefore, under

$$v^2 - u(f - \alpha R - ut\hat{}^2 - 2vt\hat{}) = 0, \tag{39}$$

or

$$v = -ut\hat{} \pm [(f - \alpha R)u]^{1/2}, \tag{40}$$

we have a *two-multiple solution*

$$b(t) = t\hat{} - [(f - \alpha R)/u]^{1/2},\ t\hat{} < t < t^*,\ b(t) = t\hat{},\ t^* \leq t. \tag{41}$$

We should still verify that the equation

$$R(t) - \int_{a(t)}^t y(s)m(s)ds = \int_{b(t)}^t [1-y(s)]m(s)ds,\ t\hat{} < t, \tag{42}$$

where $b(t)$ is given by the relation (41), has a solution $a(t)$, $0 < a(t) < t$. Let

$$h(a) = \int_a^t y(s)m(s)ds + \int_{b(t)}^t [1-y(s)]m(s)ds,\ t\hat{} < t. \tag{43}$$

Then on the strength of (41), $h(0) > R(t)$ and $h(t) < R(t)$. So, the equation (42) always has a solution $a(t)$, $0 < a(t) < t$. Thus, the system (35) under the condition (36) has a two-multiple solution.

Let us consider a case of *two-multiple repeated solutions* for the model

$$m(t) = \int_{a(t)}^t \alpha(t,s)y(s)m(s)ds,\ c(t) = \int_{b(t)}^t \beta(t,s)[1-y(s)]m(s)ds,\ 0 < t\hat{} \leq t. \tag{44}$$

It is not difficult to verify that under the condition

$$m(t) = m^\wedge = \text{constant},\ y(t) = y^\wedge,\ t \in [0, t^\wedge];\ \int_{(2k-1)t^\wedge}^{t} \alpha(t,s)ds = 1,$$

$$\beta(t,s) = 2[s-(t-t^\wedge)]/m^\wedge,\ s \in [t-t^\wedge, (2k-1)t^\wedge],\ y(t) = 1,\ c(t) = (2kt^\wedge-t)^2,$$

$$t \in [(2k-1)t^\wedge, 2kt^\wedge];\ \int_{t-t^\wedge}^{2kt^\wedge} \alpha(t,s)ds = 1,\ \beta(t,s) = 2(t-s)]/m^\wedge,$$

$$s \in [2kt^\wedge, t],\ y(t) = 0,\ c(t) = (2kt^\wedge-t)^2,\ t \in [2kt^\wedge, (2k+1)t^\wedge];\ k = 1, 2, ..., \tag{45}$$

we have *two-multiple repeated* solutions

$$b(t) = t-t^\wedge,\ m(t) = m^\wedge,\ P(t) = m^\wedge t^\wedge,\ t \in [(2k-1)t^\wedge, 2kt^\wedge],\ a(t) = t-t^\wedge,$$

$$m(t) = m^\wedge,\ P(t) = m^\wedge t^\wedge,\ t \in [2kt^\wedge, (2k+1)t^\wedge],\ k = 1, 2, ... \tag{46}$$

Note that the condition (45) is valid for

$$\alpha = 2\delta[s-(2k-1)t^\wedge],\ s \in [(2k-1)t^\wedge, 2kt^\wedge];\ \alpha = 2\delta(2kt^\wedge-s),\ s \in [2kt^\wedge, (2k+1)t^\wedge], \tag{47}$$

where δ is a Dirac generalized function.

The relations (45)--(47) can be interpreted as imitation of the process of unlimited pattern formation. From (45), it follows that to prevent the process of (46), cells should be "busy" enough with fulfilling their external functions in order for the cases of the types $y = 1$, $\beta = 0$, and $c = 0$ to be impossible.

While traditionally MM of pattern formation are connected with the events of bifurcations for unstable with unlimited growth solutions of the respective equations, in the case of MM of development, pattern formation is connected with the events of solutions branching, preassigned by regulatory processes in cells (see below).

Note that stochastic branching processes were investigated by M. Kimmel, D. Axelrod [112].

6. MM of Regulatory Processes in Cells

6.1. BIOLOGICAL BACKGROUND

That background is based on the results described in B. Alberts et al. [7], N. Campbell et al. [30], G. Cooper [36], T. Schlick [165], [167], and J. Watson [206].

6.1.1. *Proteins turn genes on or off*

Genes direct the synthesis of specific RNA and consequently protein molecules. Thus, a gene that is turned on is directing the transcription of its genetic information into RNA, and that message is being translated into specific amino acids. The overall process by which genetic information flows from genes to proteins is called *gene expression.*

The turning on and off of genes is the main way that gene expression is regulated, and it is the basis of cell specialization as a multi-cellular organism develops from a zygote. This specialization in the structure and function is called cellular *differentiation.* Cellular differentiation helps transform the cells arising from a zygote into an adult organism. Cells specialize because certain genes are expressed at certain times.

The control of gene expression makes it possible for cells to produce specific kinds of proteins when and where they are needed. The earliest understanding of gene control came from the bacterium *E. coli.* Although *E. coli* is made up of a single cell, it must change its activities from time to time to suit changes in its environment.

Remember that enzymes are proteins; their production is an example of gene expression. *E. coli* can make lactose-utilization enzymes because it has genes that code for these enzymes. In 1961, French biologists Jacob and Monod proposed a model for how an *E. coli* cell could adjust its enzyme production in response to frequent changes in its environment. The Jacob and Monod model explains how genes coding for lactose-utilization enzymes are turned off or on, depending on whether lactose is available.

E. coli uses three enzymes, whose genes are next to each other, to take up and start metabolizing lactose. The genes coding for these enzymes are regulated as a unit. Adjacent to the group of lactose enzyme genes are short sections of DNA that help control them. One stretch of nucleotides is a *promoter*, a site where the transcription enzyme, RNA polymerase, attaches and initiates transcription of all three lactose enzyme genes. Between the promoter and the enzyme genes, a DNA segment called an *operator* acts as a switch. The operator determines whether RNA polymerase can attach to the promoter and start transcribing the genes.

Such a cluster of genes with related functions, along with a promoter and an operator, is called an *operon*. The key advantage to the grouping of genes into operons is that the expression of these genes can be easily coordinated.

When *E. coli* encounters lactose, all the enzymes needed for its use are made at once, because a single switch, the operator, controls all the operon's genes. But what determines whether the operator switch is on or off?

Let the lac operon be in "off" mode, its status when no lactose is present in the cell's environment. *Transcription* is turned off by a molecule called a *repressor*, a protein that functions by binding to the operator and blocking the attachment of RNA polymerase to the promoter. A gene called a *regulatory gene*, located outside the operon, codes for the repressor. The regulatory gene is expressed continually in small amounts, so the cell always has a supply of repressor molecules.

6.1.2. *Differentiated cell has a particular pattern of genes*

The great differences among cells in an organism result from the selective expression of genes. As a developing embryo undergoes successive cell divisions, different genes must be activated in different cells at different times. Groups of cells follow diverging developmental pathways, and each group develops into a particular kind of tissue. Finally, in the mature organism, each cell type—for example, nerve or muscle—has a different pattern of turned on genes. Let us consider patterns of gene expression for a few genes in three different specialized cells. The genes for the enzymes of the metabolic pathway of glycolysis are active in all metabolizing cells, including the pancreas, embryonic eye lens, and nerve cells. On the other hand, only particular cells express genes for specialized proteins.

The specialized protein we use as an example is the transparent protein crystallin (which forms the lenses of our eyes), the hormone insulin, and the oxygen-transporting protein hemoglobin. Notice that the hemoglobin genes are not active in any of the cell types. They are turned on only in cells that are developing into red blood cells. Insulin genes are activated only in the cells of the pancreas that produce that hormone. Nerve cells express genes for other specialized protein not shown here. Mature lens cells, like mature red blood cells, achieve the final step in differentiation. After a frenzy of expressing crystallin genes and accumulating the protein products, these cells lose their nuclei, and thus all their genes. Therefore, cells become specialized because they express only certain genes.

Thus, cellular *differentiation* in multi-cellular organisms results from selective gene expression, as does the ability of bacteria to produce different enzymes as needed. Now it is time to examine the control of gene expression in cells more closely.

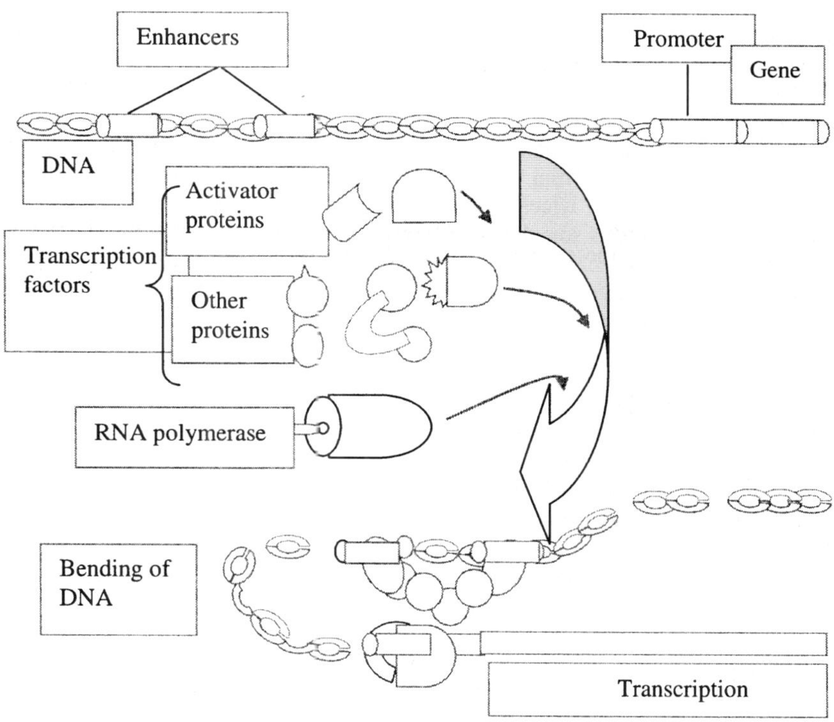

Figure 13. MM for turning on of a gene

6.1.3. *Complex proteins control transcription*

Little is known about how DNA packaging is regulated or exactly how it affects gene expression. We do know, however, that control of gene expression at the transcription stage is very important. Several species of flies, including the fruit fly, provided early evidence of transcriptional control.

At certain stages in the fly's development, DNA loops out from the chromosome at specific places, forming structures called puffs. The locations of pulling along a chromosome change as the larva develops. When the larva gets ready to shed its body covering in preparation for the next developmental stage, some puffs disappear and others form in new places. The puffs correspond to regions of intense RNA synthesis (transcription); they are visible signs of switching on and off specific genes during development.

Studying cells, biologists have found that they employ regulatory proteins that bind to DNA and turn the transcription of genes on and off. The control mechanisms involve proteins that bind to specific segments of DNA. However, the mechanisms are very complicated. Their model features an intricate array of regulatory proteins that interact with DNA and with one another to turn genes on or off. In contrast to the genes of bacterial operons, each gene usually has its own set of control sequences. Moreover, activator proteins seem to be more important in cells than repressors.

Turning on a cell gene involves regulatory *proteins* called *transcription factors* in addition to RNA polymerase. Activators are one type of transcription factor. The first step in initiating gene transcription is the binding of activators to DNA sequences called *enhancers* that are remote from the gene. DNA bends and the activators interact with other transcription factors, which then bind as a complex at the genes promoter. This large assembly of proteins somehow facilitates the correct binding of RNA polymerase to the promoter and somehow initiates transcription. Several enhancers and activators may be involved. Repressor proteins that may bind to DNA sequences are called *silencers* and function analogously to inhibit the start of transcription.

6.1.4. *RNA adds a cap and a poly-A tail and removes non-coding segments*

Cells regulate gene expression at transcription and also have additional opportunities for control. One opportunity for gene regulation after transcription arises from the fact that the DNA sequences that actually code for amino acids are usually interrupted by non-coding segments, stretches of DNA that are never translated. Most genes include long internal non-coding regions, which are called *introns*. The coding regions--the parts of a gene that are expressed--are called *exons*. Both exons and introns are transcribed from DNA into RNA. However, before RNA leaves the nucleus, the introns are removed, and the exons are joined to produce a mature mRNA molecule with a continuous coding sequence. This process is called *RNA splicing*.

Biologists have learned that RNA splicing can occur by two mechanisms. In mRNA splicing, the process is usually catalyzed by a complex of proteins and small RNA molecules. In certain other cases, splicing occurs without proteins or extra RNA molecules. In the latter case, the RNA transcript catalyzes the process.

What are the biological functions of introns? It is already clear that introns provide several possible ways for regulating gene expression. The splicing process itself may help control the flow of mRNA from nucleus to cytoplasm. Moreover, in some cases, the cell can carry out splicing in more than one way and thus generate different mRNA molecules from the same RNA transcript. In this fashion, an organism can get more than one type of polypeptide from a single gene.

Introns may also contribute to genetic diversity. Crossing over is one of the main sources of genetic variety. The presence of introns makes DNA longer, thereby increasing the number of points where crossing over can occur.

Before leaving the nucleus, mRNA undergoes other changes in addition to splicing: extra nucleotides are added to the two ends, making a small cap (a single nucleotide) and a long poly-A tail. It is suggested that the cap and the tail protect mRNA from attack by cellular enzymes and help ribosomes recognize the mRNA. The cap and the tail are not translated. The product of splicing, capping, and tailing is the cell's "final draft" of mRNA. After the “final draft” of mRNA is produced, it is transported to the cytoplasm.

Post-translational control mechanisms in cells often involve cutting cell polypeptides into smaller active final products. Another control mechanism operating after translation is the selective breakdown of proteins. For example, some of the proteins that trigger metabolic changes in cells are broken down within few minutes or hours. This regulation allows a cell to adjust the kind and amounts of its proteins in response to changes in its environment.

6.1.5. *Multiple mechanisms control gene expression*

Let us provide a review of cell gene expression and highlight the multiple control points where the process can be turned on or off, speeded up, or slowed down. Picture a series of pipes (see Fig. 13) that carry water from your local water supply, perhaps a reservoir, to a faucet in your home. At various points, valves control the flow of water.

We use this model to illustrate the flow of genetic information from a chromosome—a reservoir of genetic information—to an active protein that has been synthesized in the cell's cytoplasm.

The multiple mechanisms that control gene expression are analogous to the control valves in water pipes. Indeed, each gene-expression "valve" can be indicated by a "control knob." Note that these knobs represent possible control points. For most proteins, only few control points are probably important. The most important control point is usually the start of transcription.

Of course, this picture oversimplifies control of gene expression. What it does not show is the complicated web of control that connects genes and their products. The genes of operons in *E. coli*, for instance, are controlled by repressor or activator proteins encoded by regulatory genes on the same DNA molecule.

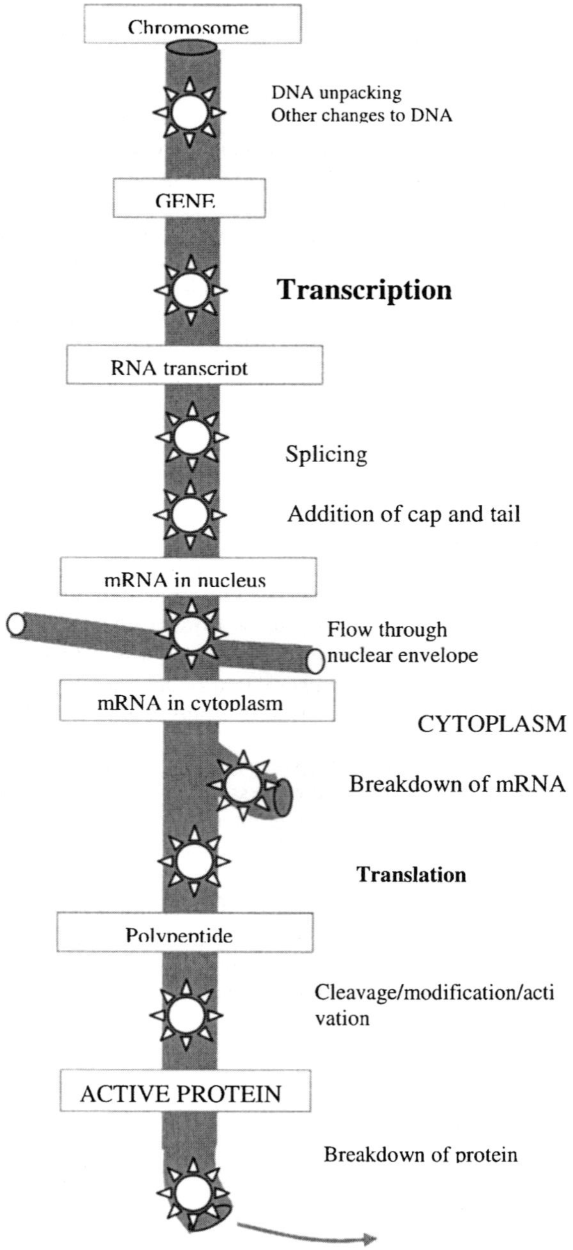

Figure 14. The gene-expression "pipe-line"

6.1.6. *Gene expressions and cell-to-cell signaling*

Among the earliest events in fruit-fly development is a cascade of gene expression and cell-to-cell signaling that determines which ends of the egg cell will become the head and the tail. These events occur in the ovaries of the mother fly and involve communication between an unfertilized egg cell and cells adjacent to it in its follicle (egg chamber). One of the first genes activated in the egg cell codes for a protein that leaves the egg cell and signals adjacent follicle cells. These follicle cells are stimulated to turn on genes for other proteins, which signal back to the egg cell. In response, the egg cell orients the microtubules of its cytoskeleton along what will become the new fly's head-to-tail axis.

Two different kinds of mRNA localize at opposite ends of the microtubules. One kind of mRNA marks the end of the egg where the fly's head will develop; the other mRNA marks where the tail will develop. With these mRNAs in place, the egg has two different poles that define a head-to-tail axis. This axis enables other axes (top to bottom and side to side) to form and ensures the proper positioning of body parts as they develop.

After the egg is fertilized, repeated mitosis transforms the zygote into a fly embryo. Translation of the "head" mRNA in the embryo produces a regulatory protein that is most concentrated at the head end. In turn, this protein gradient triggers a corresponding gradient of gene expression in the embryo's nuclei. The resulting proteins initiate several rounds of gene expression that subdivide the embryo into structural units called body segments.

Now finer details of the new fly can take shape. Protein products of some of the axis-forming genes and segment-forming genes determine the next genes to be expressed. These genes, called *homeotic genes*, in turn determine what body parts will develop from each segment. A *homeotic gene* is a *master control* gene that regulates batteries of other genes that direct the development of an organism's body plan.

Cascades of gene expression, with the protein products of one set of genes regulating the expression of a second set of genes, and so on, are a common theme in development.

6.1.7. *Signal-transduction pathways*

Cell-to-cell signaling, with proteins or other kinds of molecules carrying messages from signaling cells to receiving (target) cells, is a key mechanism in development.

In most cases, a signal molecule acts by binding to a receptor protein in the plasma membrane of the target cell and initiating a signal-transduction pathway in the cell. A *signal-transduction pathway* (see Fig. 15) is a series of molecular changes that converts a signal on a target cell's surface into a specific response inside the cell. The main elements of a signal-transduction pathway in which the target cell's response is the transcription (turning on) of a gene are the following:

1) The signaling cell first secretes the signal molecule.

2) The secreted molecule binds to a specific receptor protein embedded in the target cell's plasma membrane.

3) The binding activates the first in a series of relay proteins within the target cell. Each relay molecule activates another.

4) The last relay molecule in the series activates a transcription factor.
5) This factor triggers transcription of a specific gene.
6) Subsequently, translation produces the corresponding protein.

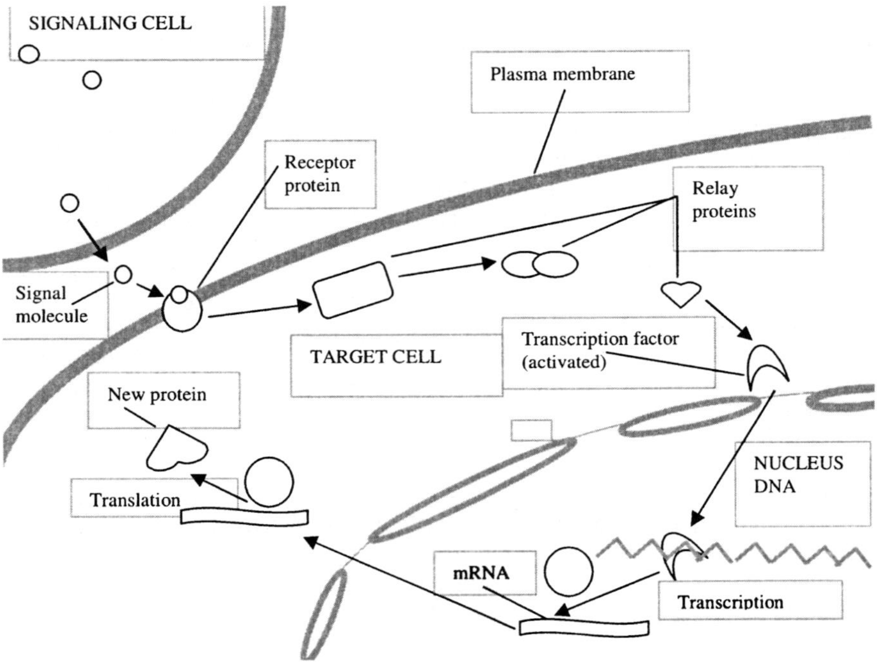

Figure 15. Ssignal-transduction pathway that turns on a gene

6.2. MM OF TURNING GENES ON OR OFF

For a bacterium, according to s. 6.1, in the system of the equations, for example,

$$m(t) = \int_{a(t)}^{t} \alpha(t,s)y(s)m(s)ds,\ c(t) = \int_{b(t)}^{t} \beta(t,s)[1-y(s)]m(s)ds,$$

$$R(t) = \int_{a(t)}^{t} y(s)m(s)ds + \int_{b(t)}^{t} [1-y(s)]m(s)ds,\ 0 < t^{\wedge} \leq t, \tag{48}$$

α and β correspond to the regulatory and utilization genes that are ready to act, m correspond to the RNA polymerase, and c to the utilization enzymes.

In addition, we have the equation

$$m^*(t) = \int_{a^*(t)}^{t} \alpha^*(t,s)y^*(s)m^*(s)ds,\ 0 < t^\wedge \leq t, \tag{49}$$

in which α^* corresponds to another regulatory gene and m^* to repressors that are binding operators (switch off). It is supposed that the last regulatory gene is expressed continuously.

While substrates $s(t)$, entering the cell, are releasing the repressors (switch on), the former genes are expressed, and the substrates and utilization enzymes start reactions, such as, for example, ATP synthesis or biosynthesis (see s. 3. and 4).

According to s. 6.1.1, turning on a eukaryotic gene involves transcription factors (among them are activators) in addition to RNA polymerase. The first step is the binding of activators to DNA enhancers. DNA bends, and activators interact with other transcription factors, which then bind as a complex to the gene's promoter.

This assembly of proteins facilitates the binding of RNA polymerase to the promoter and the initiation of transcription. Silencers play the role of repressors.

The respective MM may be the same type as (48) and (49) with those differences that we should now add several equations of (49)-type not only for silences, but also for transcription factors and for interaction of the activators A(t) with the other transcription factors TF(t).

MM of this interaction may have the form

$$\mathrm{AP}(t) = \int_{a^\wedge(t)}^{t} \alpha^\wedge(\mathrm{A},\mathrm{TF};t,s)y^\wedge(s)\mathrm{AP}(s)ds,\ 0 < t^\wedge \leq t, \tag{50}$$

where AP means assembly of proteins and $\alpha^\wedge$ is expressed by the respective gene and/or corresponds to a certain law of nature, for example,

$$\alpha^\wedge(\mathrm{A},\mathrm{TF};t,s)y^\wedge(s) = \alpha^\wedge\mathrm{A}(t)\mathrm{TF}(t),\ \alpha^\wedge = \text{constant}. \tag{51}$$

In the case of (51), we have

$$\mathrm{AP}(t) = \alpha^\wedge\mathrm{A}(t)\mathrm{TF}(t)R^\wedge(t),\ R^\wedge(t) = \int_{a^\wedge(t)}^{t} \mathrm{AP}(s)ds,\ 0 < t^\wedge \leq t. \tag{52}$$

Assuming as usual, that $R^{\wedge\prime}(t) = 0$, we have the determined system of equations with respect to AP(t) and $a^\wedge(t)$:

$$\mathrm{AP}(t) = \alpha^\wedge\mathrm{A}(t)\mathrm{TF}(t)\int_{a^\wedge(t)}^{t} \mathrm{AP}(s)ds,\ \mathrm{AP}(t) = \mathrm{AP}(a^\wedge(t))a^{\wedge\prime}(t), 0 < t^\wedge \leq t, \tag{53}$$

under the condition that AP(t) is given on the prehistory $[0, t^\wedge]$.

6.3. MM OF DNA TRANSCRIPTION AND mRNA TRANSLATION

According to s. 6.1.3. and 6.1.4., mRNA is the product of DNA transcription, splicing, capping, and tailing. MM of the respective function

$$\beta(t) = \int_{a(t)}^{t} \alpha(t,s)u(s)m(s)ds, \ \ 0 \leq u \leq 1, \tag{54}$$

where α corresponds to the DNA and m to the complex of proteins and small RNA molecules or DNA itself. In particular, $\alpha = \alpha(t\text{-}s)$, $m(s) = \alpha(s)$, and

$$\beta(t) = \int_{a(t)}^{t} \alpha(t-s)u(s)\alpha(s)ds, \ 0 \leq u \leq 1. \tag{54'}$$

The lifetime of mRNA, $t-a(t)$, is regulated likely by the redistributing factor $u(s)$.

Proteins are the product of mRNA translation by the ribosome. The respective simplest MM is

$$p(t) = \int_{b(t)}^{t} \beta(t-s)v(s)e(s)ds, \ 0 \leq v \leq 1, \tag{55}$$

where p corresponds to the proteins and e to the enzymes or enzyme-substrate complex.

After translation, selective breakdown of proteins takes place. It can be done by regulating the life span of proteins, $t\text{–}b(t)$, with the help of the function $v(s)$.

The natural conclusion that there should be genes that express mass of "controllers," MM of which may be MM of (49)-type.

6.4. MM OF SIGNAL-TRANSDUCTION PATHWAYS

A *signal-transduction pathway* is a series of molecular changes that converts a signal on a target cell's surface into a specific response inside the cell. The main elements of a signal-transduction pathway in which the target cell's response is the transcription (turning on) of a gene are 1) – 6) (see s. 1.6.7).

Let $s(t)$ correspond to the signal molecule (hormone, growth factor) from the signaling cell, $p_k(t)$, $k = 1,\ldots, l$, correspond to relay proteins within the target cell, $m(t)$ to the transcription factor, and $p(t)$ to the resulting protein.

Then the desired MM may be the following:

$$p'_k(t) = c_k\, p_{k\text{-}1}(t)e_k(t),\ p_0(t) = s(t);\ p_l(t) = \text{AP(t)},$$

$$e_k(t)(t) = \int_{a,k(t)}^{t} \alpha_k(t,s)y_k(s)e_k(t)(s)ds,\ k = 1,\ldots, l;\ 0 \leq y_k(s) \leq 1,$$

$$m(t) = \int_{a,m(t)}^{t} \alpha_m(t,s)y(s)\text{AP}(s)ds,\ 0 \leq y \leq 1,\ \beta(t) = \int_{a(t)}^{t} \alpha(t,s)u(s)m(s)ds,\ \ 0 \leq u \leq 1,$$

$$p(t) = \int_{b(t)}^{t} \beta(t-s)v(s)e(s)ds,\ 0 \leq v \leq 1,\ 0 < t\text{^} \leq t, \tag{56}$$

where $e_k(t)$ corresponds to the necessary enzymes that are expressed by the genes corresponding to $\alpha_k(t,s)$, $\alpha_m(t,s)$ corresponds to the gene that expresses the complex corresponding to $m(t)$, and the remaining magnitudes are the same as above.

There are two types of signal-transduction pathways leading to the synthesis of proteins that influence cell division. The first type occurs when the initial signal is a growth factor, and the target cell's response is the production of a protein that stimulates the cell to divide. A cell that responds to a growth factor has a specific receptor protein in its plasma membrane (cells that do not respond to growth factors lack the specific receptors). By contrast, the second type occurs when we have an inhibitory pathway, in which a growth-inhibitory factor causes the target cell to make a protein that inhibits cell division.

An oncogene protein can cause increased cell division in the absence of growth factors. A mutant tumor-suppressor protein can also affect cell division. The gene for the protein that inhibits cell division can be turned off, and cell division may occur.

6.5. MM OF MOLECULAR PUMPS

Within cells, sodium concentration is lower and potassium concentration is higher than in interstitium outside. When a nerve is stimulated, sodium ions pour into the nerve cell. The difference in concentration is restored by sodium being transported out once again. This transport requires ATP. So, there exists the ATP-degrading enzyme, *ATPase*, in the nerve membrane that is activated by sodium and potassium ions (Na^+, K^+-ATPase). *ATPase* can promote directed (vectored) transport of substances through a cell membrane, a fundamental property of all living cells.

The enzyme consists of two sub-units, alpha and beta. The first carries the enzyme's activity and the other presumably stabilizes the structure [86, 155].

Note that it corresponds to the main conception of MM under consideration (with the opposite role of the functions of α and β-types).

The enzymes are located in the cell membrane and expose their surfaces to the outside as well to the inside. Three sodium ions and ATP bind to the interior surface. A phosphate is then transferred from ATP to an amino acid in the enzyme, ADP is liberated, and sodium ions are transported to the outside. Here they are released and two potassium ions attach instead. When the phosphorus is removed the potassium ions are transported into the cell and when new ATP binds to the enzyme they are rejected.

As a result of the action of the Na^+/K^+-ATPase, the cell keeps a high concentration of potassium inside. Since the cell membrane is rather permeable to potassium ions, few of them leak out, leaving non-permeable, negative charges on the inside of the cell. Therefore, the inside of the cell membrane, as compared to the outside, becomes electrically negatively charged. The difference in potential across the membrane is necessary for a nerve stimulation to propagate along a nerve fiber or along a muscle cell. The difference in potential is also the driving force in the uptake of important nutrients, e.g. glucose and amino acids, and other ions.

Following the discovery of Na^+/K^+-ATPases, other ion pumps have been discovered with similar structures and functions. Examples include Ca^{2+}-ATPases participating in the control of muscle contraction and H^+/K^+-ATPases releasing hydrochloric acid into the stomach.

The respective MM may be as follows:

$$m(t) = \int_{a(t)}^{t} \alpha(t,s)m_1(s)ds = m_1 + m_2,$$

$$c_i(t) = \beta_i x(t)c'''(t)\int_{a(t)}^{t} m_2(s)ds,$$

$$R'(t) = m(t) - m(a)a'(t) = 0,\ 0 \le a(t) \le t,\ t > t^{\wedge} > 0, \tag{57}$$

where $m(t)$ is the rate of creation of new enzyme-substrate complexes (ATPase-ions) responsible for their own reaction and transportation of the $c_i(t)$, ions, phosphorous, and ADP; '''c is the rate of creation of ATP (see s. 3); x is a relative share of the intake of '''c; and the remained magnitudes have the usual definition as above.

Since $x(t)c'''(t)$ has a periodic character, the solutions c_i, under m_1/m = constant, also have a periodic character (see Ch. 5 and 14 for more details).

The simple model may be

$$x(t)c'''(t) = A[\sin(\omega t + \varphi) + 1]. \tag{58}$$

Chapter 7

MM OF THE CELL

1. On Cell Biology

The cell is the fundamental unit of structure and function in all-living things. This is the so-called *cell doctrine* ([26]; N. Campbell et al. [30]; R. Caret et al. [32]; H. Curtis [37]).

It has been estimated that the various organs and tissues of an adult human contain more than 10^{14} cells. All these cells are derived initially from a single cell (a fertilized *egg cell*) through growth and division. With the exceptions of muscle and nerve, replacement of old tissue cells with new ones results in a complete turnover of the body's cellular composition every few years.

A portion of the body's total mass is represented by non-cellular material that is secreted by cells. For example, most of the mass of bone and cartilage is composed of secreted calcium salts and proteins.

1.1. MOLECULAR CONSTITUENTS OF CELLS

All cells contain proteins, polysaccharides, lipids, and nucleic acids. In addition to these macromolecules, all cells contain a wide variety of smaller compounds such as water, salts, ions, vitamins, etc.

The association of water with other molecules usually takes the form of *hydrogen bonds*. These bonds have relatively low energy, yet they are vital to the molecular architecture of the cell. Salts are usually present in ionic form and many of these are associated with macromolecules. Salt ions are also important to the osmotic behavior of cells and to the buffering activities. Gases that enter the cell from the environment or are produced by cell's metabolism dissolve in the cytoplasmic water.

Some of the more important compounds are nucleoside phosphates, ligands, and vitamins. Nucleoside phosphates are important sources of energy and act to regulate specific metabolic reaction sequences in cells. Ligands are molecules that contain nitrogen or oxygen atoms capable of donating electron pairs to metals or metal ions. Vitamins are often components of coenzymes, which act in concert with an enzyme to facilitate a specific chemical reaction.

An *acid* is a proton donor and a *base* is a proton acceptor. An acid that readily donates protons to water is called a *strong acid*, while an acid that does not readily donate protons to water is termed a *weak acid*. Each acid is characterized by its tendency to dissociate; the extent of dissociation may be determined from the acid's *dissociation constant* $k' = [H^+][A^-]/[HA]$, where for a given solution, [HA] is the concentration of undissociated acid, $[A^-]$ is the concentration of conjugate base, and $[H^+]$ is the hydrogen ion concentration. To express the H^+ concentration, the *pH scale* is proposed:

$$pH = -\log_{10} [H_3O^+], pOH = -\log_{10} [OH^-].$$

At 25°C, the H_3O^+ (and OH^-) concentration of pure water is 10^{-7} *M*. At this concentration pH = 7.0. For comparison, human gastric fluid has an approximate pH of 1 and soil water has an approximate pH of 9.

A molecule is a stable union of two or more atoms. The formation of the *bonds* that exist between the atoms, comprising a molecule, requires less energy than that needed to keep the atoms apart. The breakage of the bonds that join atoms consumes energy. If a large amount of energy is required to break the bond, the bond is called a *strong bond*; otherwise, it is called a *weak bond*. Four major types of bonds may be identified: covalent bonds (the strongest), ionic bonds, hydrogen bonds, and hydrophobic bonds (the weakest).

Of the four major classes of macromolecules, the *proteins* are the most diverse and complex, being composed of one or more chains of amino acids. About 20 different amino acids occur in proteins and are covalently linked together within each chain by peptide bonds.

The specific sequence of amino acids in each of the polypeptide chains forming a protein is called the *primary structure*. Each polypeptide chain may contain one or more regions twisted to form helical structures stabilized by covalent, hydrophobic, electrostatic, and/or hydrogen bonds. This is called the *secondary* level of protein *structure*. A more compact structure may be achieved by folding the polypeptide chain in the non-helical regions; this establishes the polypeptide's *tertiary structure*. In proteins that have two or more polypeptides, the specific orientation of the chains with respect to one another is called *quaternary structure*. Why and how proteins fold the way they do remains one of the major mysteries in structural biology [178]. Scientists call this the *folding problem*.

Polysaccharides are macromolecules composed of individual sugar units called monosaccharides. From a functional standpoint, polysaccharides may be divided into nutrient and structural types. Nutrient polysaccharides such as glycogen and starch are often stored in cells as discrete granules. Polysaccharides may also occur in covalent combination with proteins, short peptides, and lipids.

Lipids are a heterogeneous collection of molecules soluble in non-polar solvents and they play two major roles in cells and tissues: (1) they are sources of reserve energy and (2) they are structural constituents of cellular membranes.

The two major nucleic acids, *DNA* and *RNA*, are composed of unbranched chains of subunits called nucleotides, each nucleotide containing a phosphate, a monosaccharide, and a nitrogenous base. DNA serves as the genetic material in all cells and in many viruses. The sperm and egg cells contain half the DNA of the diploid somatic cells.

Most RNAs are formed by a single polynucleotide twisted about itself in certain regions. The ribosomal RNAs are the most abundant and serve as functional components of the cell's ribosomes during protein synthesis. The messenger RNAs contain a sequence of nucleotides that specify the primary structures of the cell's proteins. Transfer RNAs function in the transport of amino acids to the messenger RNA-ribosome complex during protein synthesis or translation.

Cell's RNAs are produced by transcription of its DNA.

1.2. STRUCTURE AND FUNCTIONS OF CELLS

We consider only animal cells, which are characterized by a number of discrete *organelles*, especially the *plasma membrane*, *nucleus* and *nucleolus*, *mitochondria*, *Golgi bodies*, *lysosomes* and *micro bodies*, *ribosomes*, *endoplasmic reticulum* (ER), and the *cytoskeleton* (see, for example, N. Campbell et al. [30]).

The cell is delimited at its surface by the plasma membrane and the cytoplasm is furrowed by the membranes of the *endoplasmic reticulum* (ER) so that the cell volume is divided into two phases: the cytoplasm and intracisternal phase. The cell's membranes actively participate in the exchange of material between the cytoplasm and the surroundings. Both the proteins and lipids of the membrane are asymmetrically distributed across the inner and outer halves. In cells, the proteins in the inner half of the membrane are anchored in position by a cytoskeletal network consisting of filaments and microtubules. Carbohydrate chains associated with proteins and lipids in the outer membrane surface play roles in maintaining membrane organization, in transmembrane transport, in cell-to-cell adhesion, and in providing the membrane's antigenic properties. In tissues, the plasma membranes of neighboring cells exhibit specialized junctional regions that are important in maintaining tissue integrity and in regulating the passage of a substance across a tissue and from cell to cell.

According to the *signal hypothesis*, proteins that are to be either secreted from the cells, dispatched to the *lysosomes*, or incorporated into plasma membranes are encoded in mRNA by a special nucleotide sequence called a "signal." When a ribosome attaches to the mRNA in the cytoplasm and begins to translate the message, the signal sequence reorganized by the signal recognition particle (SRP) brings about a temporary halt to protein synthesis by that ribosome. Synthesis is resumed only if the SPR-ribosome complex attaches to the ER at specific sites occupied by SRP receptors called *docking proteins*. SRP is returned to the cytoplasm where it can participate in another round of signal recognition and docking (*SRP cycle*). When protein synthesis is resumed by the docked ribosome, the elongating polypeptide chain passes through the ER membrane into the intracisternal space. This process, termed *translocation*, is presumed to involve active participation of elements of the membrane. The major distinction between the synthesis of *secretory proteins* and *membrane proteins* is that the first ones are released into the lumenal phase of the ER, whereas the second proteins remain anchored in the ER.

The *Golgi bodies* consist of a parallel series of flattened vesicles or *cisternae* that form a bowl-shaped organelle. Smaller vesicles join the stack on one edge and leave the stack at the other edge. Proteins destined for secretion, insertion into the plasma membrane, or deposition into the intracellular organelles reaches the face of the Golgi apparatus from neighboring rough ER. In the cis and medial regions of the Golgi, these proteins are phosphorylated, glycosylated, sulfated, or processed in other ways. In the trans region of the Golgi apparatus, the proteins are sorted according to their final destination. Vesicles containing Golgi products destined for other cellular compartments are dispatched from the surface of the Trans Golgi. Golgi bodies also play roles in different type of cells.

The Golgi apparatus dispatches completed plasma membrane glycoproteins as small vesicles that migrate to and fuse with the plasma membrane. Synthesis of membrane lipids occurs in the plasma membrane by a similar to the proteins; that is, both membrane proteins and lipids pass from the ER to the Golgi apparatus and are later dispatched to the plasma membrane via small vesicles.

The movement of materials across cell membranes may be achieved by passive or active mechanisms. The passive movements of solutes by diffusion and of water by osmosis across the membrane occur principally as the result of concentration gradients. In some cases, diffusion through the membrane is facilitated by carrier molecules that possess the properties of enzymes. Substances can also move through the plasma membrane against their concentration gradient; this consumes cellular energy (ATP, adenosine tri-phosphate, is hydrolyzed in the process) and is termed *active transport*. For individual ions and small molecules, active transport is affected by membrane-associated enzymes that act as pumps. However, transport into and out of cells can be in bulk and is then called endocytosis (in) and exocytosis (out). *Bulk transport* involves gross movements of the plasma membrane, the three most common form of bulk transport being pinocytosis, receptor-mediated endocytosis, and phagocytosis.

Mitochondria are the "powerhouses" of the cells and the primary sites of cell oxidations. Within these organelles, elementary substrates, produced by the breakdown of carbohydrates, lipids, or nitrogenous molecules in other cell locations, are oxidized to CO_2 and water. Some of the energy of these exergonic oxidations is conserved in the phosphorylation of ADP (di-phosphate) to form ATP. All mitochondria contain two structurally and functionally different membranes—an inner and an outer membrane. A fluid-filled intermembrane space is between the two membranes. The inner membrane surrounds the mitochondrial matrix. New mitochondria form by the fission of other mitochondria. The oxidative and phosphorylation reactions occur in the inner membrane or in the matrix. The tricarboxylic acid or *Krebs cycle* reactions constitute the first phase of the oxidation of substrates such as acetate. In these reactions, the molecules are enzymatically degraded to CO_2 and water. H^+ and electrons released from the metabolic intermediates reduce NAD^+ and FAD^{2+} to NADH and $FADH_2$. Some ATP is formed at the substrate level by these reactions. Most of the ATP generated in mitochondria is via the *electron transport system* (ETS), which reoxidizes the NADH and $FADH_2$ formed in the *Krebs cycle*. The ETS functions as a multistep series of oxidation-reduction reactions, transferring electrons through a set of intermediates associated with the inner membrane and ultimately reducing molecular oxygen to water. Associated with the ETS is a mechanism for the formation of ATP called oxidative phosphorylation. For each pair of electrons transferred from NADH, three molecules of inorganic phosphate are added to three ADP to form three ATP molecules (two ATP for each FADH2 that is reoxidized). Oxidative phosphorylation is a function of the inner membrane.

Lysosomes function in the intracellular digestion of poorly functioning or superfluous organelles, as well as endocytosed materials. Lysosome engorgement is believed to be involved in the aging process. Several forms of lysosomes may be identified, including primary, secondary (e.g. heterophagic and autophagic vacuoles), and residual bodies lysosomes.

Primary lysosomes are released from the maturing or the trans face of Golgi bodies, their hydrolase content initially derived from rough ER. Fusion of primary lysosomes with endosomes forms the secondary lysosomes in which digestion occurs. Usable products of this digestive activity are transferred to the cytoplasm. Undigested or unabsorbed materials remain in the residual bodies, which may accumulate in the cell or fuse with the plasma membrane during exocytosis.

Micro bodies contain a number of flavin oxidases that produce hydrogen peroxide during their degradative activity. The potentially harmful peroxides are further degraded by peroxisomal catalase.

The *nucleus* of a cell is delimited by a pair of membranes called nuclear envelope. The fluid of the cytoplasm is continuous through the nuclear pores with the fluid of the nucleus, called the nucleoplasm. The nucleoplasm contains a number of discrete structures including one or more chromosomes, nucleoli and other structures or regions, which appear at various times depending on nuclear activity.

Chromosomes are composed of chromatin, which readily binds stains. Each species has a specific chromosome number. Human cells have 46 chromosomes. Chromosome shape and size change during the stages of nuclear divisions. Most chromosomes have two arms, one on each side of the primary constriction or centromere.

The centromere is the site of attachment of the chromosome to the microtubules of the spindle and acts as the focus of chromosome movement during the anaphase phase of division. Chromosomes that lack a centromere are said to be acentric and fail to segregate normally during division.

Secondary constrictions are associated with nucleoli and are called nucleolar organizer regions (NOR). Characteristic disposition of the tertiary constrictions helps to distinguish one chromosome from another.

The condensed chromosomes visible during mitosis are composed of an organized array of chromatin fibers. Each chromatin fiber is believed to contain one molecule of DNA. Chromatin consists of a repeating pattern of bodies called nucleosomes formed by association of DNA and specific nuclear proteins called *histones*. The histones form a core around which the 2-nm-thick molecule of DNA is wound. The DNA makes two turns around the histone core, each turn consisting of about 83 base pairs. Successive nucleosomes are connected by linker DNA. Nucleosomes can be released from chromosomes by partially digesting the chromatin with a nuclease.

During nuclear division or *mitosis*, there is a progressive change in the structure and appearance of the chromosomes. Mitosis is usually divided into four major stages: prophase, metaphase, anaphase, and telophase.

Prophase is characterized by the condensation of the chromosomes, the disappearance of nucleoli and nuclear envelope, and the formation of the microtubules of the spindle. If, prior to prophase, the cell contained a centriole, then a second centriole is formed. The two centrioles move apart as the spindle forms. The chromosomes are seen to be composed of two sister chromatids held together at the centromere. The sister chromatids are the products of replication of chromosomal DNA during interphase. Toward the end of prophase, the spindle extends between two poles positioned diametrically opposite one another in the cell and the chromosomes migrate toward the center of the spindle. In metaphase, the centromeres of each chromosome are aligned midway across the spindle on a plane called the equatorial plate. At this time, the centromeres are linked to the spindle fibers.

The centromeres are duplicated so that each chromatid becomes an independent chromosome and is attached to a spindle fiber connected to one of two poles. The onset of anaphase is characterized by the movement of the chromosomes toward opposite poles of the spindle. During anaphase, a process called cytokinesis begins and divides the cell into two halves, thereby physically separating the two complements of chromosomes. Cytokinesis is distinct from but frequently synchronized with nuclear division, occurring during the later stages of mitosis. In telophase, the chromosomes reach the poles of the spindle and begin to undergo decondensation. During telophase, nucleoli reappear, as does a new nuclear envelope enclosing the chromosomes.

The *nucleolus* is composed of RNA and proteins and is the site of formation of ribosomal constituents. NOR contain the genes that are transcribed into ribosomal RNA (rRNA).

DNA replication is semi-conservative and bi-directional. The overall growth of the two new complementary DNA strands during replication proceeds from the 5' carbon end toward the 3' carbon end. The elongation of one strand is continuous and the other grows as a series of Okazaki fragments (1000-2000 nucleotides long) that are linked together by ligase. Many enzymes are involved in the replication process.

Recombination is the exchange of genetic material between chromosomes that results in new combinations of genes in successive generations of cells. Of special significance is the use of recombinant DNA methods for transferring human genes and cloning them in bacteria.

The *ribosomes* in the microsomal fraction carry out the assembly of the polypeptide chains that make up proteins. A ribosome is made up of two subunits, each subunit containing a specific combination of proteins and rRNA. Association of these subunits with mRNA is followed by translation of the mRNA base sequence into the primary structure of polypeptides. Transfer RNA (tRNA) combines with amino acids and enters the ribosome-mRNA complex. The succession of aminoacyl-tRNA species entering the complex and donating amino acid residues to the growing polypeptide is prescribed by the mRNA base sequence. All the RNAs are transcribed from DNA and are processed (cleaved, trimmed, or otherwise modified) prior to becoming functionally active. Ribosomes may be free in the cytoplasm or attached to the ER, each variety functioning in the assembly of different proteins. Many substances, including a number of antibiotics, act as inhibitors of protein synthesis.

The *cytoskeleton* or cytoplasmic matrix is composed of cytoplasmic filaments, microtubules, microtrabeculae, and the intertrabecular space. These structures give shape to the cell, position and move the intracellular organelles, and move the cell as a whole. Cytoplasmic filaments are formed by a linear array of proteins subunits, whereas microtubules consist of an array of protein dimers forming a hollow cylinder. Cytoplasmic filaments and microtubules interact with each other during a number of cellular activities including muscle contraction and cytokinesis.

Cell-to-cell signaling, with proteins or other kinds of molecules carrying messages from signaling cells to receiving (target) cells, is a key mechanism in development. In most cases, a signal molecule acts by binding to a receptor protein in the plasma membrane of the target cell and initiating a signal-transduction pathway in the cell.

1.3. ON CELL METABOLISM

Common reactions in cells are reactions involving two or more reactants (substrates) and/or two or more products.

Nearly all reactions that occur in cells are catalyzed by a class of proteins called *enzymes*. Enzymes increase the likelihood that potentially reacting molecules (*substrates*) will encounter each other with the necessary orientation in space and activate their substrates by imposing strains on certain bonds. Although enzymes greatly increase reaction rates, they do not alter equilibrium concentrations. In the presence of increasing amounts of substrate, reactions approach a limiting velocity, the kinetics of which may be used to determine the enzyme's *Michaelis constant* K_m, a measure of the tendency of the enzyme and substrate to combine with each other. The greater affinity of an enzyme for its substrate, the lower the K_m value. This is because the K_m value is numerically equal to the substrate concentration at which half of the enzyme molecules are associated with substrate (in the enzyme-substrate, or ES form).

The catalytic properties of enzymes are intimately related to the enzyme's primary, secondary, tertiary, and quaternary structure. Change in structure is often accompanied by loss of catalytic activity. Some enzymes require a non-protein component called a *cofactor* to be catalytically active. In addition to the active site, some enzymes possess an allosteric or regulatory site. The binding of a positive effector to the allosteric site increases the activity of the enzyme, whereas the binding of a negative effector decreases the activity of the enzyme.

The metabolism of a cell can be subdivided into two categories: *catabolism* (the breakdown of molecules into simpler forms, accompanied by the release of energy) and *anabolism* (the synthesis of complex molecules from simpler ones, accompanied by the consumption of energy).

Bioenergetics is the study of all the energy changes that take place in living systems. Metabolic reactions that yield energy are called *exergonic reactions* and reactions that consume energy are called *endergonic reactions*. In most instances the energy of exergonic reactions is used to attach phosphate to ADP, thereby forming ATP.

Energy changes in cells obey the first two *laws of thermodynamics* (see [180] for more detail). The first law states that energy cannot be created or destroyed but can be converted from one form into another. The second law states that in all processes involving energy changes within a system, the entropy of the system increases until equilibrium is achieved.

Conversion of energy from one form to another is called *transduction*. Light energy occurs in units called quanta. Light energy is converted to chemical energy in such mechanisms as photosynthesis and vision.

The reaction sequences, called *metabolic pathways*, are frequently associated with specific organelles, the enzymes that catalyze the reactions being compartmentalized between or within the membranes of the organelle. The products of catabolism in one organelle may be transported to other organelles for further catabolism or anabolism.

The most common metabolic pathways in cells are:

Glycolysis - the catabolism of monosaccharides to pyruvate.

Fermentation – the catabolism of monosaccharides to products such as ethanol and CO_2 in the absence of air (anaerobic conditions).
Tricarboxylic acid (Krebs) cycle – the oxidation of pyruvate and its catabolism to CO_2 and water in the presence of oxygen.
Gluconeogenesis – the synthesis of glucose from simple organic acids.
Glycogen and starch synthesis – the incorporation of sugars into polysaccharides.
Fatty acid synthesis – the pathway leading to the formation of fatty acids and triglycerides from acetyl-CoA and glycerol.
β-oxidation – the pathway for breakdown of fatty acids.
Amino acid synthesis – the mechanism for formation of amino acids to be incorporated into proteins.

1.4. ON CONTROL OF GENE EXPRESSION

Cascades of gene expression, with the protein products of one set of genes regulating the expression of a second set of genes, and so on, are a common theme in development.

Differences among cells in an organism result from the selective expression of genes. As a developing embryo undergoes successive cell divisions, different genes must be activated in different cells at different times. Groups of cells follow diverging developmental pathways, and each group develops into a particular kind of tissue. Finally, in the mature organism, each cell type—nerve or muscle, for instance—has a different pattern of turned-on genes.

We have already considered the biological background, the problems and MM of gene expression in Ch. 6, s. 6.1 – 6.5. Here we present certain additional aspects of gene expression.

The lifetime of mRNA molecules is one important factor regulating the amounts of various proteins that are produced in the cell. Long-lived mRNAs are generally translated into much more protein than short-lived ones.

This is one reason bacteria can change their proteins relatively quickly in response to environmental changes. In contrast, mRNAs of eukaryotes can have lifetimes of hours or even weeks.

The process of translation also offers opportunities for regulation.

The final opportunities for the control of gene expression occur after translation. Post-translational control mechanisms in cells often involve cutting cell polypeptides into smaller, active final products.

Another control mechanism operating after translation is the selective breakdown of proteins. This regulation allows a cell to adjust the kind and amounts of its proteins in response to changes in its environment.

The genes of one cell can produce chemicals such as hormones that induce a neighboring or distant cell to develop along a certain pathway.

1.5. SUMMING UP

As was described briefly above, the *generalized structure* of interactions of different component of a *cell* has the following form (**f** is a flux into a cell, **g**, **c** is a flux out of a cell, **g** is waste products).

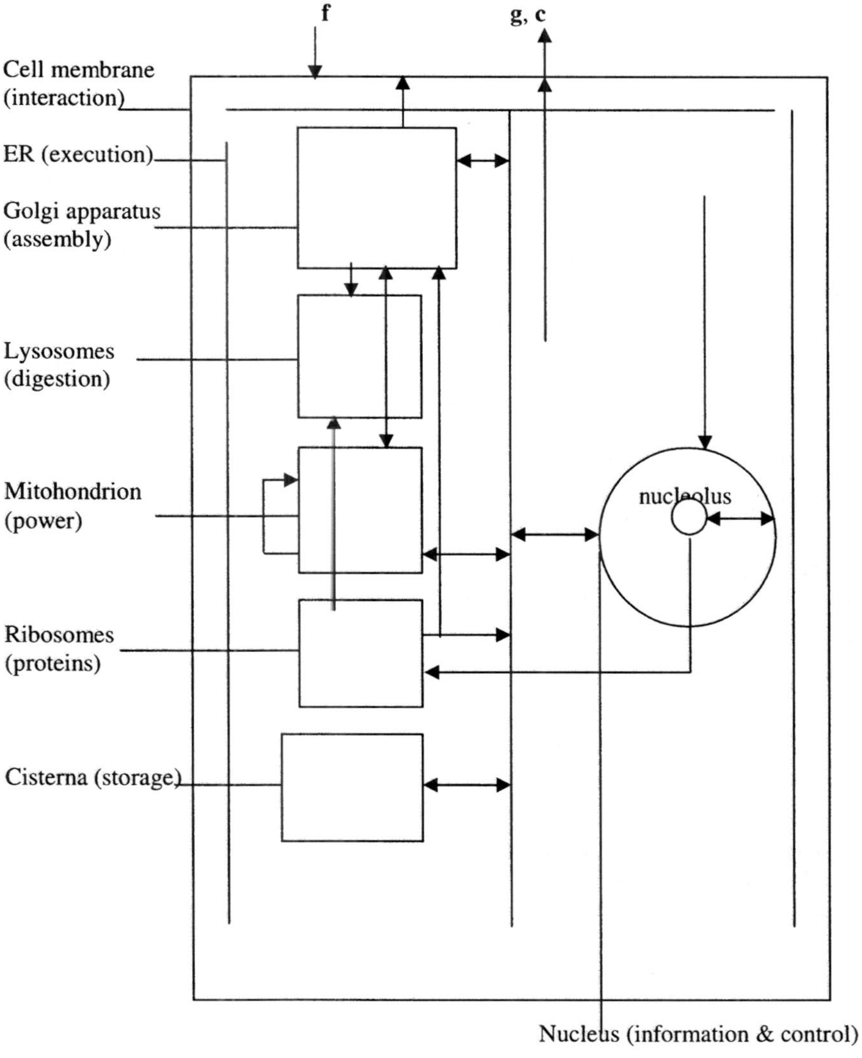

Figure 16. Generalized structure of the cell

2. Base MM of a Cell

This MM has the form

$$\boldsymbol{m}(t) = \int_{a(t)}^{t} \alpha(t,s)\boldsymbol{y}(s)\boldsymbol{m}(s)ds = (m^1(t), m^2(t), m^3(t)), \boldsymbol{M}(t) = \int_0^t \boldsymbol{m}(s)ds,$$

$$\boldsymbol{c}(t) = \int_{b(t)}^{t} \beta(t,s)[\boldsymbol{I}-\boldsymbol{y}(s)]\boldsymbol{m}(s)ds = (c^1(t), \ldots, c^P(t)), \boldsymbol{C}(t) = \int_{t^\wedge}^{t} \boldsymbol{c}(s)ds,$$

$$\boldsymbol{R}(t) = \int_{a(t)}^{t} \boldsymbol{y}(s)\boldsymbol{m}(s)ds + \int_{b(t)}^{t} [\boldsymbol{I}-\boldsymbol{y}(s)]\boldsymbol{m}(s)ds, \boldsymbol{G}(t) = \boldsymbol{M}(t) - \boldsymbol{R}(t) > 0,$$

$$m^1(t) + m^2(t) + m^3(t) = m(t), c^1(t) + \ldots + c^P(t) = c(t),$$

$$f(t) = m(t) + c(t), F(t) = \int_0^t f(s)ds, \boldsymbol{y}(t) = (y^1(t), y^2(t), y^3(t)),$$

$$\boldsymbol{a}(t) = (a^1(t), a^2(t), a^3(t)), \boldsymbol{b}(t) = (b^1(t), b^2(t), b^3(t)),$$

$$0 \le \boldsymbol{y}(t) \le 1, 0 \le \boldsymbol{a}(t), \boldsymbol{b}(t) \le t, t > t^\wedge > 0; \qquad (1)$$

where $m^1(t)$ is the rate of creation of new enzyme-substrate complexes responsible for recreation of themselves and the cell bio-mass components, $m^2(t)$ is the rate of creation of new enzyme-substrate complexes responsible for maintenance and recreation of themselves and cell energy components, $m^3(t)$ is the rate of creation of new enzyme-substrate complexes responsible for maintenance and recreation of themselves and cell information components, $c^i(t)$ is the rate of creation of new ith internal or external product of cell, and the remaining magnitudes have the usual definition as above.

If DNA of the cell and/or the respective natural laws (i.e., the functions of α and β-types), the control factors $\boldsymbol{y}$, the function f, c or the condition $f' = 0$, $c' = 0$, and all the functions on the prehistory $[0, t^\wedge]$ are known, then the system (1) is determined with respect to all the remaining magnitudes. The control factors $\boldsymbol{y}$ are usually determined with the help of some of optimization problems, unless they are already given due to the cell development prehistory.

The condition $f' = 0$, $c' = 0$, obviously have a place for stationary functioning of the cell.

2.1. MM OF CELL STRUCTURE AND FUNCTIONS

MM in question has the form

$$m(t,X) = \int_{a(t)}^{t} \int_V \alpha(t,s,e(X \cup Y))m(^1)(s,Y)dYds,\ M(t,X) = \int_0^t m(s,X)ds,$$

$$M(t) = \int_V M(t,X)dX,\ c(t,X) = \int_{\cdot(t)}^{t} \int_V \beta(t,s,e(X \cup Y))m(^2)(s,Y)dYds,$$

$$c(t,Z) = \int_{a(t)}^{t} \int_V \beta(t,s,e(Z \cup Y))m(^2)(s,Y)dYds,\ C(t,Z) = \int_{t\wedge}^{t} c(s,Z)ds,$$

$$R(t) = \int_{a(t)}^{t} \int_V m(^1)(s,Y)dYds + \int_{a(t)}^{t} \int_V m(^2)(s,Y)dYds,\ m(^1)(t,X) + m(^2)(t,X) = m(t,X),$$

$$G(t) = M(t) - R(t), f(t,X,Z) = m(t,X) + c(t,Z), f(t) = \int_{R,3} f(t,X,Z)dXdZ = k(f)\int_V m(t,X)dX,$$

$$X(t) = \int_{a(t)}^{t} \int_{O}^{X(t)} \alpha(t,s,e(X(t) \cup Y))m(^1)(s,Y)dYds,$$

$$F(t) = \int_0^t f(s)ds,\ 0 \le a, b \le t,\ t > t\wedge > 0,\ O, X \in V, Z \in R,3\backslash V, \qquad (2)$$

where $V = V(t)$ is the domain that a cell occupies in $R^3 = R,3$: $X = (x_1, x_2, x_3)$; $m(t,X)$ are the new component numbers (or concentrations) of cell's enzyme-substrate complex at the instant t in its point X per unit of time (this means that X is simultaneously notation for the products $m(t,X)$ of X-type and by the same token the various products of a cell are distributed in the space); $M(t)$ are the total created enzyme-substrate complex quantities of a cell; $R(t)$ are the total functioning enzyme-substrate complex quantities of a cell; $G(t)$ are the total obsolete enzyme-substrate complex quantities of a cell; $c(t,X)$ and $c(t,Z)$ are the new additional resource quantities in the unit of time in the point X and Z that are necessary for fulfillment of the internal and external functions of a cell; $\alpha(t,s,e(X \cup Y))$ are the efficiency indices of a cell functioning along the channel $m(^1)(s,Y)$ -- $m(t,X)$, corresponding to genes of a cell, etc.; $\alpha(t,s,X)$ and $\beta(t,s,Y))$, $X \cup Y \in V^*$, where V^* is the domain for the nucleus, corresponding to DNA of a cell, and the set function e: $R,3 \to V^*$ corresponds to the mapping, realized by RNAs of a cell.

If α, β, e, $m(^1)/m$, and c, R or the conditions $c'_t = R'_t = 0$ are given, then we have the determined system of the equations (2), whenever all elements are known on the prehistory $[0, t\wedge]$.

Instead of c, the function f or the condition $f'_t = 0$ may be given.

2.2. MM OF A CELL WITH REGARD TO ITS GIVEN SPATIAL STRUCTURE

According to s. 1.5, we can transform MM (2) into a more specific one. Indeed, let

$$V = \cup V{,}j \ (j = 1, \ldots, 10), \tag{3}$$

where $V{,}j$ are the domains of the respective parts of a cell's location: the nucleus, $j = 1$; the cytoplasm, $j = 2$; the mitochondria, $j = 3$; the ribosome, $j = 4$; the ER, $j = 5$; the Golgi apparatus, $j = 6$; the plasma membrane, $j = 7$; the lysosomes and peroxisomes, $j = 8$; the cisterna, $j = 9$; and the cytoskeleton, $j = 10$.

Then in connection with the structure and functions of the particular parts of a cell

$$\boldsymbol{m}(t,X) = \Sigma \int_{a(t)}^{t} \int_{V,j} \alpha(t,s,\boldsymbol{e}(X \cup Y))\boldsymbol{m}(^{1})(s,Y)dYds \ (j = 1,\ldots,5),\ X \in V{,}1,\ V{,}2,\ V{,}3,\ V{,}6;$$

$$\boldsymbol{m}(t,X) = \Sigma \int_{a(t)}^{t} \int_{V,j} \alpha(t,s,\boldsymbol{e}(X \cup Y))\, \boldsymbol{m}(^{1})(s,Y)dYds \ (j = 1,3,5,6),\ X \in V{,}4;$$

$$\boldsymbol{m}(t,X) = \Sigma \int_{a(t)}^{t} \int_{V,j} \alpha(t,s,\boldsymbol{e}(X \cup Y))\boldsymbol{m}(^{1})(s,Y)dYds \ (j = 1,3,4,7),\ X \in V{,}5;$$

$$\boldsymbol{m}(t,X) = \Sigma \int_{a(t)}^{t} \int_{V,j} \alpha(t,s,\boldsymbol{e}(X \cup Y))\boldsymbol{m}(^{1})(s,Y)dYds \ (j = 1,\ldots,6),\ X \in V{,}7;$$

$$\boldsymbol{m}(t,X) = \Sigma \int_{a(t)}^{t} \int_{V,j} \alpha(t,s,\boldsymbol{e}(X \cup Y))\boldsymbol{m}(^{1})(s,Y)dYds \ (j = 1,\ldots,7),\ X \in V{,}8;$$

$$\boldsymbol{m}(t,X) = \Sigma \int_{a(t)}^{t} \int_{V,j} \alpha(t,s,\boldsymbol{e}(X \cup Y))\boldsymbol{m}(^{1})(s,Y)dYds \ (j = 1,\ldots,8),\ X \in V{,}9;$$

$$\boldsymbol{atp}(t,X) = \int_{b(t)}^{t} \int_{V,3} \beta(t,s,\boldsymbol{e}(X \cup Y))\boldsymbol{m}(^{2})(s,Y)dYds,$$

$$\boldsymbol{pr1}(t,X) = \int_{b(t)}^{t} \int_{V,4 \cup V,5} \beta(t,s,\boldsymbol{e}(X \cup Y))\boldsymbol{m}(^{2})(s,Y)dYds,$$

$$\boldsymbol{pr2}(t,X) = \int_{b(t)}^{t} \int_{V,6} \beta(t,s,\boldsymbol{e}(X \cup Y))\, \boldsymbol{m}(^{2})(s,Y)dYds,$$

$$\boldsymbol{r}(t,X) = \int_{b(t)}^{t} \int_{V} \beta(t,s,\boldsymbol{e}(X \cup Y))\boldsymbol{m}(^{2})(s,Y)dYds,\ X \in V;$$

$$\boldsymbol{c}(t,Z) = \int_{b(t)}^{t} \int_{V} \beta(t,s,\boldsymbol{e}(Z \cup Y))\boldsymbol{m}(^{2})(s,Y)dYds,\ Z \in V\hat{}\subset R{,}3, \tag{4}$$

where ***atp***, ***pr1***, ***pr2***, and ***r*** are respectively the rate of creation of ATP by the mitochondria, of proteins by the ribosomes and Golgi apparatus, of the remaining components of a cell including the cytoskeleton fibers, the surface membrane, etc.; the matrices α and β may depend on all the other elements.

We divide all functions α and β-types on three classes: the first that corresponds to the given procedures by the genome, the second that corresponds to the known laws of the nature, and the third that corresponds to combination of two former classes. Among the known laws are the mass action law and Ohm's law that are used frequently under modeling of cells and CAO. In MM (4), for $X \in V,1, V,2$, the functions α and β-types relate probably to the first class, for $X \in V,3, V,4, V,5, V,6, V,7$ to the third class, and for $X \in V,8, V,9$ to the second class. Particular confirmation comes from MM of (1)-(3) and from many results above and below.

In the case of the first class, the respective control may come from the control of gene expression, from gene therapy. In the case of the third and especially the second class, the respective control may come from the control of the known laws at the expense of their certain parameters and variables.

In particular, for the mass action law, we have a certain dependence of the reaction rate on cell pH, temperature T, and enthalpy (see Ch. 2, s. 5).

Note a survey of laws in [118]; V. Ivanov et al. [93], pp. 128-148; and A. Samarskii, A. Mikhailov [164].

2.3. MM OF DEVELOPING CELL

The MM in question has the form

$$\boldsymbol{\alpha}'(t,X) = \sum \int_{a(t)}^{t} \int_{V,j} \boldsymbol{\alpha}(t,s,\boldsymbol{e}(X \cup Y))\boldsymbol{m}(^{1})(s,Y)dYds,\ (j = 1,\ldots,5),\ X \in V',1;$$

$$\boldsymbol{m}'(t,X) = \int_{a'(t)}^{t} \int_{O}^{X} \boldsymbol{\alpha}'(t-s,\boldsymbol{e}'(Y))\boldsymbol{m}'(^{1})(s,Y)dYds,\ \boldsymbol{M}'(t,X) = \int_{0}^{t} \boldsymbol{m}'(s,X)ds,$$

$$\boldsymbol{M}'(t) = \int_{V'} \boldsymbol{M}'(t,X)dX,\ \boldsymbol{\beta}'(t,X) = \sum \int_{a(t)}^{t} \int_{V,j} \boldsymbol{\beta}(t,s,\boldsymbol{e}(X \cup Y))\boldsymbol{m}(^{2})(s,Y)dYds,\ (j = 1,\ldots,5),$$

$$X \in V',1,\ \boldsymbol{c}'(t,Z) = \int_{b'(t)}^{t} \int_{V'} \boldsymbol{\beta}'(t-s,\boldsymbol{e}'(Z \cup Y))m'(^{2})(s,Y)dYds,$$

$$\boldsymbol{C}'(t,Z) = \int_{t^{\wedge}}^{t} \boldsymbol{c}'(s,Z)ds,\ \boldsymbol{C}'(t) = \int_{V^{\wedge}} \boldsymbol{C}'(t,Z)dZ,$$

$$X = X(t) = \int_{a(t)}^{t} \int_{V} \boldsymbol{\alpha}(t,s,\boldsymbol{e}(X \cup Y))\boldsymbol{m}(^{1})(s,Y)dYds,$$

$$Z = Z(t) = \int_{b(t)}^{t} \int_V \beta(t,s,e(Z \cup Y))\boldsymbol{m}(^2)(s,Y)dYds,$$

$$\boldsymbol{R}'(t,X) = \int_{a'(t)}^{t} \int_{O}^{X} m'(^1)(s,Y)dYds + \int_{b'(t)}^{t} \int_{V'} m'(^2)(s,Y)dYds,$$

$$R'(t) = \int_{V'} R'(t,X)dX,\ m'(^1)(t,X) + m'(^2)(t,X) = m'(t,X),$$

$$G'(t,X) = M'(t,X) - R'(t,X) > 0,\ G'(t) = M'(t) - R'(t),$$

$$f'(t) = k'(f')\int_{V'} m'(t,X)dX,\ F'(t) = \int_0^t f'(s)ds,$$

$$0 \le a', b' \le t,\ t > t'\wedge > 0,\ O, X \in V', Z \in V\wedge' \subset R^3, \tag{5}$$

where the definitions of all functions are the same as in (1) -- (4), excluding X and Z regularities of development that may also be given by DNA and natural laws.

The domains V' and $V\wedge'$ don't have exactly the same shape as V and $V\wedge$, and after $V'$and $V\wedge'$ settle, the developing cell separates from the mother cell.

The reason for the start of development of a new cell is the process of signal transduction (Ch. 6, s. 6.1.7). The final act of division can be explained by the critical osmotic state of the cell membrane.

We can consider the models (5) as MM of a cell that is in the state of proliferation or differentiation.

Thus, we have another mechanism for explanation of proliferation and differentiation.

It can also be shown that in a certain case of the oscillating functions α and β-types, the equations (5)-type itself and different optimization problems for the models have non-unique solutions (see Ch. 6 and Ch. 14, 15).

Assuming, that cells' development pursues the solution of determined optimization problems, e.g., cells' duplication maximizes reliability of their functioning, and cells' differentiation maximizes efficiency of fulfillment of given functions, we may explain cells' duplication and differentiation by the non-uniqueness of solutions of the appropriate optimization problems.

2.4. MM OF CELL BIO-FIELD

Let V be the domain in R^3 for a cell, and $R^3\backslash V$ be the complement of V to R^3. We accept that the external bio-field with respect to a cell will be decided if the vector function

$$c(t,Z) = \{c_1(t,Z), c_2(t,Z), c_3(t,Z)\}, Z \in R^3\backslash V, \tag{6}$$

is given.

Here t is the time, (z_1, z_2, z_3) are Cartesian coordinates of the point Z; c_1 is the rate of creation along t, Z (i.e., along t, z_1 z_2, and z_3) of the (generalized) new material product of t, Z-type in the point Z at the instant t; c_2 is the rate of creation along t, Z of the (generalized) new energy product of t, Z-type in the point Z at the instant t; c_3 is the rate of creation along t, Z of the information product of t,Z-type in the point Z at the instant t.

As the *MM of a cell bio-field*, we assume the system

$$m(t,X) = \int_{a(t,X)}^{t} \int_V \alpha(t,X,s,Y) m(^1)(s,Y)dYds, X \in V,$$

$$c(t,Z) = \int_{b(t,Z)}^{t} \int_V \beta(t,Z,s,Y) m(^2)(s,Y)dYds, Z \in R^3,$$

$$R(t,X) = \int_{a(t,X)}^{t} \int_V m(s,Y)dYds, f(t,X,Z) = m(t,X) + c(t,Z),$$

$$m(^1) + m(^2) = m(t,X) = \{m_1(t,X), m_2(t,X), m_3(t,X)\}, X \in V, t \geq t\text{^} > 0, \tag{7}$$

where $m_i(t,X)$ are the rates of creation along t, X of the (generalized) internal products of t, X-type respectively bio-mass ($i = 1$), energy ($i = 2$), and information ($i = 3$) in the point X at the instant t; α and β are matrices of the order three whose components are indices of efficiencies of functioning of a cell along the corresponding channels and definitions of all the other values are similar to the previous ones.

Note that the functions $m_i(t,X)$ can be different from 0 under the same t, X. In addition, we can also consider the Gibbous free energy as

$$g(t,X) = m_2(t,X) - m_3(t,X). \tag{8}$$

It is natural to call V with (7) and (8) as a bio-field of the internal sphere of a cell or the internal cell bio-field and $R^3\backslash V$ with (6) and (7) as the external cell bio-field.

We will come back to these MM when considering cancer cells.

3. On the Protein Folding Problem

The problem of spatial structure and holding DNA in small volume as well as the problem of protein folding can be modeled by introducing the *simplest spatial variant* of the model (1)-type:

$$m(t,X) = \int_{a(t)}^{t\wedge} \alpha(t-s,r(X))y\wedge(s)m\wedge(s,X)ds + y\int_{t\wedge}^{t} \alpha(t-s,r(X))m(s,X)ds,$$

$$pr(t,X) = \int_{a(t)}^{t\wedge} \alpha(t-s,r(X))z\wedge(s)m\wedge(s,X)ds + z\int_{t\wedge}^{t} \alpha(t-s,r(X))m(s,X)ds,$$

$$R(t,X) = \int_{a(t)}^{t} m(s,X)ds,\ y,\ z,\ y + z \le 1,\ X \in V,\ t \ge t\wedge \ge a(t) \ge 0, \qquad (9)$$

where V is the domain a cell occupied, r: $V \to V\wedge$, where $V\wedge$ is the domain in the nucleus the gene of DNA for respective protein occupied and the mapping r is realized by the cell RNA; m is the concentration of new component of enzyme-substrate complex at point X; pr is the concentration of new component of protein at point X; $y\wedge(s)$, $z\wedge(s)$, $m\wedge(s,X)$ are the given functions on the prehistory $[0, t\wedge]$; and y, z are positive constants.

We assume that the functions $\alpha(u,r(X))$ and $m(u,X)$ are periodic over u for $u > t\wedge$ with the period T, which is small enough such that we can consider y and z as constants.

It should be emphasized that $m(t,X)$ and $pr(t,X)$ in (9) for most X in V are equal to 0, in the case of separate m and pr, and if $m(t,X) \neq 0$, then $pr(t,X) = 0$, and vice versa.

It is well known that DNA code is about 10^6 times longer than the diameter of the nucleus. However, this code can be embedded in the nucleus due to DNA helical structure and its partial tight wrapping, which has also been represented by MM (9).

It is easy to see that (9) implies

$$pr(t,X) - pr\wedge(t,X) = z/y[m(t,X) - m\wedge(t,X)] = z\int_{t\wedge}^{t} \alpha(t-s,r(X))m(s,X)ds,$$

$$pr\wedge(t,X) = \int_{a(t)}^{t\wedge} \alpha(t-s,r(X))z\wedge(s)m\wedge(s,X)ds\ ,$$

$$m\wedge(t,X) = \int_{a(t)}^{t\wedge} \alpha(t-s,r(X))y\wedge(s)m\wedge(s,X)ds. \qquad (10)$$

It means that the structure of the enzyme-substrate complex and the respective protein may be about equally complicated, which in turn is not less complicated than the structure of DNA.

Denoting the Laplace transform by $F(p) = \int_0^\infty e^{-pt}f(t)dt$ for any function f, on the strength of (10) we have

$$PR(p,X) = PR\wedge(p,X) + z/y[M(p,X) - M\wedge(p,X)],$$

$$M(p,X) = M\wedge(p,X)/[1 - yA(p,r(X))], \qquad (11)$$

from where under certain conditions (V. Ivanov [89]; G. Korn, T. Korn [115])

$$pr(t,X) = pr^\wedge(t,X) + z/(2\pi i)\lim \int_{[s-ir,\, s+ir]} e^{pt} M^\wedge(p,X) A(p,r(X))/[1 - yA(p,r(X))]dp \ (r \to \infty). \quad (12)$$

and also

$$PR(X) = \int_{t^\wedge}^{t^\wedge+T} pr(t,X)dt. \quad (13)$$

Since we have a certain physical model of DNA (see Ch. 1, s. 4.4), the relations (9)–(13) may probably be used for the solution of the protein folding problem. It is more likely so if the plausible hypothesis is valid that maximal entropy of the DNA gene corresponds physically to maximal entropy of the respective protein.

Yet note that under

$$a(t) = at,\ 0 < a < 1,\ P'_t(t,X) = m(t,X) - am(at,X) = 0, \quad (14)$$

we have

$$pr(t,X) = pr^\wedge(t,X) + z/y[m(t,X) - m^\wedge(t,X)],$$

$$m(t,X) = a^n m^\wedge(a^n t,X),\ t \in [t^\wedge/a^{n-1}, t^\wedge/a^n],\ n = 1, 2, \ldots, \quad (15)$$

where $m^\wedge$ is a given function on the prehistory $[0, t^\wedge]$.

If the above mentioned hypothesis is not valid, then using the equations of the cell biofield (7) we can consider the relation of (13)-type for determination of the starting values of the Gibbous free energy $G(X)$ of the protein.

One of further steps of modeling may then be a representation of the desired solution as the harmonic function in the form

$$G(Y) = 1/(4\pi)\int_S 1/r\partial G(X)/\partial n dS - 1/(4\pi)\int_S \partial(1/r)/\partial n G(X)dS,\ Y \in F, \quad (16)$$

where S is the initial surface and F is the desired surface on which $G(Y)$ is minimal; r is the distance between dS and Y; and n is the direction of the exterior normal to S.

Let $Y = (y_1, y_2, y_3)$. Then, the condition of minimum for each (y_1, y_2) is

$$\partial G(Y)/\partial y_3 = 0. \quad (17)$$

Solving this equation, we have the desired surface F as given by

$$y_3 = f(y_1, y_2). \quad (18)$$

Note that the equation (17) may have zero, one, or more solutions for different (y_1, y_2). Also note that the numerical realization of (12) and (17) is not trivial because we have to deal here with the so-called ill-posed problems (see Ch. 16).

4. MM of Cell Cycle, II

4.1. BIOLOGICAL BACKGROUND

As a biological background, we describe a certain part of the press release on the Nobel Prize 2001 in Physiology and Medicine (see www.nobel.se).

Summary: All organisms consist of cells that multiply through cell division. An adult human being has approximately 100 000 billion cells, all originating from a single cell, the fertilized egg cell. In adults, there is also an enormous number of continuously dividing cells replacing those dying. Before a cell can divide it has to grow in size, duplicate its chromosomes, and separate the chromosomes for exact distribution between the two daughter cells. These different processes are coordinated in the cell cycle.

The Nobel Laureates of 2001 have made seminal discoveries concerning the control of the cell cycle. They have identified key molecules that regulate the cell cycle in all eukaryotic organisms, including yeasts, plants, animals and humans. They have discovered that CDK-molecules and cyclins drive the cell from one phase of the cell cycle to the next. The CDK-molecules can be compared with an engine and the cyclins with a gearbox controlling whether the engine will run in the idling state or drive the cell forward in the cell cycle. These fundamental discoveries have a great impact on all aspects of cell growth.

Leland Hartwell, Fred Hutchinson Cancer Research Center, Seattle, USA, is awarded for his discoveries of a specific class of genes that control the cell cycle. One of these genes called "start" was found to have a central role in controlling the first step of cell cycle. Paul Nurse, Imperial Cancer Research Fund, London, identified, cloned, and characterized with genetic and molecular methods, one of the key regulators of the cell cycle, CDK (cyclin dependent kinase). He showed that the function of CDK was highly conserved during evolution. CDK drives the cell through the cell cycle by chemical modifications (phosphorylation) of other proteins. Timothy Hunt, Imperial Cancer Research Fund, London, is awarded for his discovery of cyclins, proteins that regulate the CDK function. He showed that cyclins are degraded periodically at each cell division.

The phases of the cell cycle: The cell cycle consists of several phases (see Fig. 16). In the first phase (G_1), the cell grows and becomes larger. When it has reached a certain size it enters the next phase (S), in which DNA synthesis takes place. The cell duplicates its hereditary material (DNA-replication) and a copy of each chromosome is formed. During the next phase (G_2), the cell checks that DNA-replication is complete and prepares for cell division. Chromosomes are separated in the next phase (mitosis, M) and the cell divides into two daughter cells. After division, the cells are back in G_1.

The duration of the cell cycle varies between different cell types. In most mammalian cells it lasts between 10 and 30 hours. Cells in the first cell cycle phase (G_1) do not always continue through the cycle. Instead they can exit from the cell cycle and enter a resting stage (G_0).

Cell cycle control: For all living eukaryotic organisms it is essential that the different phases of the cell cycle are precisely coordinated. The phases must follow in correct order, and one phase must be completed before the next phase can begin. Errors in this coordination may lead to chromosomal alterations. Chromosomes or parts of chromosomes may be lost, rearranged, or distributed unequally between the two daughter cells. This type of chromosome alteration is often seen in cancer cells.

It is of central importance in the fields of biology and medicine to understand how the cell cycle is controlled.

Cell cycle genes in yeast cells: Leland Hartwell isolated yeast cells in which genes controlling the cell cycle were altered (mutated). Through mutation he succeeded to identify more than one hundred genes specifically involved in cell cycle control, so called CDC-genes (cell division cycle genes). One of these genes, designated CDC28, controls the first step in the progression through the G_1-phase of the cell cycle, and was therefore also called "start".

In addition, Hartwell introduced the concept of checkpoint, which means that the cell cycle is arrested when DNA is damaged. The purpose of this is to allow time for DNA repair before the cell continues to the next phase of the cycle. Later Hartwell extended the checkpoint concept to include controls ensuring a correct order between the cell cycle phases.

A general principle: Paul Nurse used a different type of yeast, *Schizosaccharomyces pombe*, as a model organism, and discovered the gene cdc2 in *S. pombe*. He showed that this gene had a key function in the control of cell division (transition from G_2 to mitosis, M). Later, he found that cdc2 had a more general function. It was identical to the "start) gene controlling the transition from G_1 to S. This gene (cdc2) was thus found to regulate different phases of the cell cycle. In 1987, Paul Nurse isolated the corresponding gene in humans, and it was later given the name CDK1 (cyclin dependent kinase 1). The gene encodes a protein that is a member of a family called cyclin dependent kinases, CDK. Nurse showed that activation of CDK is dependent on reversible phosphorylation, i.e. those phosphated groups are linked to or removed from proteins. On the basis of these findings, half a dozen different CDK molecules have been found in humans.

The discovery of the first cyclin: Tim Hunt discovered the first cyclin molecule in the early 1980s. Cyclins are proteins formed and degraded during each cell cycle. They were named cyclins because the levels of these proteins vary periodically during the cell cycle. The cyclins bind to the CDK molecules, thereby regulating the CDK activity and selecting the proteins to be phosphorylated. Today about ten different cyclins have been found in humans.

The engine and the gearbox of the cell cycle: The three Nobel Laureates have discovered molecular mechanisms that regulate the cell cycle. The amount of CDK-molecules is constant during the cell cycle, but their activities vary because of the regulatory function of the cyclins. CDK and cyclin together drive the cell from one cell cycle phase to the next. The CDK-molecules can be compared with an engine and the cyclins with a gearbox controlling whether the engine will run in the idling state or drive the cell forward in the cell cycle.

A great impact of the discoveries: The discoveries are important in understanding how chromosomal instability develops in cancer cells, i.e. how parts of chromosomes are rearranged, lost or distributed unequally between daughter cells. It is likely that such chromosome alterations are the result of defective cell cycle control. It has been shown that genes for CDK-molecules and cyclins can function as oncogenes. CDK-molecules and cyclins also collaborate with the products of tumor suppressor genes (e.g., p53 and Rb) during the cell cycle. The findings in the cell cycle field are about to be applied to tumor diagnostics. Increased levels of CDK-molecules and cyclins are sometimes found in human tumors, such as breast cancer and brain tumors. The discoveries may, in the long run, open new principles for cancer therapy. Already clinical trials are in progress using inhibitors of CDK-molecules.

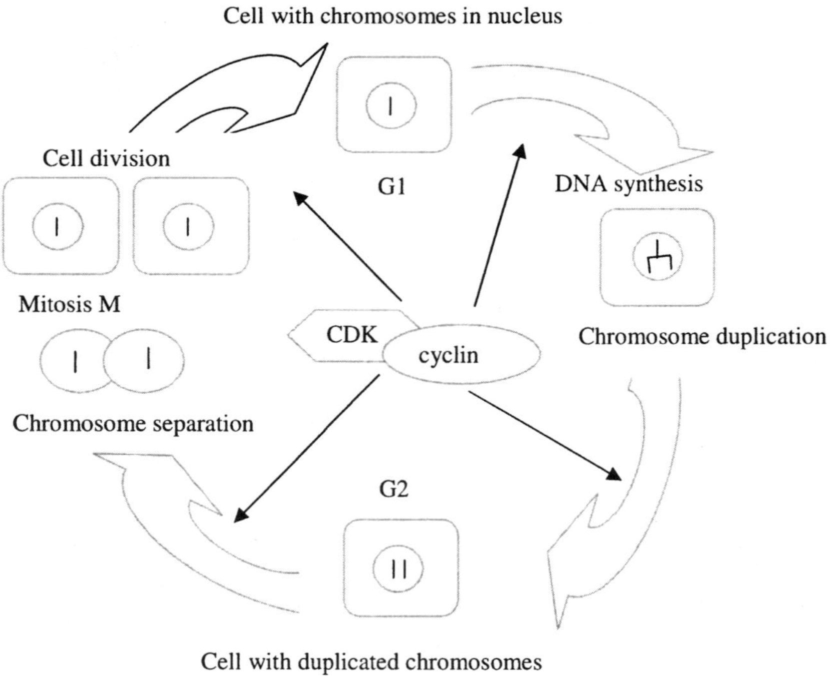

Figure 17. The cell cycle

4.2. THE RESPECTIVE MM

Let us consider the following variety of MM (3) and (9):

$$`\alpha'(t,X') = \Sigma_{j=1}^{5} \int_{a(t)}^{t} \int_{V,j} `\alpha(s,e'(X' \cup Y))x(s)m'(s,Y)dYds =$$

$$\Sigma_{j=1}^{5} [\int_{a(t)}^{t\wedge} \int_{V,j} \alpha(s,e'(X' \cup Y))x(s)m'(s,Y)dYds + \int_{t\wedge}^{t} \int_{V,j} `\alpha(s,e'(X' \cup Y))x(s)m'(s,Y)dYds],$$

$$`\alpha'(t\wedge,X') = \Sigma_{j=1}^{5} \int_{0}^{t\wedge} \int_{V,j} \alpha(s,e\,(X' \cup Y))x(s)m'(s,Y)dYds,\ X' \in V';$$

$$m'(t,X) = \int_{a(t)}^{t} \int_{O}^{X} \alpha(t - s,e(Y))y(s)m(s,Y)dYds,$$

$$m(t,X) = \int_{a(t)}^{t} \int_{O}^{X} \alpha(t - s,e(Y))[1 - x(s) - y(s)]m(s,Y)dYds,$$

$$0 \le a \le t\wedge,\ t > t\wedge > 0,\ O,\ X \in V, \tag{19}$$

where $`\alpha'(t,X) = \partial'\alpha(t,X)/\partial t$; $\alpha$ and $`\alpha$ correspond to DNA of the initial and daughter cell; $\alpha(s,e'(X' \cup Y))$, $s \in [a(t), t\wedge]$, corresponds to the "start" gene; m' to the CDK molecules; and x to the cyclin.

It is not difficult to see that due to non-negativity of all the elements of MM (19),

$$`\alpha'(t,X') > 0, \tag{20}$$

if x becomes greater than 0, which guarantees the further development of $`\alpha'$.

If we differentiate by t the first equality in (19) one time, we find

$$`\alpha''(t,X') = \Sigma_{j=1}^{5} \int_{V,j} `\alpha(t,e'(X' \cup Y))x(t)m'(t,Y)dY + d(t,X'),$$

$$`\alpha''(t\wedge,X') = \Sigma_{j=1}^{5} \int_{V,j} \alpha(t\wedge,e'(X' \cup Y))x(t\wedge)m'(t\wedge,Y)dY + d(t\wedge,X'),$$

$$`\alpha'(t\wedge,X') = \Sigma_{j=1}^{5} \int_{0}^{t\wedge} \int_{V,j} \alpha(s,e'(X' \cup Y))x(s)m'(s,Y)dYds, \tag{21}$$

where $d(t,X')$ can be the known function.

In particular, the relations (21) can be reduced to the Caushy problem for ordinary differential equation of the order two. This problem and a related problem of optimization of certain functionals at the expense of x and y are considered in more detail in Ch. 15.

5. MM of Morphogenesis

5.1. BIOLOGICAL BACKGROUND

As a biological background, we describe a certain part of the press release on Nobel Prize 2002 in Physiology and Medicine (see www.nobel.se).

Summary: The human body consists of hundreds of cell types, all originating from the fertilized egg. During the embryonic and fetal periods, the number of cells increases dramatically. The cells mature and become specialized to form the various tissues and organs of the body. Large numbers of cells are also formed in the adult body. In parallel with this generation of new cells, cell death is a normal process, both in the fetus and in the adult, to maintain an appropriate number of cells in the tissues. This delicate, controlled elimination of cells is called *programmed cell death.*

The Nobel Laureates of 2002 in Physiology or Medicine have made seminal discoveries concerning the genetic regulation of organ development and programmed cell death. Establishing and using the nematode *Caenorhabditis elegans* as an experimental model system opened possibilities to follow cell division and differentiation from the fertilized egg to the adult. The Laureates have identified key genes regulating organ development and programmed cell death, and have shown that corresponding genes exist in higher species, including man. The discoveries are important for medical research and have shed new light on the pathogenesis of many diseases. Sydney Brenner, Berkeley, CA, USA, established *C. elegans* as a novel experimental model organism. This provided a unique opportunity to link genetic analysis to cell division, differentiation and organ development, and to follow these processes under the microscope. John Sulston, Cambridge, England, mapped a cell lineage where every cell division and differentiation could be followed in the development of a tissue in *C. elegans.* He showed that specific cells undergo programmed cell death as an integral part of the normal differentiation process, and he identified the first mutation of a gene participating in the cell death process. Robert Horvitz, Cambridge, MA, USA, has discovered and characterized key genes controlling cell death in *C. elegans.* He has shown how these genes interact with each other in the cell death process and has shown that corresponding genes exist in humans.

Cell lineage – from egg to adult: All cells in our body are descendents from the fertilized egg cell. Their relationship can be referred to as a cellular pedigree or *cell lineage*. Cells differentiate and specialize to form various tissues and organs such as muscle, blood, heart, and the nervous system.

The human body consists of several hundreds of cell types, and the cooperation between specialized cells makes the body function as an integrated unit. To maintain the appropriate number of cells in the tissues, a fine-tuned balance between cell division and cell death is required. Cells have to differentiate in a correct manner and at the right time during development in order to generate the correct cell type.

In unicellular model organisms (e.g., bacteria and yeast) organ development and the interplay between different cells cannot be studied. Mammals, on the other hand, are too complex for these basic studies, as they are composed of an enormous number of cells. The nematode *C. elegans,* being multi-cellular, yet relatively simple, was therefore chosen as the most appropriate model system, which has then led to characterization of these processes in humans.

Programmed cell death: Normal life requires cell division to generate new cells but also requires cell death, so that a balance is maintained in our organs. In an adult human being, more than a thousand billion cells are created every day. At the same time, an equal number of cells die through a controlled "suicide process", referred to as *programmed cell death.* Developmental biologists first described programmed cell death. They noted that cell death was necessary for embryonic development, for example, when tadpoles undergo metamorphosis to become adult frogs. In the human fetus, the interdigital mesoderm initially formed between fingers and toes is removed by programmed cell death. The vast excess of neuronal cells present during the early stages of brain development is also eliminated by the same mechanism.

The Nobel Laureates mentioned above discovered that specific genes control the programmed cell death in the nematode *C. elegans.* Detailed studies in this simple model organism demonstrated that 131 of total 1090 cells die reproducibly during development, and that this natural cell death is controlled by a unique set of genes.

The model organism C. elegans: Sydney Brenner realized that fundamental questions regarding cell differentiation and organ development were hard to tackle in higher animals. Therefore, a genetically amenable and multicellular model organism that was simpler than mammals was required. The ideal solution proved to be the nematode *Caenorhabditis elegans.* This worm, approximately 1 mm long, has a short generation time and is transparent, which made it possible to follow cell division directly under the microscope. Brenner provided the basis, in which he broke new ground by demonstrating that specific gene mutations could be induced in the genome of *C. elegans* by the chemical compound EMS (ethyl methane sulphonate). Different mutations could be linked to specific genes and effects on organ development. This combination of genetic analysis and visualization of cell divisions observed under the microscope initiated the discoveries that were awarded the Nobel Prize.

Mapping cell lineage: John Sulston extended Brenner's work with *C. elegans* and developed techniques to study all cell divisions in the nematode, from the fertilized egg to the 959 cells in the adult organism. Sulston described the cell lineage for a part of the developing nervous system. He showed that the cell lineage is invariant, i.e. every nematode underwent exactly the same program of cell division and differentiation.

As a result of these findings, Sulston made the seminal discovery that specific cells in the cell lineage always die through programmed cell death and that this could be monitored in the living organism. He described the visible steps in the cellular death process and demonstrated the first mutations of genes participating in programmed cell death, including the *nuc-1* gene. Sulston also showed that the protein encoded by the *nuc-1* gene is required for degradation of the DNA of the dead cell.

Identification of "death genes": Robert Horvitz continued Brenner's and Sulston's work on the genetics and cell lineage of *C. elegans*. In a series of elegant experiments, Horvitz used *C. elegans* to investigate whether there was a genetic program controlling cell death. He identified the first two bona fide "death genes", *ced*-3 and *ced*-4. He showed that functional *ced*-3 and *ced*-4 genes were a prerequisite for cell death to be executed. Later, Horvitz showed that another gene, *ced*-9, protects against cell death by interacting with *ced*-4 and *ced*-3. He also identified a number of genes that direct how the dead cell is eliminated. Horvitz showed that the human genome contains a *ced*-3-like gene.

Of importance for many research disciplines: The development of *C. elegans* as a novel experimental model system has proven valuable for many research disciplines. The characterization of genes controlling programmed cell death in *C. elegans* soon made it possible to identify related genes with similar functions in humans. It is now clear that one of the signaling pathways in humans leading to cell death is evolutionarily well conserved. In this pathway *ced*-3-, *ced*-4- and *ced*-9-like molecules participate.

Disease and programmed cell death: Knowledge of programmed cell death has helped us to understand the mechanisms by which some viruses and bacteria invade our cells. We also know that in AIDS, neurodegenerative diseases, stroke, and myocardial infarction, cells are lost as a result of excessive cell death. Other diseases, like autoimmune conditions and cancer, are characterized by a reduction in cell death, leading to the survival of cells normally destined to die.

Research on programmed cell death is intense, including in the field of cancer. Many treatment strategies are based on stimulation of the cellular "suicide program". This is, for the future, a most interesting and challenging task to further explore in order to reach a more refined manner to induce cell death in cancer cells.

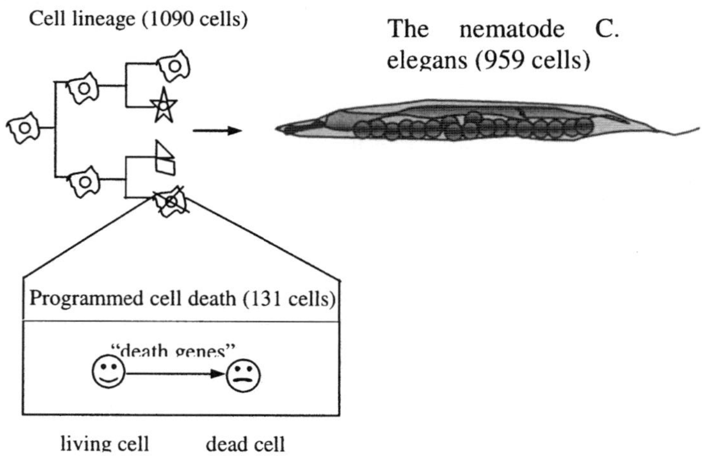

Figure 18. Development of the worm

5.2. THE RESPECTIVE MM

These MM should combine at least the following simplest MM of cell death (22), cell cycles (23), and morphogenesis (24) and (25):

$$\boldsymbol{m}(t) = \int_{a(t)}^{t} \alpha(t,s)\boldsymbol{y}(s)\boldsymbol{m}(s)ds = (m^1(t), m^2(t), m^3(t)), \boldsymbol{M}(t) = \int_0^t \boldsymbol{m}(s)ds,$$

$$\boldsymbol{c}(t) = \int_{b(t)}^{t} \beta(t,s)[\boldsymbol{I}-\boldsymbol{y}(s)]\boldsymbol{m}(s)ds = (c^1(t), \ldots, c^p(t)), \boldsymbol{C}(t) = \int_{t^\wedge}^{t} \boldsymbol{c}(s)ds,$$

$$\boldsymbol{R}(t) = \int_{a(t)}^{t} \boldsymbol{y}(s)\boldsymbol{m}(s)ds + \int_{b(t)}^{t} [\boldsymbol{I}-\boldsymbol{y}(s)]\boldsymbol{m}(s)ds, \boldsymbol{G}(t) = \boldsymbol{M}(t) - \boldsymbol{R}(t) > 0,$$

$$m^1(t) + m^2(t) + m^3(t) = m(t), c^1(t) + \ldots + c^p(t) = c(t),$$

$$f(t) = m(t) + c(t), F(t) = \int_0^t f(s)ds, \boldsymbol{y}(t) = (y^1(t), y^2(t), y^3(t)),$$

$$\boldsymbol{a}(t) = (a^1(t), a^2(t), a^3(t)), \boldsymbol{b}(t) = (b^1(t), b^2(t), b^3(t)),$$

$$0 \le \boldsymbol{y}(t) \le 1, 0 \le \boldsymbol{a}(t), \boldsymbol{b}(t) \le t, t > t^\wedge > 0; \qquad (22)$$

$$`\alpha'(t,X') = \sum_{j=1}^{5} \int_{a(t)}^{t} \int_{V,j} `\alpha(s,e'(X' \cup Y))x(s)m'(s,Y)dYds =$$

$$\sum_{j=1}^{5} \int_{a(t)}^{t^\wedge} \int_{V,j} \alpha(s,e'(X' \cup Y))x(s)m'(s,Y)dYds +$$

$$\sum_{j=1}^{5} \int_{t^\wedge}^{t} \int_{V,j} `\alpha(s,e'(X' \cup Y))x(s)m'(s,Y)dYds,$$

$$`\alpha'(t^\wedge,X') = \sum_{j=1}^{5} \int_0^{t^\wedge} \int_{V,j} \alpha(s,e'(X' \cup Y))x(s)m'(s,Y)dYds, X' \in V';$$

$$m'(t,X) = \int_{a(t)}^{t} \int_O^X \alpha(t-s,e(Y))y(s)m(s,Y)dYds,$$

$$m(t,X) = \int_{a(t)}^{t} \int_O^X \alpha(t-s,e(Y))[1 - x(s) - y(s)]m(s,Y)dYds,$$

$$0 \le a \le t^\wedge, t > t^\wedge > 0, O, X \in V; \qquad (23)$$

$$m(t) = \int_{a(t)}^{t} \alpha(t,s)y(s)m(s)ds, c(t) = \int_{b(t)}^{t} \beta(t,s)[1-y(s)]m(s)ds,$$

$$R(t) = \int_{a(t)}^{t} y(s)m(s)ds + \int_{b(t)}^{t} [1-y(s)]m(s)ds, 0 < t^\wedge \le t, \qquad (24)$$

where

$$m(t) = m^\wedge = \text{constant},\ y(t) = y^\wedge,\ t \in [0, t^\wedge];$$

$$\int_{(2k-1)t^\wedge}^{t} \alpha(t,s)ds = 1,\ \beta(t,s) = 2[s-(t-t^\wedge)]/m^\wedge,\ s \in [t-t^\wedge, (2k-1)t^\wedge],$$

$$y(t) = 1,\ c(t) = (2kt^\wedge - t)^2,\ t \in [(2k-1)t^\wedge, 2kt^\wedge];$$

$$\int_{t-t^\wedge}^{2kt^\wedge} \alpha(t,s)ds = 1,\ \beta(t,s) = 2(t-s)]/m^\wedge,\ s \in [2kt^\wedge, t],$$

$$y(t) = 0,\ c(t) = (2kt^\wedge - t)^2,\ t \in [2kt^\wedge, (2k+1)t^\wedge];\ k = 1, 2, ...,$$

$$b(t) = t\text{-}t^\wedge,\ m(t) = m^\wedge,\ P(t) = m^\wedge t^\wedge,\ \ t \in [(2k-1)t^\wedge, 2kt^\wedge],\ a(t) = t-t^\wedge,$$

$$m(t) = m^\wedge,\ P(t) = m^\wedge t^\wedge,\ t \in [2kt^\wedge, (2k+1)t^\wedge],\ k = 1, 2, ...$$

$$\alpha(t,s) = 2\delta[s-(2k-1)t^\wedge],\ s \in [(2k-1)t^\wedge, 2kt^\wedge];$$

$$\alpha(t,s) = 2\delta(2kt^\wedge - s),\ s \in [2kt^\wedge, (2k+1)t^\wedge], \tag{25}$$

δ is Dirac generalized function.

It follows from (22) that for realization of the programmed cell death, there are, at least, three possibilities: elimination of $\boldsymbol{m}$ with the help of (i) $\boldsymbol{y}$ decrease (by probable influence of the chemical compound EMS), (ii) $\boldsymbol{\alpha}$ decrease (by mutation of the respective *ced*-3-like gene), and (iii) $\boldsymbol{y}$ and $\boldsymbol{\alpha}$ decrease. To stop this process, $\boldsymbol{y}$ and $\boldsymbol{\alpha}$ (by probable interaction of gene *ced*-9 with *ced*-4 and *ced*-3 genes) have to be restored to the norm.

The process of cell proliferation and differentiation through MM (23) of the cell cycle can be realized at the expense of the same factors: the "start" gene, CDK molecules, and the cyclin as above (see s. 4 and the relations (19) – (21)).

Supposing that the "start" gene switches on and off by the special program in DNA and that the process of cell proliferation and differentiation accompanied by the respective process of reproduction of all other chemicals organism needed through MM (24) and (25), we can complete the picture of the morphogenesis modeling.

Chapter 8

MM OF THE IMMUNE NETWORK

The new models of the *immune network* (IN) are constructed and investigated. Most assumptions, biological in nature that follow from the known experimental data on idio-type anti-idio-typical interactions of various sub-populations of cells and antibodies of IN and from the foundation of MM of IN, are considered. Such assumptions, as the refractoriness of cells towards the action of antigen (Ag) during their proliferation stimulated by Ag and their passage out of proliferation according to a certain law after which the cells can be stimulated again, lead to MM that are essentially different from well-known ones. The *immune network MM* are important here not only by themselves, but also as verifying phase of validity and quality of the models of CAO developed in this book. We show below that the immunology models following from the biological data and observations entirely correspond to MM of development. And what is more, MM of CAO allow us to deepen our understanding of IN dynamics. We also compare our MM with well-known MM.

1. Introduction

This Ch. is aimed at development of the models of functioning of *IN* (N. Jerne [107]) and *immune system* (IS) (A. Abbas et al. [2]; [84]; [85]; [179]) regarding available experimental data on the idio-type anti-idio-typical (Id-AId) interactions among distinctive sub-populations of lymphocytes (L. Fontalin, L. Pevnitskiy [49]; R. Mohler et al. [140]). This Ch. discusses tolerance and the rapid immune response. The role of the macrophages that examine foreign particles and then present their findings to other cells is omitted here because their number is usually not limiting.

The postulates or assumptions underlying the models were mainly developed by the Moscow immunologists Dr. L. N. Fontalin and Dr. N. G. Nesterenko. These assumptions lead to provide rather sophisticated differences and continuous integro-functional MM different from the well-known ones (V. Ivanov et al. [94]). In particular, they are essentially different from the models of the so-called +, - - theory of IN (G. Hoffman [79]) and the second-generation theory of IN (F. Vareba, A. Continto [190]. Among the models reviewed by G. Rowe [163], the models by A. Perelson [147], arising from certain extreme principles of IS functioning, remind us of some important features of the models under consideration.

The comparison of those MM with the novel class of models of CAO and also with the models of cell-to-cell interactions, considered in Ch. 3, can be found in s. 4. Some results of application to the immune-therapy of AIDS and cancers are given later in the next Ch. 10 and 11. Since MM of IN have been described in detail in the book by V. Ivanov [92], we dwell only on a continuous variant of models and make some corrections of previous MM.

2. Cells of the Immune System

2.1. ON THE IMMUNE SYSTEM

The *immune system* (*IS*) is the body's system of defense against viruses, bacteria, and other pathogenic agents and also serves as the principal line of defense against cancer. The IS responds to infection in two ways:

(1) the *cellular immune response*, in which there is direct interaction between lymphocytes and invading pathogens; and

(2) the *humoral immune response*, which acts primarily against extracellular phases of infection and in which antibodies are secreted by lymphocytes into the body fluids.

Lymphocytes (Lc) are white blood cells that spend most of their time in the tissue and organs of the lymphatic system. They originate from stem cells in the bone marrow (see the Fig. 19). Some immature Lc continue development in the bone marrow and become *B cells*. Other immature Lc are carried by the blood from the bone marrow to the *thymus*, a gland in the upper chest region. There the Lc become specialized as *T cells*. Both B cells and T cells make their way via the blood to the lymph nodes and other lymphatic organs.

Two types of Lc mediate the immune response: *T cells* (which mediate cellular immunity) and *B cells* (which mediate humoral immunity).

In response to the presence of foreign substances called *antigens* (Ag), B cells give rise to antibody-secreting *plasma cells* and T cells give rise to *helper cells* and cytotoxic *effector* and *supressor* T *cells*.

Humoral immunity defends primarily against bacteria and viruses present in body fluids. Antibodies secreted by plasma cells protect the body by precipitating free Ag and agglutinating foreign cells. Antibodies are carried in the blood to sites of infection wherever they occur in the body. T cells circulate in the blood, attacking body cells that have been infected with bacteria or viruses.

When a B cell or T cell develops, certain genes in the cell are turned on, and the cell synthesizes molecules of a specific protein and builds them into its plasma membrane. The molecules are Ag receptors, capable of binding one specific type of Ag.

These antibody-like proteins present in the plasma membranes of differentiated T cells endow T cells with their capacity to interact directly with the body's own virus-infected cells and with tissue of macrophages. This interaction is founded on the presence in the membranes of target cells of foreign antigens and proteins of the major histocompatibility complex. Both B cells and T cells also produce *memory cells*, which are reserved by the body and respond during later exposures to the same source of antigen.

Hybrid cells called *hybridomas* can be produced *in vitro* by fusing B cells with tumor cells. These cells can be grown in culture, where they provide a continuous source of monoclonal antibodies.

Details of the so-called hybridoma technology can be found in modern books on cell biology and in book by J. Rifkin [160].

There are several levels (about fore) of interactions of B and T cells that constitute the immune network (IN). Properties of the IN cells considered in detail below.

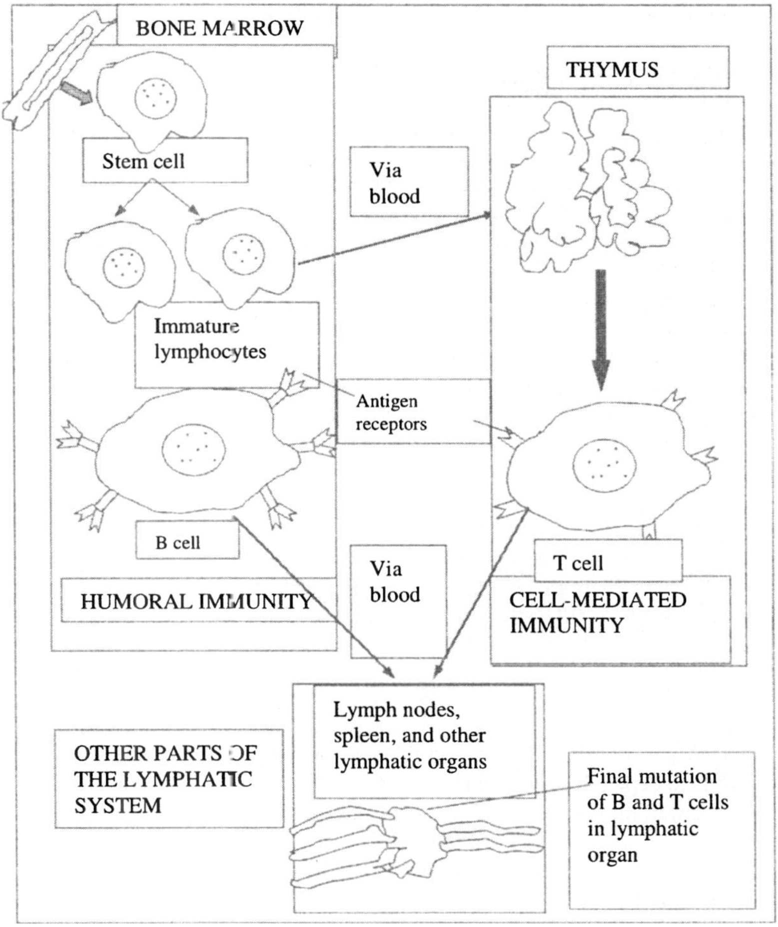

Figure 19. The development of B cells and T cells

2.2. PROPERTIES OF THE IMMUNE NETWORK CELLS

The following postulates or assumptions underlie the properties of *lymphocytes*:

1) Receptors of lymphocytes (*Lc*) and immunoglobulin (*Ig*) that recognize the *antigen* (*Ag*) have centers of fixation of Ag, the para-tops, and individual antigenic determinants, the idio-tops, located in the variable region of the Lc receptors and Ig.

2) Various levels of immune network (IN) are distinguished. The first level consists of idio-type-positive (Id) *T Lc* (T helpers - Th, T suppressors - Ts, T effectors of hypersensitivity and T killers - Tef), *B Lc* - B cells and their products (idio-type-positive immune-globulin - Ig). The second level consists of anti-idio-type (AId) Lc (Th, Ts, Tef, B) and Ig. The third level consists of anti-anti-idio-type (AAId) Lc (Th, Ts, Tef, B) and Ig, etc.

3) Behavior of IN essentially depends on the presence of the epi-tops of Ag (immune-gene, tolerant-gene) or on their absence. The situation is discussed when the elements of the first level react with the molecular-disperse Ag of the protein nature. For the higher order levels the antigens are the cells and Ig of the previous level.

4) A supplement of the population of Lc due to the differentiation of stem cells as well as a natural loss in the population of lymphoid cells caused by aging is ignored.

5) For Th of every level, the molecular-disperse Ag (for the cells of the second, third and succeeding levels such Ag are the Ig of the first, second and the higher levels, respectively) can induce three states according to its concentration: suppression by small doses, stimulation by average doses, and suppression by large doses. For Ts, Tef, and B, the molecular-disperse Ag results in two states: stimulation by average doses and suppression by large doses. The cells of every level as Ag for cells of the higher level induce stimulation.

6) Among the B cells of every level we distinguish, first, the B_2 cells (90% of all B cells) which are launching into proliferation and differentiation only by Th Lc and, secondly, the B_1 cells (10% of B cells) proliferating and differentiating without Th.

7) A part of Th Lc that are in the state of suppression due to small doses of Ag is determined depending on the concentration of Ag ($10^{11.5}$ - 10^{12} molecules) by the truncated normal law of the decimal logarithm of Ag concentration with the mean $x_1 = 11.75$ and mean square deviation $y_1 = 0.0833$ (dispersion is y_1^2).

8) The part of the cells that are stimulated by mean doses of Ag and suppressed by large doses can also be determined using the truncated normal law. The concentrations of Ag inducing the stimulation of Th, Ts, and Tef are in the interval 10^{12} - 10^{15} molecules ($x_2 = 13.5$, $y_2 = 0.500$), for B_1 the range is 10^{13} - 10^{15} molecules ($x_3 = 14$, $y_3 = 0.333$) and for B_2 the range is $10^{12.5}$- 10^{15} molecules ($x_4 = 13.5$, $y_4 = 0.417$).

9) The concentrations of Ag inducing for all the cells suppression with large doses lay within the range 10^{15} - 10^{16} molecules ($x_5 = 15.5$, $y_5 = 0.167$).

10) The part of Id, AId, AAId, etc. cells eliminated by AId-, AAId-, AAAId-Ig, etc. is determined according to the concentration of Ig $10^{12.5}$-$10^{14.5}$ by the truncated normal law of the decimal logarithm of Ig concentration with $x_6 = 13.5$ and $y_6 = 0.333$.

11) The Lc of adult mice can recognize about 10^6 different epi-topes of Ag. Therefore, before the introduction of Ag, on each level (regarding the relative part of the cells of each type) the number of Th, Ts, Tef is set to be equal to 2×10^2, the number of B_1 equal to 0.4×10^2, the number of B_2 equal to 3.6×10^2, and the number of molecules Ig equal to 10^{11}.

Here the concentration of Ag and cells is indicated in calculation per 1 mm^3 of blood.

12) The period of half-decay of Ig (and of the protein Ag) is assumed to be 50 – 100 hours and the time of the mitotic cycle of cells is assumed to be 8 – 10 hours.

13) The stimulation of Th, Ts, and Tef causes their proliferation in 12 hours and the proliferation of B in 18 hours. B cells produce Ig in 30 hours after stimulation.

14) B_2 cells stimulated by Ag become ready to accept a signal of stimulated Th during 18 hours. If contact of B_2 cell with Th cell takes place, B_2 cell will start to proliferate in 12 hours, otherwise B_2 cell will go to the rest (*r*) phase.

15) The part of X target-cells that are in one of the states: stimulation, suppression or elimination, due to the interaction with the respective Y control (or effector)-cells is determined by the truncated normal law for the decimal logarithm of Y with the mean $x_7 = 5$ and deviation $y_7 = 0.667$ and anyway do not exceed $k^*\min(X, Y)$, where k^* is the average productivity of Y-cells. Thus, we accept the approach for the cell-cell interactions similar to the Ag (Ig)-cell interactions.

16) The following properties of Lc are employed:

a) The start of the proliferation of Lc is impossible without their stimulation;

b) The proliferating Lc are characterized by refractoriness with respect to the stimulus, i.e. they proliferate by themselves;

c) The Lc that began to proliferate stopped their proliferation spontaneously by the truncated normal law for the number of hours ($x_8 = 114$, $y_8 = 10$);

d) The Lc that stopped to proliferate can be stimulated again and then, in 6 hours, they become mature (*m*) cells, which cannot proliferate;

e) The period of half-decay of the mature cells is assumed to be 48 – 96 hours.

17) The interaction of cells within each and between each level is caused by Ag and Id-AId bridges that realize the functional potentials of Lc of the appropriate sub-populations. Therefore, IdTh help IdB_2 and $AIdB_2$; IdTs suppress IdTh, IdTef, IdB, AIdTh, AIdTef, and AIdB; IdTef eliminate IdTh, IdTs, IdTef, IdB, AIdTh, AIdTs, AIdTef, and AIdB; IdB produce IdIg; IdIg eliminate Ag; AIdTh help IdB_2, $AIdB_2$, and $AAIdB_2$; AIdTs suppress IdTh, IdTef, IdB, AIdTh, AIdTef, AIdB, AAIdTh, AAIdTef, and AAIdB; AIdTef eliminate IdTh, IdTs, IdTef, IdB, AIdTh, AIdTs, AIdTef, AIdB, AAIdTh, AAIdTs, AAIdTef, and AAIdB; AIdB produce AIdIg; AIdIg eliminates IdIg and kills Id cells. Behavior of the elements of the 3rd and 4th levels is similar to behavior of the elements of the second level.

18) Ts suppress the cell-targets within their level at the instant of their stimulation (through the antigen bridge). Ts suppress the cell-targets of the lower and higher levels (due to Id-AId interactions) except the plasma B cells because they do not have Ig receptors. Because of the interactions between Ts and the cell-targets, the Ts are inactivated and the cell-targets are suppressed throughout the reactions being considered.

19) Tef eliminate the cell-targets except for the plasma B cells.

20) Every control-cell can affect only one target-cell during three hours. The general productivity of Ts, Tef, and Th can make up for the proliferating cells from 2 to 4 targets and for the mature cells from 6 to 10 targets.

21) The mean rate of Ig production by one proliferating B cell is 10^7 molecules per hour. The mature (plasma) B cell gives 10^8 molecules per hour.

22) One molecule of Ig eliminates on the average one molecule of Ag when large concentrations of Ag and Ig are considered.

23) Within all the interactions Ig have priority over cells.

24) Ag are subdivided into T-dependent and T-independent Ag. T-dependent Ag stimulate all the sub-populations of T cells and B_2 cells. T-independent Ag stimulate Ts and B_1 cells. Ig of any level are related to T-dependent Ag.

25) Two classes of Ig are introduced: IgM and IgG. Half-life of IgM is 50 hours and that of IgG (75% of all Ig) is 100 hours. They are similar in all the other respects.

26) During the primary response, IgG are derived only by the plasma B_2 cells. All the other B cells (proliferating B_2 and proliferating and plasma B_1) produce IgM. During the secondary response, IgG are derived by B_2 cells and IgM by B_1 cells.

27) There are priorities in action of different T cells on the same target: Tef $\succ$ Ts $\succ$ Th, i.e., first Tef acts, then Ts, and at last Th, and also $\mathrm{Ti}^m \succ \mathrm{Ti}^p$, where m means mature cells and p means proliferating cells. Tef and Ts of the next, higher level have the priorities in action in comparison with Tef and Ts of the antecedent level. On the contrary, Th of the higher level acts after Th of the antecedent level.

28) All the cells except the proliferating and mature Tef, Ts, Th, and plasma cells enter the quantity of cells of the given level that stimulate the cells of the next level.

29) The resting Lc, memory cells are not reactive to the action of cells of the same level.

30) The following *MM* is accepted here as the basis of *cell-cell interactions*:

$$z = kxy/Y = k\text{^}xy,\ k = k(Y) = 1/3F_7(\lg Y),\ k\text{^} = k/Y, \tag{1}$$

where $3k$^ is probability of two cells of a control-type and target-type contact; Y is the total number of control cells (regarding priorities in action) at the time instant t for x; x is any part of the total number of target cells $X = X(Y)$ at the time instant t for Y; y is any part of the total number of control cells Y; z is the result of x-cells and y-cells interactions during an hour. Interactions, while the time is Δt, result in $z\Delta t$ instead of z.

Thus, we actually accept the "acting mass" law, where k^ depends on t, but not on x, y. Under large concentrations of Y when $F_7(\lg Y)$ is 1, we have the expression (1) usually used. From (1), it follows that under fixed t the results of the cell-cell interactions are additive over all x or y for which X and Y are fixed values. Thus, the very important property of mass law also takes place in our more general law (1). It should be noted that k^ may also depend on pH, temperature, enthalpy and some other factors.

31) All the cells of the immune network must be subjected to the *laws* of *affinity* to Ag (Ig). In particular, the capability of each cell to react to a certain dose of Ag is kept by its offsprings. And also, we have to employ the following relations

$$F_i(v')F_i(v) = F_i(v')cF_i(v) = max\,[F_i(v'), F_i(v)];\ F_i(v')(1\text{-}F_i(v)) =$$

$$F_i(v')\text{-min}\,[F_i(v'), F_i(v)];\ (1\text{-}F_i(v'))(1\text{-}F_i(v)) = 1 - F_i(v')(1 - F_i(v)). \tag{2}$$

where F_i, $i = 1, 2,\ldots, 6$, are the functions defined above.

3. Continuous MM

Let M_{5n+1}, M_{5n+2}, M_{5n+3}, M_{5n+4}, and M_{5n+5} be the numbers of Th, Ts, Tef, B, and Ig, respectively, at the instant t at the level $n+1$, where $n = 0, 1, 2, \ldots$; $M_{5n+4,1}$ and $M_{5n+4,2}$ are the numbers of the cells B_1 and B_2 – type, respectively, and let $M_i^{\,r}$, $M_i^{\,s}$, $M_i^{\,s'}$; $M_i^{\,p}$, $M_i^{\,ss}$, $M_i^{\,ss'}$; and $M_i^{\,m}$ be the numbers of the resting (memory) cells stimulated by Ag, stimulated by Ag and Th-cells (for B_2-cells), proliferating, second time stimulated by Ag, second time stimulated by Ag and Th-cells (for B_2-cells), and mature (plasma for B) cells, respectively, all from M_i,

$$M_i = M_i^{\,r} + M_i^{\,s} + M_i^{\,s'} + M_i^{\,p} + M_i^{\,ss} + M_i^{\,ss'} + M_i^{\,m}. \tag{3}$$

Let M_0 be the number of Ag molecules at the instant t and $M_{i\,,0}$ be the numbers of the cells and molecules at the instant $t = 0$ of the introduction of Ag.

We denote

$$y_0 = \lg M_0,\ D_0 = M_0,\ D_i = M_{5i} + 10^7 M_{5i-1}^{\ \ p,1} + 10^8 M_{5i-1}^{\ \ m},\ i = 1, 2, \ldots, \tag{4}$$

where $M_{5i-1}^{\ \ p,1}$ is the number of the cells from $M_{5i-1}^{\ \ p}$ that produce Ig, and

$$\Phi_s(v) = 1/(2\pi y_s^2)^{1/2} \int_{-\infty}^{v} \exp\,[-(u-x_s)^2/(2y_s^2)]du;\ F_s(v) = \Phi_s(v),\ v \in [x_s\text{-}3y_s,\ x_s\text{+}3y_s],$$

$$F_s(v) = 0,\ v < x_s\text{-}3y_s\ \ F_s(v) = 1,\ v > x_s\text{+}3y_s,\ s = 1, 2, \ldots, 8. \tag{5}$$

If $D_{i-1} - D_i \geq 0$, $i = 1, 2, \ldots$, then (see the assumptions 12), 22), s. 2.2)

$$M_0(t+1) = [D_0 - D_1]0.99,\ M_5(t+1) = 0,\ M_{10}(t+1) = [D_2 - D_3]0.99,$$

$$M_{15}(t+1) = 0,\ \text{etc.} \tag{6}$$

If $D_0 - D_1 < 0$, $D_1 - D_0 - D_2 \geq 0$, $D_{i-1} - D_i \geq 0$, $i > 2$, then

$$M_0(t+1) = 0,\ M_5(t+1) = [D_1 - D_0 - D_2]0.99,\ M_{10}(t+1) = 0,$$

$$M_{15}(t+1) = [D_3 - D_4]0.99,\ \text{etc.} \tag{7}$$

If $D_0 - D_1 < 0$, $D_1 - D_0 - D_2 < 0$, $D_2 - D_1 + D_0 - D_3 \geq 0$, then

$$M_0(t+1) = M_5(t+1) = 0,\ M_{10}(t+1) = [D_2 - D_1 + D_0 - D_3]0.99,$$

$$M_{15}(t+1) = 0,\ M_{20}(t+1) = [D_4 - D_5]0.99,\ \text{etc.} \tag{8}$$

and so on.

According to the assumptions 16), 17), 20), and 30), we have the following cases

$$k_i^j = k(M_i^j) = 1/3F_7[\lg (M_i^j)],\ k\hat{}_i^j = k_i^j/M_i^j,\ i = 1, 2, 3, 6, 7, 8, \ldots, j = m, p;$$

$$k_i = 1/3F_7[\lg (\Sigma_i)],\ k\hat{}_i = k_i/\Sigma_i,\ i = 0, 5, \ldots \quad (9)$$

where (see assumption 28))

$$\Sigma_i = \Sigma_{s=1}^{4} (M_{s+i} - M_{s+i}^{p} - M_{s+i}^{m}) + M_{4+i}^{p}, \quad (10)$$

M_{s+i} is given by (1) with the replacement of i by $i+s$. We denote, in addition,

$$[1\text{-}k_{i1,\ldots,ij}^{m,p}] = [1\text{-}k_{i1}^{m}][1\text{-}k_{i1}^{p}]\ldots[1\text{-}k_{ij}^{m}][1\text{-}k_{ij}^{p}],\ i1, \ldots, ij = 1, 2, 3, 6, 7, 8, \ldots,$$

$$U_1(t) = U_1 = U_3(t) = U_3 = U = [1\text{-}F_6(y_5)][1\text{-}k_{8,7,3,2}^{m,p}],\ U_2(t) = U_2 = [1\text{-}F_6(y_5)][1\text{-}k_{8,3}^{m,p}],$$

$$U\hat{}_1(t) = U\hat{}_1 = U\hat{}_3(t) = U\hat{}_3 = U\hat{} = [1\text{-}F_6(y_5)][1\text{-}k_{8,7}^{m,p}],\ U\hat{}_2(t) = U\hat{}_2 = [1\text{-}F_6(y_5)][1\text{-}k_8^{m,p}],$$

$$V(t) = V = [1\text{-}F_6(y_{10})][1\text{-}k_{13,12,8,7,3,2}^{m,p}],$$

$$W(t) = W = k_1^{m} + k_1^{p}(1\text{-}k_1^{m}) + k_6^{m}(1\text{-}k_1^{m,p}) + k_6^{p}(1\text{-}k_6^{m})(1\text{-}k_1^{m,p}), \quad (11)$$

Let

$$y_i = y_i(t) = \lg M_i,\ z = z(t) = \lg [M_1+M_2+M_3+M_4 - \Sigma M_i^j\ (i = 1, 2, 3, 4; j = p, m) + M_4^{p}],$$

$$V_6(t) = V_6 = [1\text{-}F_1(y_5)][1\text{-}F_5(y_5)][1\text{-}F_6(y_{15})][1\text{-}k_{13,12,8,7,3,2}^{m,p}],\ V_7(t) = V_7 =$$
$$[1\text{-}F_5(y_5)][1\text{-}F_6(y_{15})][1\text{-}k_{13,8,3}^{m,p}],\ V_8(t) = V_8 = [1\text{-}F_5(y_5)][1\text{-}F_6(y_{15})][1\text{-}k_{13,12,8,7,3,2}^{m,p}],$$

$$V\hat{}_6(t) = V\hat{}_6 = [1\text{-}F_1(y_5)]\,[1\text{-}F_5(y_5)][1\text{-}F_6(y_{15})][1\text{-}k_{13,12,3,2}^{m,p}],\ V\hat{}_7(t) = V\hat{}_7 =$$
$$[1\text{-}F_5(y_5)][1\text{-}F_6(y_{15})][1\text{-}k_{13,3}^{m,p}],\ V\hat{}_8(t) = V\hat{}_8 = [1\text{-}F_5(y_5)][1\text{-}F_6(y_{15})][1\text{-}k_{13,12,3,2}^{m,p}],$$

$$X(t) = X = k_1^{m} + k_1^{p}(1\text{-}k_1^{m}) + k_6^{m}(1\text{-}k_1^{m,p}) + k_6^{p}(1\text{-}k_6^{m})(1\text{-}k_1^{m,p}) +$$
$$k_{11}^{m}(1\text{-}k_{6,1}^{m,p}) + k_{11}^{p}(1\text{-}k_{11}^{m})(1\text{-}k_{6,1}^{m,p}),$$

$$u = \Sigma\, k_i^j\ (i = 2, 3, 7, 8; j = m, p),\ u\hat{} = \Sigma\, k_i^j\ (i = 7, 8; j = m, p),$$

$$f(t,s) = \lim \prod [1\text{-}F_5(y_0(k\Delta t))][1\text{-}F_6(y_{10}(k\Delta t))]\ (k\Delta t = s, \ldots, t),\ \Delta t \to 0. \quad (12)$$

Discrete MM of IN (see V. Ivanov [92]) can be obtained directly on the strength of the accepted postulates. The respective continuous analogy of those models can be obtained on the strength of the ordinary procedure of passage to the limit. As a result, we have

$$M_i^p = M_i^p(12)+\int_{12}^{t} m_i^p(s)ds = M_i^p(12)f(t,12)\exp\left(\int_{12}^{t} [0.08-u(s)-c_i^p(s)]ds\right)[1- F_8(t-12)],$$

$$M_i^p(12) = M_{i,0}F_2(y_0(0)),\ t > 12,\ i = 1, 2, 3,$$

$$c_1^p = 1/3k\hat{}_1{}^p(1-k_1{}^m)[U\sum M_i^j (i = 4,2; j = s, s,s) + V\sum M_i^j (i = 9,2; j = s, s,s)],$$

$$c_2^p = 1/3k\hat{}_2{}^p/(1-k_2^p)[U\sum M_i^j (i = 1, 3, j \neq r; i = 4, j \neq r, m) + V\sum M_i^j (i = 6, 8; i = 9, j \neq m)]\},$$

$$c_3^p = 1/3k\hat{}_3{}^p/(1-k_3{}^p)[U\sum M_i^j (i = 1, 2, 3, j \neq r; i = 4, j \neq r, m) +$$

$$V\sum M_i^j (i = 6, 7, 8; i = 9, j \neq m)\};$$

$$m_i^m(t) = -(u+0.01+c_i{}^m)M_i{}^m+M_i^p(t-6)F'_8(t-18)F_2(y_0(t-6))f(t,t-6)\exp\left[-\int_{t-6}^{t} u(s)ds\right],$$

$$M_i^m(18) = 0,\ t \geq 18,\ i = 1, 2, 3;$$

$$m_i^r(t) = - u\hat{}M_i{}^r+N_i^pF'_8(t-12)[1-F_2(y_0)],\ t \geq 12,\ M_i^r(12) = M_{i,0}[1-F_2(y_0(0))],\ i = 1, 2, 3,$$

$$c_1{}^m = 1/8k\hat{}_1{}^m[U\sum M_i^j (i = 4,2; j = s,\ s,s) + V\sum M_i^j (i = 9,2; j = s, s,s)]Ue^{-0.01},$$

$$c_2{}^m = 1/8k\hat{}_2{}^m/(1-k_2{}^{m,p})[U\sum M_i^j\ (i =1, 3, j \neq r; i = 4, j \neq r, m) +$$

$$V\sum M_i^j (i = 6, 8; i = 9, j \neq m)]\}U_2e^{-0.01},$$

$$c_3{}^m =1/8k\hat{}_3{}^m/(1-k_3{}^{m,p})[U\sum M_i^j\ (i =1, 2, 3, j \neq r; i = 4, j \neq r, m) +$$

$$V\sum M_i^j\ (i = 6, 7, 8; i = 9, j \neq m)]Ue^{-0.01};$$

$$m_{4,1}{}^m(t) = -0.01M_{4,1}{}^m+M_{4,1}{}^p(t-6)F'_8(t-24)f(t,18)F_3(y_0(t-6)\exp\left[-\int_{t-6}^{t} u(s)ds\right],$$

$$M_{4,1}{}^m(24) = 0,\ t \geq 24;$$

$$m_{4,1}{}^r(t) = -u\hat{}M_{4,1}{}^r+ M_{4,1}{}^pF'_8(t-13)[1-F_3(y_0)],\ M_{4,1}{}^r(18) = M_{4,1,0}[1-F_3(y_0(0))],\ t \leq 18;$$

$$M_{4,2}{}^p(r+12,r) = M_{4,2}{}^{s'}(r)f(r+12,r)\exp\left[-\int_{r}^{r+12} u(s)ds\right],\ r > 0,$$

$$M_{4,2}{}^p(t,r) = M_{4,2}{}^p(r+12,r)f(t,r+12)\exp\left(\int_{r+12}^{t} [0.08 - u(s)]ds\right)[1- F_8(t-u-12)],\ t > r+12,$$

$$m_{4,2}{}^{s'}(t) = -u[1-F_6(y_5)]M_{4,2}{}^{s'}+M_{4,2}{}^{s}UW-M_{4,2}{}^{s'}(t-12)f(t,t-12)\exp\,[-\textstyle\int_{t-12}^{t} u(s)ds],$$

$$m_{4,2}{}^{s}(t) = M_{4,2,0}f(t,0)\exp\,\{-\textstyle\int_0^t [u(s)+W(s)]ds\}F_4(y_0(18k)),\ 18k < t \le 18(k+1),$$

$$k = 0, 1, \ldots, M_{4,2}{}^{p}(t) = \textstyle\int_0^{t-12} M_{4,2}{}^{p}(t,r)dr,\ t > 12;$$

$$m_{4,2}{}^{m}(t,r) = -\,0.01M_{4,2}{}^{m}(t,r)+dM_{4,2}{}^{s,s'}(t-6)f(t,t-6)\exp\,[-\textstyle\int_{t-6}^{t} u(s)ds],$$

$$dM_{4,2}{}^{s,s'} = [M_{4,2}{}^{p}(t,r)F'_8(t-r-12)+dM]F_4(y_0)W,\ t > r+12,$$

$$dM = \{\hat{}(dM_{4,2}{}^{s,s}) + \hat{}(\hat{}(dM_{4,2}{}^{s,s})F_4(y_0)U(1-W)) +$$

$$\hat{}(\hat{}(\hat{}(dM_{4,2}{}^{s,s})F_4(y_0)U(1-W))F_4(y_0)U(1-W)) +\ldots\},$$

$$\hat{}(h) = h(t-18)f(t,t-18)\exp\,(-\textstyle\int_{t-18}^{t} [u(s)+W(s)]ds]),\ m_{4,2}{}^{m}(t) = \textstyle\int_0^t m_{4,2}{}^{m}(t,r)dr,\ t > 0;$$

$$m_{4,2}{}^{r}(t,\tau) = -u\hat{}[1-F_6(y_5)]M_{4,2}{}^{r}(t,\tau) + dM_{4,2}{}^{r},\ dM_{4,2}{}^{r} = [M_{4,2}{}^{p}(t,\tau)F'_8(t-\tau-12) + dM][1-F_4(y_0)],$$

$$M\hat{}_{4,2}{}^{r}(t) = M_{4,2,0}f(t,0)\exp\,\{[-\textstyle\int_0^t [W(s)+u\hat{}(s)]ds\}[1-F_4(y_0(18k))],\ 18k < t \le 18(k+1),$$

$$k = 0, 1, \ldots, m_{4,2}{}^{r}(t) = [\,M\hat{}_{4,2}{}^{r}(t)]' + \textstyle\int_0^t m_{4,2}{}^{r}(t,\tau)d\tau,\ t > 0. \qquad (13)$$

In the case of the relations for the second level of IN, let us put

$$v = v_6 = v_8 = \textstyle\sum k_i{}^{j}\ (i = 2, 3, 7, 8, 12, 13; j = m, p),$$

$$v_7 = \textstyle\sum k_i{}^{j}\ (i = 3, 8, 13; j = m, p), v\hat{} = v\hat{}_6 = v\hat{}_8 = \textstyle\sum k_i{}^{j}\ (i = 2, 3, 12, 13; j = m, p),$$

$$v\hat{}_7 = \textstyle\sum k_i{}^{j}\ (i = 3, 13; j = m, p),$$

$$f_6(t,s) = \lim \textstyle\prod\{[1-F_1(y_5(k\Delta t))][1-F_5(y_5(k\Delta t)][1-F_6(y_{15}(k\Delta t)]\},$$

$$f_7(t,s) = f_8(t,s) = \lim \textstyle\prod\{[1-F_5(y_5(k\Delta t)][1-F_6(y_{15}(k\Delta t)]\},$$

$$f_{6,1}(t,s) = \lim \textstyle\prod\{[1-F_1(y_5(k\Delta t))][1-F_2(y_5(k\Delta t))][1-F_5(y_5(k\Delta t)][1-F_6(y_{15}(k\Delta t)]\},$$

$$f_{7,1}(t,s) = f_{8,1}(t,s) = \lim \textstyle\prod\{[1-F_2(y_5(k\Delta t))][1-F_5(y_5(k\Delta t)][1-F_6(y_{15}(k\Delta t)],$$

$$f_{9,1,1}(t,s) = \lim \textstyle\prod\{[1-F_3(y_5(k\Delta t))][1-F_5(y_5(k\Delta t)][1-F_6(y_{15}(k\Delta t)]\},$$

$$f_{9,2,1}(t,s) = \lim \textstyle\prod\{[1-F_4(y_5(k\Delta t))][1-F_5(y_5(k\Delta t)][1-F_6(y_{15}(k\Delta t)]\},$$

$f_{6,2}(t,s) = \lim \prod\{[1-F_1(y_5(k\Delta t))]F_2(y_5(k\Delta t))[1-F_5(y_5(k\Delta t))][1-F_6(y_{15}(k\Delta t))]\}$,

$f_{7,2}(t,s) = f_{8,2}(t,s) = \lim \prod\{F_2(y_5(k\Delta t))[1-F_5(y_5(k\Delta t))][1-F_6(y_{15}(k\Delta t))]\}$,

$f_{9,1,2}(t,s) = \lim \prod\{F_3(y_5(k\Delta t))[1-F_5(y_5(k\Delta t))][1-F_6(y_{15}(k\Delta t))]\}$,

$f_{9,2,2}(t,s) = \lim \prod\{F_4(y_5(k\Delta t))[1-F_5(y_5(k\Delta t))][1-F_6(y_{15}(k\Delta t))]\}$, $(k\Delta t = s, \ldots, t)$, $\Delta t \to 0$, (14)

where F_i is one of the functions in (5). Then instead of (13), we have

$M_i{}^P(t,u) = M_i{}^P(u+12,u)f_i(t,u+12)\exp(\int_{u+12}^{t} [0.08 - v_i(s)-c_i{}^P]ds)[1-F_8(t-u-12)]$,

$M_i{}^P(u+12,u) = m_i{}^s(u)f_i(u+12,u)\exp[-\int_u^{u+12} v_i(s)]ds]$, $u > 0$,

$m_i{}^s(u) = M_i{}^s(u)-M_{i,0}\{(f_{i,2}(u,0)\exp[-\int_0^u v_i(s)ds])'+(f_{i,1}(u,0)\exp[-\int_0^u (v_i(s) +$

$F_7(z(s)))ds])'\}F_7(z)$. $M_i{}^P(t) = \int_0^{t-12} M_i{}^P(t,u)du$, $t > 12$;

$M_i{}^m(t,u) = -(v_i+0.01+c_i{}^m)M_i{}^m(t,u) + dM_i{}^{s,s}(t-6,u)f_i(t,t-6)\exp[-\int_{t-6}^{t} v_i(s)ds]$,

$DM_i{}^{s,s}(t,u) = M_i{}^P(t,u)F'_8(t-u-18)\{F_2(y_5)+[1-F_2(y_5)]F_7(z)\}$,

$M_i{}^m(u,u) = 0$, $t > u > 0$, $m_i{}^m(t) = \int^t{}_0 m_i{}^m(t,u)d$, $t > 0$;

$M\hat{}_i{}^r = M_{i,0}f_{i,2}(t,0)\exp(-\int^t{}_0 [F_7(z(u))+v\hat{}_i(u)]du)$, $t > 0$,

$m_i{}^r(t,u) = -v\hat{}M_i{}^r(t,u)+dM_i{}^r$, $M_i{}^r(u,u) = 0$,

$dM_i{}^r = M_i{}^P(t,u)F'_8(t-u-12)[1-F_2(y_5)][1-F_7(z)]$, $t > u > 0$,

$M_i{}^r = M\hat{}_i{}^r+\int^t{}_0 M_i{}^r(t,u)du$, $t > 0$, $i = 6, 7, 8$;

$M_{9,1}{}^P(t,u) = M_{9,1}{}^P(u+18,u)f_7(t,u+18)\exp(\int_{u+18}^{t} [0.08-v(s)]ds)[1-F_8(t-u-18)]$,

$M_{9,1}{}^P(u+18,u) = dM_{9,1}{}^s(u)f_7(u+18,u)\exp[-\int_u^{u+18} v(s)]ds]$, $u > 0$,

$dM_{9,1}{}^s(u) = M_{9,1,0}\{(f_{9,1,2}(u,0)\exp[-\int^u{}_0 v(s)ds])'+(f_{9,1,1}(u,0)\exp[-\int^u{}_0(v_i(s)+F_7(z(s)))ds])'F_7(z)\}$,

$M_{9,1}{}^P(t) = \int_0^{t-18} M_{9,1}{}^P(t,u)du$, $t > 18$; $M_{9,1}{}^m(t,u) = 0$, $0 < t-u < 109$,

$m_{9,1}{}^{m}(t,u) = -0.01M_{9,1}{}^{m}(t,u) + dM_{9,1}{}^{s,s}(t\text{-}6,u\text{-}6)\exp[-\int_{t-6}^{t} v(s)ds, 107+u < t < 360,$

$dM_{9,1}{}^{s,s} = M_{9,1}{}^{P}(t,u)F'_8(t\text{-}u\text{-}18)\{F_3(y_5)+[1\text{-}F_3(y_5)]F_7(z)\}V, 101+u < t < 360,$

$$M_{9,1}{}^{m} = \int_0^t M_{9,1}{}^{,m}(t,u)du, 0 < t < 360;$$

$$M\hat{}_{9,1}{}^{r} = M_{9,1,0}f_{9,1,1}(t,0)\exp(-\int_0^t [F_7(z(u))+v\hat{}(u))du), t > 0,$$

$\hat{}M_{9,1}{}^{r}(t,u) = 0, 0 < t\text{-}u < 97, \hat{}m_{9,1}{}^{r}(t,u) = -v\hat{}\ \hat{}M_{9,1}{}^{,r}(t,u)+dM_{9,1}{}^{r},$

$dM_{9,1}{}^{r} = M_{9,1}{}^{P}(t,u)F'_8(t\text{-}u\text{-}18)[1\text{-}F_3(y_5)][1\text{-}F_7(z)]V\hat{}, 95+u < t < 360,$

$\hat{}M_{9,1}{}^{,r} = \int_0^t \hat{}M_{9,1}{}^{,r}(t,u)du, M_{9,1}{}^{r} = M\hat{}_{9.1}{}^{r}+\hat{}M_{9,1}{}^{r}, 0 < t < 360,$

$$dM_{9,2}{}^{s} = M_{9,2,0}\{(f_{9,2,2}(u,0)\exp[-\int_0^u v(s)ds])' +$$

$(f_{9,2,1}(u,0)\}\exp[-\int_0^u (v(s)+ F_7(z(u))ds])'(1\text{-}X), 0 < t < 360;$

$dM_{9,2}{}^{s'} = dM_{9,2}{}^{s}X+g(dM_{9,2}{}^{s})[F_4(y_5)+[1\text{-}F_4(y_5)]F_7(z)]VX,$

$g(dM_{9,2}{}^{s}) = \hat{}(dM_{9,2}{}^{s})+\hat{}(\hat{}(dM_{9,2}{}^{s})[F_4(y_5)+[1\text{-}F_4(y_5)]F_7(z)](1\text{-}X))+$

$\hat{}(\hat{}(\hat{}(dM_{9,2}{}^{s})[F_4(y_5)+[1\text{-}F_4(y_5)]F_7(z)](1\text{-}X))[F_4(y_5)+[1\text{-}F_4(y_5)]F_7(z)](1\text{-}X))+\ldots\},$

$\hat{}(f) = f(t\text{-}18)f_{9,2,1}(t,t\text{-}18)\exp(-\int_{t-18}^{t} [v(s)+X(s)]ds]), 0 < t < 360;$

$M_{9,2}{}^{P}(t,u) = 0, 0 < t\text{-}u < 12, M_{9,2}{}^{P}(t,u) = M_{9,2}{}^{P}(u+12,u)\exp[\int_u^t (0.08\text{-}v(s))ds][1\text{-}F_8(t\text{-}u\text{-}12)],$

$M_{9,2}{}^{P}(u+12,u) = dM_6{}^{s'}(u)\exp[-\int_u^{u+12} v(s))ds], 12+u < t < 360,$

$$M_{9,2}{}^{P} = \int_0^{t-12} M_{9,2}{}^{,P}(t,u)du, 0 < t < 360;$$

$dM_{9,2}{}^{s,s} = M_{9,2}{}^{P}(t,u)F'_8(t\text{-}u\text{-}12)\{F_4(y_5)+[1\text{-}F_4(y_5)]F_7(z)\}V(1\text{-}X), 95+u < t < 360,$

$dM_{9,2}{}^{s,s'} = \{M_{9,2}{}^{P}(t,u)F'_8(t\text{-}u\text{-}12)+g(dM_{9,2}{}^{s,s})\}[F_4(y_5)+[1\text{-}F_4(y_5)]F_7(z)]VX,$

$95+u < t < 360, 0 < t < 360; M_{9,2}{}^{m}(t,u) = 0, 0 < t\text{-}u < 103,$

$m_{9,2}{}^{m}(t,u) = -0.01M_{9,2}{}^{m}(t,u)+dM_{9,2}{}^{s,s'}(t\text{-}6,u\text{-}6)f_{9,2,1}(t,t\text{-}6)\exp[-\int_{t-6}^{t} v(s)ds, 101+u < t < 360,$

$$M_{9,2}{}^{m} = \int^{t}_{0} M_{9,2}{}'^{m}(t,u)du,\ 0 < t < 360;$$

$$\wedge M_{9,2}{}^{r}(t,u) = 0,\ 0 < t\text{-}u < 97,\ m_{9,2}{}^{r}(t,u) = dM_{9,2}{}^{r} = \{M_{9,2}{}^{p}(t,u)F'_{8}(t\text{-}u\text{-}12) +$$

$$g(dM_{9,2}{}^{s,s})\}[1\text{-}F_4(y_5)][1\text{-}F_7(z)]V\wedge,\ 95{+}u < t < 360,\ \wedge M_{9,2}{}^{r} = \int^{t}_{0}\wedge M_{9,2}{}'^{r}(t,u)du),\ 0 < t < 360,$$

$$M\wedge_{9,2}{}^{r}(t,u) = M_{9,2,0}f_{9,2,1}(t,u)\exp\{[-\int^{t}_{u}[X(s)+v\wedge(s)]ds\}[1\text{-}F_4(y_5(18k))],$$

$$18k < t\text{-}u \le 18(k+1),\ k \ge 0,$$

$$M_{9,2}{}^{r} = M\wedge_{9,2}{}^{r} + \wedge M_{9,2}{}^{r},\ m_{9,2}{}^{r}(t) = [M\wedge_{9,2}{}^{r}(t)]' + \int^{t}_{0} m_{9,2}{}^{r}(t,\tau)d\tau,\ t > 0. \quad (15)$$

Here $c_i^{p,m}$ are the same expressions as $c_i^{p,m}$ of the first level with replacement of i by i+5. The equations of the model for the third level can be obtained by adding 5 to all the subscripts of the previous level. The following levels can be obtained in the same way.

Note at last that we have not considered the models corresponding to the assumptions 24)-26). These models can be constructed in a similar way.

4. On the Classical and the New Approach

Let us show that the proposed models belong to the class of MM in the general theory of development. Indeed, the base equations of the class are the equation of restoration and guarantee of external functions,

$$m_i(t) = \int^{t}_{0} \alpha_i(t,s)\lambda_i(t,s)y_i(s)m_i(s)ds,\ R_i(t) = \int^{t}_{0} \lambda_i(t,s)m_i(s)ds,\ G_i(t) = \int^{t}_{0} [1-\lambda_i(t,s)]m_i(s)ds,$$

$$\lambda_i(t,s) = \mu_i(t,s),\ c_i(t) = \int^{t}_{0} \beta_i(t,s)\mu_i(t,s)[1-y(s)]m_i(s)ds,\ f_i(t) = m_i(t)+c_i(t), \quad (16)$$

and the equation of realization of the external functions,

$$c_i{}^{j}(t) = \int^{t}_{0} \beta_i{}^{j}(t,s)\mu_i{}^{j}(t,s)z_i{}^{j}(s)m_i(s)ds,\ j = 1, \ldots, n_i,$$

$$0 \le y_i,\ z_i^j \le 1,\ y_i + \sum z_i{}^{j}\ (j = 1, \ldots, n_i) = 1,\ t \ge 0,\ i = 1, \ldots, n. \quad (17)$$

In our case for every level of IN we have 5 equations of (16)-type relative to $m_i{}^{p}$, i=5n+1, 5n+2, 5n+3, 5n+4,1, 5n+4,2, and 16+6+1 = 23 equations of (17)-type relative to $m_i{}^{j}$, j = s, s', s,s, s,s', m, r, and m_{n+5} (the number of Ig per unit of the time).

Comparing (16) and (13), i = 1, we have

$$R_1 = \lambda_1(t)[M_1{}^{p}(12)+\int^{t}_{12} m_1(s)ds] = \lambda_1(t)M_1 = M_1{}^{p} =$$

$$M_1{}^P(12)\exp\{\int_{12}^t [0.08-u(s)-c_1{}^P(s)]ds\}[1-F_8(t-12)],\ M_1 = R_1+G_1,$$

$$G_1 = [1-\lambda_1(t)]\int_0^t m_1(s)ds = \int_{12}^t M_1{}^P(s)[u(s)+c_1{}^P(s)]ds;$$

$$m_1(t) = M_1{}^P/M_1\int_{12}^t \alpha_1(t,s)y_1(s)m_1(s)ds = (R_1+G_1)' = M_1{}^P[0.08-d(t)],$$

$$d(t) = d = F'_8(t-12)/[1-F_8(t-12)],\ m_1(t) = 0,\ t \geq t^*,$$

$$d(t^*) = 0.08,\ t^* \approx 126,\ \alpha_1(t,s) = [1+25\delta(s-12)]\alpha(t,s);$$

$$c_1(t) = M_1{}^P/M_1\int_{12}^t \beta_1(t,s)[1-y_1(s)]m_1(s)ds = M_1{}^P d,\ c_1(t) = 0,\ t < 96,\ t > 156,$$

$$\beta_1(t,s) = [1+2M_1(96)d(96)/\{[1-y_1(96)]m_1(96)\delta(s-96)]\}\beta(t,s);$$

$$f_1(t) = m_1(t)+c_1(t) = M_1{}^P 0.08,\ f_1(t) = 0,\ t > 156, \tag{18}$$

where δ is Dirac δ-function. It is natural to suppose that

$$\alpha(t,s) = 0.08,\ \beta(t,s) = z(t)/z(s), \tag{19}$$

Then

$$y_1(t) = 1,\ M_1(12)\alpha(t,12) + \int_{12}^t \alpha(t,s)m_1(s)ds = M_1 0.08,\ 12 < s < t < 96,$$

$$y_1(t) = 0,\ [M_1(12)+\int_{12}^{96} m_1(s)ds]x(t) = M_1[0.08\text{-}d(t)],\ x(t) = M_1/M_1(96)[0.08\text{-}d(t)],\ 96 \leq t < t^*;$$

$$c_1(t) = 0,\ 12 < t < 96,\ M_1(96)d(96) + \int_{96}^t m_1(s)/z(s)ds = m_1(t)/z(t)M_1(t)/m_1(t)d(t) =$$

$$M_1(t)/z(t)d(t) = M_1(96)d(96)\exp\{\int_{96}^t m_1(s)/[M_1(s)d(s)]ds\},$$

$$z(t) = M_1(t)d(t)/[M_1(96)d(96)]\exp\{-\int_{96}^t m_1(s)/[M_1(s)d(s)]ds\},\ 96 \leq t \leq 156. \tag{20}$$

Note that in (16)--(18), $m_1(t)$ is the increment of new bio-mass of the cells arisen from the proliferating cells, $f_1(t)$ is the resource that is necessary for new bio-mass and for the external product output, and

$$\int_{12}^t M_1{}^P(s)u(s)ds,\ \int_{12}^t M_1{}^P(s)F'_8(s-12)/[1-F_8(s-12)]ds,\ \int_{12}^t M_1{}^P(s)c_1{}^P(s)ds \tag{21}$$

are respectively the suppressed cells, stopped proliferating cells, and stimulated B_2-cells of the first and the second level (with the help of $M_1{}^P$ cells).

It is not difficult to see that in our case the description by the new approach is likely more complicated than by the classical one, but in return the new approach gives us more information and proves useful for accounting new, likely more profound, assumptions. Probable examples are different assumptions about properties of biomass of the cells depending on their age. If immunologists confirm that older target-cells are open to control-cells more than younger ones, then $\lambda_1(t,s)$ can make sense instead of $\lambda_1(t)$, in particular,

$$\lambda_1(t,s) = 0,\ 0 < s < a(t),\ \lambda_1(t,s) = 1,\ a(t) < s < t, \qquad (22)$$

can make sense. With (20), we would have the typical expression

$$M_1{}^P = \int_{a(t)}^{t} m_1(s)ds,\ \ G_1 = \int_0^{a(t)} m_1(s)ds,\ \ f_1(t) = m_1(t)+c_1(t),$$

$$m_1(t) = \int_{a(t)}^{t} \alpha_1(t-s)y_1(s)m_1(s)ds = (M_1{}^P+G_1)',\ c_1(t) = \int_{a(t)}^{t} \beta_1(t-s)[1-y_1(s)]m_1(s)\mathrm{d}s, \qquad (23)$$

which cannot be obtained easily by the classical approach.

It should be noted that knowing $M_1{}^P$, G_1, and $f_1(t)$ in (23) and finding y from an experiment and/or a certain extreme principle, we can find the others.

Comparing (17) and (13), $i = 1$, we have

$$c_1{}^1(t) = \int_{12}^{t} \beta_1{}^1(t,s)\mu_1{}^1(t,s)m_1(s)z_1{}^1(s)ds = m_1{}^{s,s}(t) =$$

$$c_1 - c_1(t\text{-}6)\exp\left[-\int_{t-6}^{t} u(s)ds\right],\ c_1{}^2(t) = \int_{18}^{t} \beta_1{}^2(t,s)\mu_1{}^2(t,s)m_l(s)z_1{}^2(s)ds = m_1{}^m(t) =$$

$$-[u+0.01+c_1{}^m]M_1{}^m + c_1(t\text{-}6)F_2(y_0(t\text{-}6))f(t,t\text{-}6)\exp\left[-\int_{t-6}^{t} u(s)ds\right],\ M_1{}^m(18) = 0,\ t > 18,$$

$$c_1{}^3(t) = m_1{}^r(t) = \int_{12}^{t} \beta_1{}^3(t,s)\mu_1{}^3(t,s)m_l(s)z_1{}^3(s)ds = -u^\wedge M_1{}^r + c_1[1-F_2(y_0)],$$

$$M_1{}^r(12) = M_{1,0}[1-F_2(y_0(0))],\ t > 12,\ \ z_1{}^1 = z_1{}^2 = z_1{}^3 = 1. \qquad (24)$$

Solving the equations in the relations (24) with respect to $m_1{}^m(t)$ and $m_1{}^r(t)$, we can find the kernels $\beta_1{}^2(t,s)\mu_1{}^2(t,s)$ and $\beta_1{}^3(t,s)\mu_1{}^3(t,s)$.

Here and in many other cases, we need to use the well-known formulae of the type

$$x'(t) + g(t)x(t) = f(t),\ x(t) = x(t^\wedge)\exp\left[-\int_{t^\wedge}^{t} g(v)dv\right] + \int_{t^\wedge}^{t} f(s)\exp\left[-\int_{s}^{t} g(v)dv\right]ds,$$

$$x'(t) = g(t)\left[-x(t^\wedge)\exp\left[-\int_{t^\wedge}^{t} g(v)dv\right] + \int_{t^\wedge}^{t} [f(t)/g(t)-f(s)/g(s)]\exp\left[-\int_{s}^{t} g(v)dv\right]g(s)\mathrm{d}s.\right. \qquad (25)$$

It should also be noted that

$$m_1{}^{g} = c_1 - m_1{}^{s,s} - m_1{}^{r} - m_1{}^{m} + M_1{}^{p}(u+c_1{}^{p}) \tag{26}$$

is the increment of the general "obsolete" product under the output of $m_1{}^{p} + m_1{}^{s,s} + m_1{}^{r} + m_1{}^{m}$, but the part of it, which is $M_1{}^{p}c_1{}^{p}+M_1{}^{m}c_1{}^{m}$, turns out to be stimulated by B_2 cells of the first and second level.

Let us consider now the comparison of our models with the models of the kinetic cellular theory by N. Bellomo, L. Preziosi [20] (see also Ch. 3). It is clear that in our notations M_i^j the superscript j, j = s, s', s,s, s,s', p, m, and r, may be corresponded to the variable u in [20], which called there the state of the cell activity. However, we do not summarize M_i^j over j because for each j we have the specific kind of activity that uses the specific way later. Instead, we summarize each given kind of activity by our variable u, which plays the role of the time preceding the current time t. Another kind of activity, connecting by the affinity of Ag and Ig for cells, has been introduced in the explicit form.

So, our models of IN cell populations are essentially different from the models of the kinetic cellular theory [20]. These last models essentially different from the novel class models considered in this book, as our models, are Volterra-type non-linear relations with the prehistory and feedback. Of course, the main effort in [20] (see Ch. 3) is devoted to interaction of IN and cancers. So, we will come back to the comparison when we consider the application of our models to cancers in Ch. 11.

5. Some Results of Investigation

It is not difficult to see that the equations (4)--(13) have the unique nonnegative boundary solutions. Boundaries of these solutions follow from those of $M_i{}^{r}(t)$ and $M_i{}^{P}(t)$, which, in turn, follow from the postulates in 16).

The following experimental facts are known (which are the results of multilevel interaction) about the first level cells:

1. The number of antibodies generating B cells (AGC) is maximal within the 4th or 6th day after the primary or within the 4th or 5th day after the secondary immunization.

2. The number of $B_2{}^{P}$ cells is maximal within the 4th or 6th day and the number of plasma cells is maximal within the 6th or 8th day after the primary immunization. After the secondary immunization the total number of AGC is maximal within the 4th or 6th day.

3. The total number of AGC of the given specificity is maximal within the 4th or 6th day.

4. The number of memory cells of the given specificity is less than 10^4.

5. The number of memory B cells must be less than the initial number of stimulated B cells (the so-called exhausting differentiation) over the range that is less or equal to 0.3 of the decimal logarithm of the Ag doses (on the right edge of the dose curve, before the tolerance concentration).

6. The "plateau" of the curvature for the maximal number of AGC depending on the dose of Ag should be no wider than 0.6 of the Ag dose logarithm.

7. Free Ig should be under the condition that the logarithm of number Ag is not less than

one logarithm of the stimulating Ag doses.

8. Contents of free Ig at the top of the primary immune response should not be below the initial Ig dose if only in some intervals (no less than 0.6 of logarithm) of the dose range.

9. Contents of free Ig at the top of the secondary immune response should exceed the summary Ag dose by no less than 0.5-1 order, if only over some intervals of the dose range.

10. The total number of the observed levels of IN does not exceed 3-4.

<u>Guess 1</u>. *All the experimental facts* 1 – 10 *are contained in the properties of the MM of IN* (*under some certain subintervals of IN parameters falling within the possible intervals*). *In addition, the solutions of the MM equations have, as a rule* (*due to the assumption* 17), *namely,* Tef *eliminate themselves*), *a fluctuating nature* (*effect of "cannibalism"*) *and only their averaged values correspond to the usually expected values.*

<u>Guess 2</u>. *The solutions of the model have the structural stability* (G. Hoffman, [80]), *that is provided by a dominating idio-type on each level* (*under the strong asymmetry of the matrix of IN interactions*). *Here, the evolutionary principle economizing of the interactions' number and structure under guarantee of the required functions manifests itself.*

The attempt (V. Ivanov et al. [94]) of qualitative investigation of this kind of models confirms that some of the properties 1 – 10 follow from the models. However, the proof of the guesses to a complete extent is still ahead of us.

The attempt of evaluation of the respective software for the models was done by programmers V.A. Dul'sky and T.A. Dyn'ko in Glushkov Institute of Cybernetics, Kiev, Ukraine. However, it was not finished. Some of the numerical results of modeling on a computer confirming the observations of IN can be taken from [94].

It should be emphasized here that this is by no means a completed work. Besides the above-mentioned open problems, there are many others, in particular, the problems of the control with the help of changing of the initial data of the type $M_{i,0}$. The latter are possible due to the biotechnology of the production of *hybridomas* that are derived from the fusion of cancerous Lc and antibody-producing cells.

<u>Guess 3</u>. *Each possible steady behavior of IN and, in general, of the immune system can be achieved, as a rule, by a tremendous number of controls. It is one of the manifestations of immunology mobile principle* (R. Petrov, R. Stepanenko [149]).

6. Conclusion

On the whole, we discuss the IN models constructed for the first time in a complete form according to the 31 detailed *postulates* or assumptions of *IN*.

It is well known that 7 simple postulates of Euclidean geometry have generated numerous important results both in theory and practice. Therefore, we believe that above-mentioned 31 immunology postulates will also generate numerous significant results. However, the respective investigation exceeds the limits of this book.

The IN models are also important here due to the main goals of the book, especially in proving the validity of CAO models for the wide spectrum of CAO.

What has one learned by using this application of the general theory in Ch. 3 and 4 to IN modeling? We have showed that all the equations of the IN models belong to the class of the models in question. It has been important to receive explicit expressions for the functions of α, β-types, which are usually not entirely determined indices of efficiency functioning.

One can see that the distribution function y-type takes in IN the value of 1 or 0 that corresponds to the optimal strategy of control of the general theory. It follows that the control of y-type has been likely selected in natural systems. Therefore, it is important to study more profound *factorization* of initial dynamic characteristics of natural systems in comparison with artificial systems. This factorization has been given in Ch. 5. Besides, we have showed that in contrast to artificial systems the "obsolete products" in IN turn out to be useful for its external functions. It should be an important goal of bionic system.

At the same time, based on the general laws energy conservation, the functioning of IN cannot be without some waste. A certain possibility to regard this fact and, by the same token, deepen the models of IN follows from the general theory of development modeling, and has also been shown in s. 5.

On the whole, we have much more reasons now to believe that the novel class of MM of the cell and CAO is very useful for mathematical modeling of a wide spectrum of biological objects.

Chapter 9

MM OF SOME OTHER CAO

1. Functions of Special Cells

Structural and biochemical changes in the cells of multi-cellular organisms leading to specialized cell roles constitute cell *differentiation*. The potential for cellular differentiation depends on the conditions for the expression (i.e., transcription and translation) of the cell's genetic information. Interactions between the cytoplasm and the nucleus (*intrinsic factors*) and environmental factors outside the cell (*extrinsic factors*) influence the course of differentiation.

Muscle cells, nerve cells, and blood cells are excellent examples of the degree of specialization that may accompany cell development.

Muscle cells develop a complex mechanism for contraction. Myofibrils containing actin and myosin filaments are geometrically differentiated parallel to the long axis to the cell and are spatially arranged about one another with a number of cross-bridges formed between them. When stimulation is applied or ATP or Mg^{2+} are added, the cross-bridges between the filaments progressively break and reform as the filaments slide past one another, and the myofibril (and cell) shortens.

Nerve cells or neurons develop processes much longer than the cell body. Impulses are conducted between remote parts of the body along these processes and are transmitted from one cell to the next as a result of specialized properties of the nerve cell membrane.

Stem cells in the bone marrow may give rise to red or white *blood cells*. In the case of red blood cells, increasing level of hormone erythropoietin travelling in bloodstream to the bone marrow may cause the stem cells to form mRNAs for the globin chains of hemoglobin.

The cells become committed to be red blood cells, as 95% of all protein synthesis is directed to hemoglobin. The buildup of hemoglobin is then followed by the dissolution or loss of cell structures such as the nucleus and the residual nucleic acids. The cells irreversibly differentiate into red blood cells. When primitive stem cells undergo division, one of the daughter cells remains undifferentiated and pluripotent, so that depletion of marrow stem cells does not normally occur.

Complete differentiation and maturation from stem cells to the erythrocytes takes about 7 to 10 days. The mature erythrocyte is incapable of further proliferation. In humans, its average life span is 120 days. As there are 5 billion red blood cells in each cubic centimeter of blood, a few simple calculations (regarding that 8% of body weight is blood) reveal that in a typical adult the differentiation and maturation of 3 million erythrocytes is completed each second. White blood cells or Lc are capable of secreting thousands of different *immunoglobulins* (*antibodies*), each of which is specific to a single antigenic determinant. Liver cell loss stimulates liver cell proliferation. The same occurs with the kidney cells.

Endothelial cells that form the lining of blood vessels have a remarkable capacity to adjust their numbers and arrangement by simple duplication to suit local requirements.

Some other cells (intestine, skin, blood, etc.) are renewed by means of *stem cells*.

The defining properties of a *stem cell* are as follows: It is not itself terminally differentiated. It can divide without limit. When it divides, each daughter has a choice: it can either remain a stem cell, or it can embark on a course leading irreversibly to terminal differentiation.

Many cells (photoreceptors, heart muscle cells, and nerve cells) are unable to divide, but they all metabolically active and capable not only of synthesizing RNA and protein, but also of altering their size and/or structure. Altogether, there are about 210 varieties of *special cells* (one can find the catalog and the classification of those cells in W. Becker et al. [19]).

2. MM of Stem Cells

The *stem cells* is the right example for confirmation of the philosophy of the novel class of MM because it is a widely known fact that they recreate themselves and give propagation of different specialized cells. We can consider MM of stem cells in detail on the basis of the relations (127)-(130)-types, Ch. 5. But, unfortunately, there is not enough evidence about what determines their proliferation or differentiation. So, we consider here the simpler *MM of stem cells*. As the reasonable simple hypotheses, we accept

$$\alpha_1(t,s) = \alpha_1(s)\exp\,[-c_1(t-s)],\ \beta_1(t,s) = \beta_1(s)\exp\,[-d_1(t-s)]; \tag{1}$$

or

$$\alpha_2(t,s) = \alpha_2(t-s)\exp\,(c_2 t),\ \beta_2(t,s) = \beta_2(t-s)\exp\,(d_2 t), \tag{2}$$

where c_i and d_i are given constants, $c_i \geq 0$.

Then reasoning about the structure of stem cells as in the general case of CAO results in

$$\alpha_\mu^{\nu}(t) = \int_{a(t)}^{t} \alpha_\mu(t,s)x(s)m(s)ds,\ \nu = 0,\ 1(\equiv');\ m(t) = \int_{a(t)}^{t} \alpha_\mu(t,s)y(s)m(s)ds,\ M(t) = \int_0^t m(s)ds;$$

$$\beta_\kappa(t) = \int_{a(t)}^{t} \alpha_\mu(t,s)z(s)m(s)ds,\ \mu,\ \kappa = 1,\ 2\ (\mu = 1,\ \nu \neq 0);$$

$$c(t) = \int_{a(t)}^{t} \beta_\kappa(t,s)[1-x(s)-y(s)-z(s)]m(s)ds;\ R(t) = \int_{a(t)}^{t} m(s)ds,$$

$$G(t) = \int_0^{a(t)} m(s)ds = M(t)-R(t);\ f(t) = \alpha_\mu^{\nu}(t)+m(t)+\beta_\kappa(t)+c(t);$$

$$0 \leq x,\ y,\ z,\ 1-x-y-z \leq 1;\ 0 \leq a(t) \leq t,\ T \geq t \geq t^{\wedge} \geq 0, \tag{3}$$

where $\alpha_\mu^{\nu}(t)$ is νth derivative of $\alpha_\mu(t)$, $\alpha_\mu^{0} = \alpha_\mu$, $\alpha_\mu(t^{\wedge}) = \alpha_\mu^{\wedge}$, $x(s)m(s)$--$\alpha_\mu^{\nu}(t)$ and $z(s)m(s)$--$\beta_\kappa(t)$ are the channels of creation of the new technology quantities in the respective sub-systems (for stem cells, these subsystems are likely to be parts of the genome), $c(t)$ is the necessary germ initial value of the specialized cell-type, and the remaining notations were previously noted.

Clearly $\alpha_\mu(t)$ is always increasing in the case when $\nu = 1$, since $\alpha_\mu'(t) > 0$. For stem cells, x, y, z, and G (or m) may be given. The distributions x, y, and z usually can be found from the solution of some extreme problems. For stem cells, we may probably count on the fact that those extreme problems have been solved by nature during evolution.

The functions α and β depend on control factors u and v respectively that relate to influence of different hormones, interferon, interleukins, means of chemotherapy, etc. The simplest form of this dependence may be

$$\alpha = u\alpha\hat{}, \ \beta = v\beta\hat{}, \tag{4}$$

where, in turn, u and v depend on deviation of m and c from their average stationary normal values m^* and c^*, for example,

$$u = k|m - m^*|, \ v = l|c - c^*|. \tag{5}$$

Here k and l are coefficients subjected to identification; the sign of absolute value means that we actually have two models for control with the help, for example, of the so-called antagonistic hormones. In addition, quantitative and qualitative composition of u and v determine the character of germ c. In the case of stem cells of bone marrow, for example, c can be the precursors (the secondary stem cells) of leukopoiesis or red blood cells and platelets (blood clotting) In turn, this c may play the role of m in the model of (3)-type, where the previous c may play the role of the next germ of the "poiesis", and so on.

It should be noted that researchers have recently grown *in vitro* human cells, embryonic stem cells, which have the potential of becoming virtually any kind of body tissue. Researchers already have demonstrated that these cells can be converted into other types of cells, including bone, muscle, neural, and gut cells. Since these stem cells do not age, they could be potentially used to generate a limitless supply of tissues for transplantation.

Among the diseases that can potentially be treated are Parkinson's disease, hart disease, diabetes, and cancers. However, researchers have not yet deciphered the programming required to order the cultured cells to become the pre-assigned cells or tissues.

Obviously, the Human Genome Project may help researchers to learn how to control the genetically based process by which the stem cells develop into various types of tissues and organs. It is most likely that mathematical modeling, especially MM under consideration, may help researchers in this direction because these MM focus on the connection among the genome-based functions of α and β-types, the control functions of y, a and u, v-types, the enzyme function of m-type, and the product function of c-type. Note here that one of the values of our MM is the reduction of the number of elements and number of times subject to experiments.

Having in mind the perspective of biotech century (see J. Rifkin [160]), we may go further assuming that this class of MM may help solve not only various problems of analysis, but also problems of synthesis of tissues and organs with pre-assigned properties.

3. MM of Thymus and Spleen

Certain details about the structure and properties of the thymus, that is an organ of both the immune and endocrine system, as well as the results and analysis of various experiments with influence of hydrocortisone on thymuses of white mongrel rats can be found in V. Glushkov, V. Ivanov et al. [68].

Here, we discuss only the following assumptions that were used under construction of MM of the thymus: 1. In the thymus, there is a permanent process of proliferation and differentiation of lymphoid cells. 2. Proliferation takes place in the cortex of thymus' fractions and differentiation in its cerebral substance. 3. The rate of proliferation depends on the level of hydrocortisone and other factors. 4. In the process of selection, many thymus cortex cells are destroyed. 5. Most mature cells migrate to other lymphoid organs, but mainly to the spleen.

Based on these assumptions and on the theory of the new class of models, we have the simplest, base *MM of the thymus and spleen* in the form

$$m(t) = \int_{a(t)}^{t} \alpha(t,s)y(s)m(s)ds + \Delta m(t),\ M(t) = \int_{a(t^\wedge)}^{t} m(s)ds,$$

$$c(t) = \int_{a(t)}^{t} \beta(t,s)z(s)m(s)ds,\ C(t) = \int_{t^\wedge}^{t} c(s)ds,\ d(t) = \int_{a(t)}^{t} \gamma(t,s)[1-y(s)-z(s)]m(s)ds,$$

$$D(t) = \int_{t^\wedge}^{t} d(s)ds,\ R(t) = \int_{a(t)}^{t} m(s)ds,\ G(t) = \int_{a(t^\wedge)}^{a(t)} m(s)ds,\ F(t) = \int_{t^\wedge}^{t} f(s)ds,$$

$$F(t) = M(t) + C(t),\ 0 \le y,\ z,\ 1-y-z \le 1,\ 0 < t^\wedge \le t, \tag{6}$$

where $t^\wedge$ is the start of modeling, m is the number of the new thymus cells in the unit of time remaining in the thymus, Δm is the cells inflow from outside (e.g. from bone marrow) to the thymus, c is the number of the new thymus cells in the unit of time leaving the thymus, d is the number of the new destroyed cells in the unit of time (products of their destruction can be used to create new cells), α is the efficiency index for the cortex cells proliferation, β is the efficiency index for the cerebral substance cells differentiation, γ is the efficiency index for the cortex cells selection and destruction, and the remaining magnitudes are usual.

If the functions α, β, γ, and P (or F) are given and y, z are found from a certain optimization problem, we will have the determined system (6) with respect to m, a, c, and d. Assuming that z is selected from requirement of destruction of all useless cells and y from the condition of maximization of the total number of useful cells $C(T)$, we come again to the optimization problem investigated earlier and in the Ch. 15.

The respective properties of this optimization in connection with the thymus were named by the *principle of adaptable redistribution and regenerative accumulation* of the thymus cells (*ARRA-principle*) [63-69].

The ARRA- principle can be formulated as follows.

Under stress actions influencing reactions of the hypothalamus-adrenal gland system and resulting in increasing of hydrocortisone and other factors, the thymus "record" for small T-t^ will be achieved under maximal possible z (adaptable redistribution); while for large T-t^ the "record" is possible under z essentially less than 1, *i.e., under the condition that an essential part of all available resources going to the internal needs of the thymus (regenerative accumulation).*

Experimental confirmation of this principle, using 132 white mongrel rats, was demonstrated in the work [68]. In the same work, certain properties and analytical forms of the functions of α and β-types can be found.

Note that instead of the distribution functions of y, z-types, we can use the control functions u, v-types (see Ch. 5, s. 5.2.4. and (6) here) with the same qualitative results as before. Besides, MM of (6)-type allows us to introduce additional control factors, for example, with the help of the so-called antagonistic hormones.

We come back to MM of the thymus and the spleen when we discuss the MM of AIDS.

4. MM of the Immune and Blood System Cells

According to s. 6.1.6 and 6.1.7, Ch. 6, let $s(t)$ correspond to the signal molecule (hormone) from the signaling cell, $p_k(t)$, $k = 1,\ldots, l$, correspond to relay proteins within the target cell, $m(t)$ to the transcription factor, and $p(t)$ to the resulting protein. Then the desired *MM of the immune and blood system cells* may be the following:

$$p'_k(t) = c_k\, p_{k-1}(t)e_k(t),\ p_0(t) = s(t);\ p_l(t) = \mathrm{AP}(t),$$

$$e_k(t) = \int_{a,k(t)}^{t} \alpha_k(t,s)y_k(s)e_k(t)(s)ds,\ k = 1,\ldots, l;\ 0 \le y_k(s) \le 1,$$

$$m(t) = \int_{a,m(t)}^{t} \alpha_m(t,s)y(s)\mathrm{AP}(s)ds,\ 0 \le y \le 1,\ \beta(t) = \int_{a(t)}^{t} \alpha(t,s)u(s)m(s)ds,\ \ 0 \le u \le 1,$$

$$p(t) = \int_{b(t)}^{t} \beta(t-s)v(s)e(s)ds,\ 0 \le v \le 1,\ 0 < t^{\wedge} \le t, \tag{7}$$

where $e_k(t)$ correspond to the necessary enzymes that are expressed by the genes corresponding to $\alpha_k(t,s)$, $\alpha_m(t,s)$ corresponds to the gene that expresses the complex corresponding to $m(t)$, and the remaining magnitudes are the same as above.

There are two types of signal-transduction pathways leading to the synthesis of proteins that influence cell division. The first type occurs when the initial signal is a growth factor and the target cells ultimate response is the production of the protein that stimulates the cell to divide. By contrast, the second type occurs when we have an inhibitory pathway, in which a growth-inhibitory factor causes the target cell to make the protein that inhibits cell division. In both cases, the proteins function by interacting with components of the cell-cycle system.

What can happen when a proto-oncogene in the target cell has become an oncogene? An oncogene protein can cause increased cell division even in the absence of a growth factor. A mutant tumor-suppressor protein can also affect cell division. The gene for the protein that inhibits cell division can be remained turned off, and inappropriate cell division may occur.

Complex *MM of the immune and blood system cells* are considered by V. Mikhalevich, V. Ivanov et al. [137]. For the *j*th germ of the system, $j > 1$ ($j = 1$ corresponds to the stem cells), we have

$$m_j(t) = \int_{a,j(t)}^{t} \alpha_j(t,s)y_j(s)m_j(s)ds + \Delta c_{j-1}(t),$$

$$c_j(t) = \int_{a,j(t)}^{t} \beta_j(t,s)[1-y_j(s)]m_j(s)ds,$$

$$R_j(t) = \int_{a,j(t)}^{t} m_j(s)ds,\ G_j(t) = \int_{a,j(t\wedge)}^{a,j(t)} m_j(s)ds,$$

$$f_j(t) = m_j(t) + c_j(t),\ 0 \le y_j \le 1,\ 0 < t\wedge \le t, \tag{8}$$

where all notations have the same sense as above and by analogy with (4)

$$\alpha_j = u_j\alpha\wedge_j,\ \beta_j = v_j\beta\wedge_j, \tag{9}$$

where, in turn, u_j and v_j depend on deviation of m_j and c_j from their average stationary normal values m^*_j and c^*_j (see (5)). It should still be noted that the first equation from above in (8) does not correspond obligatory to cell proliferation, but also or only to the process of cell regeneration, which always take place.

Beside (8) and (9), the relations

$$R_j'(t) = m_j(t) - m_j(a_j(t))a_j'(t) = 0, \tag{10}$$

meaning the observance of *j*th germ's homeostasis, may be useful.

Along with the optimization problems investigated below (and in Ch. 15) and the principle of ARRA, some other problems have been considered in [137].

If stress-like effects on the immune and blood system happen too often, then this system (in particular the stem cells of the bone marrow) may be exhausted and the phase of RA (i.e., regenerative accumulation of the cells) may be impossible.

To evade this, stress-realizing factors can be inhibited and stem cells' regenerative factors can be induced.

There is also a possibility of the contradiction [137] between the requirements to "accumulate" and "attend" to the cells that can be the reason for the excess of immature cells of certain germs (one of the reasons for leukemia).

5. MM of Endothelial Cells

It is well known (see W. Becker et al. [19]) that the endothelial cells' main functions are self-proliferation and secretion of various products for the endothelium.

So, *MM of the endothelial cell population* has the form

$$m(t) = \int_{a(t)}^{t} \alpha(t,s)y(s)m(s)ds + \Delta m(t),\ c(t) = \int_{a(t)}^{t} \beta(t,s)[1-y(s)]m(s)ds,$$

$$R(t) = \int_{a(t)}^{t} m(s)ds,\ G(t) = \int_{a(t\text{^})}^{a(t)} m(s)ds,$$

$$f(t) = m(t) + c(t),\ 0 \le y \le 1,\ 0 < t\text{^} \le t, \tag{11}$$

where all notations have the same sense as above and by analogy with (4) and (9) $\alpha = u\alpha\text{^}$, $\beta = v\beta\text{^}$, where, in turn, u and v depend on deviation of m and c from their average stationary normal values m^* and c^*.

We may also consider (11) as *MM of an endothelial cell* with the respective correction of the definition of the notations. Note that the functions $\alpha\text{^}$ and $\beta\text{^}$ are likely be given by the genome and the function c means, in particular, the new concentration of the gas nitric oxide, acting as a messenger molecule that tells blood vessels to relax and dilate, thus lowering blood pressure. Professor R. Furchgott, 1998 Nobel Laureate in Medicine, made this discovery.

We are especially interested in the endothelial cells because of their exceptional role in development of different kinds of cancers. This aspect will be considered more in Ch. 11.

6. Conclusion

The base, simplest MM of cells of the rest of the connective tissues, the epithelial tissues, the muscle tissues, the nervous tissues, the multiple tissues of different organs, and organ systems themselves may be expressed by the same relations (9)-type and, in addition, by the relations (10)-type, in which the functions α and β-types are probably given by the genome and/or certain natural laws, the function m means the rate of proliferation, differentiation, or renovation of the new respective cells, and the function c means the rate of secretion of their new external product.

Consideration of all those cells in detail is beyond the limits of this book. However, this consideration is, of course, very important, especially in connection with temporal sequence of events of the whole organism development.

The respective additional results related to the properties of different kinds of cells in connection with the application of MM in question to the problems of treatment of certain diseases can be found in the subsequent Ch. 10 and 11.

Part III:

INTRODUCTION TO APPLICATIONS

There are potentially limitless opportunities for application of MM under consideration due to limitless numbers of various CAO and the respective problems. We demonstrate here the capacity of the new approach through the examples of the problems of modeling and treatment of such diseases as AIDS and cancers.

We also consider such an important problem as life longevity.

Part III furthermore demonstrates possibilities of application of this class of MM to modeling of organism subsystems (OS) and to a wide class of their diseases.

This part also applies to a medical business, whose external product should be the cure (i.e., the number of people who a doctor cured).

Chapter 10

AIDS

1. Introduction

Let us remind the reader of the following abbreviations: immune system of an organism (IS), immune status of an organism (ISO), immune network of an organism (IN), immune competent cells (ICC), antigens (Ag), immune-globulin (Ig), the specific rates (SR), human immune deficiency viruses (HIV), acquired immune-deficiency syndrome (AIDS), mathematical models (MM), cell-associated objects (CAO), and evolutionary systems (ES).

The general situation about AIDS has been described in [5]; A. Levin [120]; [208].

The first part of this section is devoted to a formalized description of ISO, which is a crucial point of various viral effects on an organism. In this connection, we dwell on a collection of ISO parameters that are necessary for the description of IN (see this description in more detail in Ch. 8).

The central part is mainly devoted to a formalized description of AIDS and the joint mathematical models of a patient with HIV infection. All in all, this part is based on the extension and development of MM obtained earlier (Ch. 5-8). Qualitative investigation of these MM is given.

The final part is devoted to the preliminary consideration of an applied medical problem. It consists of a selection of immune and other correctors, including correctors of gene technology that help the organism to neutralize HIV and to induce the ISO to the norm.

2. Immune Status of an Organism

One can evaluate the *immune status of an organism* (*ISO*) as a collection of parameters such that the appropriate MM describes IS adequately (i.e., with the assigned accuracy). ISO includes the parameters for IN (see Ch. 8) and for different organs of IS [84], namely marrow, thymus, spleen, lymphatic glands, and so on.

In addition to the parameters, ISO can include various characteristics (grounded with the help of parameters), important instances of which can be given by the indices of efficiency of functioning of α and β-types (Ch. 5-8) for some subsystems of IS. In special cases, α and β have the meaning of SR for the processes to be described by the systems of differential equations.

Keeping in mind the complexity of ISO, it is not difficult to understand the presence of the vast variety of IS' behavior and also the fact that a lot of the same IS' behavior can be realized under various combinations of the parameters. Here, we dwell on the collection of ISO parameters that are necessary for description of IN.

According to Ch. 8, the parameters in question are:

1/. The initial numbers $M_{6j+i,0}$ of ICC and Ig for the level j ($j = 0, 1, 2, 3$): T helpers (Th, $i = 1$), T suppressors (Ts, $i = 2$), T effectors (Tef, $i = 3$), B_1 cells ($i = 4$), B_2 cells ($i = 5$) and Ig ($i = 6$). Taking into account that the number has the exponent 10^6 for various Ag, different levels can amount to 3-5, the main species of ICC and Ig to 6, and essential gradation of the cells by avidity (*affinity*) to Ag about 20-40, the total number of parameters has the exponent

$$N_1 = 10^6 \times 4 \times 6 \times 30 = 0.72 \times 10^9. \tag{1}$$

2/. SR (α_i) for the proliferating cells (8-10 hours).

3/. SR for decay of Ig (50-100 hours).

4/. SR for decay of the mature cells (2-4 days).

5/. SR (β_i) for realization of the following control functions: 10^7 Ig per hour for the proliferating B cells, 10^8 Ig per hour for the plasma cells, 2-4 target cells (one cell per 3 hours) for the proliferating T cells and 6-0 target cells for the mature T cells.

6/. Phases of duration for some kind of cell activation (6-30 hours).

7/. x_i, averages and y_i, variances of the truncated normal distributions F_i for the cells' output from proliferation and also for various interactions among the cells of different levels and between the cells and Ig, $i = 1, \ldots, 8$ (see Ch. 8).

Assuming that the cells and Ig have the same parameters at all the levels for any molecular-disperse Ag or Ig, we have

$$N_2 = 1 + 1 + 1 + 4 + 2 + 16 = 25 \tag{2}$$

as the number of the main parameters of ISO in conformity with IN. In Ch. 8, MM of IN are constructed, regarding the 31 immunology hypotheses. We select some of those hypotheses in order to exhibit the fact that the decline of Th-cells results in the immune disruption.

1) The first level of IN consists of the idio-type-positive (Id) lymphocytes and Ag. The second level of IN consists of the anti-idio-type (AId) lymphocytes and Ig. The third level consists of the anti-anti-idio-type (AAId) lymphocytes and Ig, etc.

2) For the higher order level the Ag are the certain cells and Ig of the previous level.

3) For lymphocytes of every level, Ag can induce two states: stimulation by the mean doses and suppression by the large doses. The part of the cells stimulated by the mean doses of Ag (10^{14} molecules) and suppressed by the large doses ($10^{15.5}$ molecules) can be determined according to the concentration of Ag using the normal law for its decimal logarithm.

4) The following properties of the lymphocytes are employed:

a) The start of proliferation of the lymphocytes is impossible without their stimulation.
b) The proliferating lymphocytes characterized by refractoriness with respect to the stimulus.
c) The lymphocytes that began to proliferate stopped their proliferation spontaneously in 2-5 days after the stimulation.
d) The lymphocytes that stopped to proliferate can be stimulated again.

5) Among the B cells of every level, we distinguish the B_2 cells (90% of all B cells), which are lunching into proliferation and differentiation only by Th lymphocytes.

6) Ag and Id-AId bridges cause the interaction of the cells within each level and between levels. Therefore, IdTh help IdB and AIdB; IdTs suppress IdTh, IdTef, IdB, AIdTh, AIdTef, and AIdB; IdTef eliminate IdTh, IdTs, IdTef, IdB, AIdTh, AIdTs, AIdTef, and AIdB; AIdTh help IdB, AIdB, and AAIdB; AIdTs suppress IdTh, IdTef, IdB, AIdTh, AIdTef, AIdB, AAIdTh, AAIdTef, AAIdB, and so on.

7) One molecule of Ig eliminates, on the average, one molecule of Ag.

Regarding these hypotheses, under the condition of the decline of Th cells, Ag-determinants of HIV can increase up to large doses that induce the suppression of ICC of a certain spectrum. In general, the decline of Th cells perhaps results in the tolerance to any Ag from the respective spectrum and a cellular immune response to suppressed ICC. The latter can result in different diseases, including the autoimmune diseases.

3. Viruses

3.1. MAIN PROPERTIES OF VIRUSES

Viruses are particles that consist of a DNA or RNA molecule packaged in a protein capsid. This capsid allows viruses to move and penetrate through cells. A virus can multiply only inside a host cell. Both the structure of the viral genome and its mode of replication differ widely between viruses.

The usual outcome of a viral infection is the lysis of the infected cell, with the release of viral particles. Some viruses can instead become integrated into the host cell. Because viruses sometimes carry host DNA sequences from one species to another, they make possible an occasional mixing of different gene pools during evolution. Thus, viruses can be called "mobile genes" and even "mobile chromosomes".

The reproductive cycle of an enveloped RNA virus consists of the following steps:

1. The envelope fuses with the cell's membrane, allowing the protein-coated RNA to enter the cytoplasm.
2. Enzymes then remove the protein coat.
3. The viral RNA then serves as a template for making complementary strands of RNA.
4. The new strands serve as mRNA for the synthesis of new viral proteins and as templates for synthesizing new viral RNA.
5. The new coat proteins assemble around the new viral RNA.
6. The viruses leave the cell by coating themselves in the membrane.

3.2. ON RETROVIRUSES AND HIV

Retrovirus is an RNA virus that reproduces by means of a DNA molecule. It reverses the usual DNA $\rightarrow$ RNA flow of genetic information. *Human immunodeficiency virus* (HIV) is an example of retroviruses.

The reproductive cycle of HIV, after HIV RNA is uncoated in the cytoplasm of a host cell, consists of the following steps:

1. The reverse transcriptase uses the RNA as a template to make a DNA strand and then adds a second, complementary DNA strand.
2. The resulting double-stranded DNA enters the host cell nucleus and inserts itself into the chromosomal DNA, becoming a pro-virus.
3. Occasionally the pro-virus is transcribed into RNA and translated into viral proteins.
4. New viruses assembled from these components leave the cell and can infect cells.

According to the well-known data on HIV (see, for example [5; D. Kirscchner [113]):

The size of viral grains is 85-200nm. The size of thorns (receptor-binding proteins) extending from the external surface area of HIV membranes and presenting apparently the main medium of infection is 15 nm in diameter and 9 nm in height.

HIVs coincide with retrovirus structure by their main morphological marks.

Pro-viral DNA consists of 9213 nucleotides and contains at least 9 genes, namely, *gag, pol, env, tat, rev, nef, vif, vpr,* and *vpu.* As a whole, the genes coding structural proteins are concentrated to the 5'-end of the genome, and the control genes (*tat, rev, nef,* (*vpr, vpu*)) are adjacent to the 3'-end. The *gag* gene encodes a protein that is cleaved by the viral protease into four entities. The *pol* gene encodes the viral protease (pro), reverse transcriptase (RT), and integrase.

RT enzyme takes up a key position in HIV biology. RT is magnesium-dependent, has molecular mass of 53-100 kD and is highly immunogenic (98% of serum of AIDS patients contains Ig for RT).

The *env* gene produces a protein cleaved into two glycoproteins, gp120 and gp41. The protein gp120 on the surface of the virus has a high affinity for the CD4 protein on the surface of the Th cell and monocytes.

Binding takes place, and the protein gp41 promotes fusion of the viral lipid envelope and the host-cell plasma membrane into one continuous layer of membrane, and contents of the HIV is injected into the cytoplasm.

RT copies the viral RNA into DNA, which then integrates into the cellular DNA. The provirus produces mRNAs for translation into the viral proteins.

The control mechanism of *nef* and *rev* includes a negative feedback and that of the *tat* includes a positive feedback that can enlarge a level of gene expression approximately 1000 times. At the same time, *rev* promotes more viral structural proteins in an infected cell. The functions of two additional genes are presently unclear.

Strong resistance of HIV to a drug due to HIV's variation, in particular of genes, is the most important factor of HIV pathogenesis.

Th cells of an organism, more exactly the so-called CD4+ T cells, are the main HIV targets. After the host cell has duplicated the DNA of the virus, it is reassembled and new virus particles bud from the surface of the host cell. This budding can take place slowly, sparing the host cells, or rapidly, bursting and killing the host cells.

Death rate of infected T cell population is 0.5--1.0 per day. Death rate of uninfected T cell population is 0.02 per day. Number of the free virus produced by the bursting infected cell is 1000. Source of new Th cells from thymus, $c^*(t)$, is about 1000/24.

3.3. PATTERNS OF HIV INFCTION

The duration of the asymptomatic carrier state of HIV infection can be seven to nine years or longer. It has been suggested that environmental and/or perhaps genetic factors can accelerate virus production. For example, the addition of mitogens (agents that stimulate active division of T cells) and/or foreign antigens to HIV infected cells in culture will result in dramatic increases in virus production and cyto-pathology.

The infected Th cell or macrophage begins to produce HIV, and the virus buds from the plasma membrane. Inserted into the membrane of infected cells are gp120 and gp41 proteins, which protrude from the surface of these cells. When an infected Th cell is close to an uninfected Th cell or macrophage, the gp120 on the infected cell binds to the uninfected cell's CD4 protein, just as a virus particle might bind to this receptor. The gp41 protein then helps to promote the fusion of the two cells. The plasma membranes merge, and one cell with two nuclei is formed. This process is repeated, forming a giant cell called a *syncytium*. In effect, normal cells take in an infected sibling, and the virus kills them all.

As time goes by, many individuals with HIV infection develop AIDS, that is, a symptomatic latent period followed by clinical symptoms characterized by a severely impaired IS and, consequently, by infections and cancer. Thus, the main test of AIDS is the impaired ISO, following from the killing of too many Th cells. Once the diagnosis of AIDS is made, usually 50% of individuals do not survive for more than one year.

A decreased number of Th cells in the blood and a falling ratio of Th cells to Tef cells are common indicators of the progressing disease. The Th cells recognize foreign Ag in the body, help B cells make antibodies, and regulate numbers of Tef and Ts cells of the given level through the interaction of Ag, cells, and Ig of IN. The loss of the Th cells "thus compromises both arm of the immune system – antibodies and killer T cells." (A. Levine [120], page 141).

At about a quarter of Th cells from the normal number, individuals with HIV infection are at risk for Pneumocystis pneumonia. Some of the more common cancers associated with immunosupression occur in patients with about 1/10 of the Th cells from the norm.

We do not dwell on the patterns of HIV infection connecting to its origin and difference for different groups of people at risk. Certain details of this can be found in [5, 208].

4. MM of Viruses

4.1. BASE MM

Let the *base MM* of a *virus* have the form

$$m(t) = \int_{a(t)}^{t} \alpha(t,s)u(s)y(s)m(s)ds, \; M(t) = \int_{0}^{t} m(s)ds$$

$$c(t) = \int_{b(t)}^{t} \beta(t,s)v(s)[1-y(s)]m(s)ds, \; C(t) = \int_{t^\wedge}^{t} c(s)ds,$$

$$R(t) = \int_{a(t)}^{t} y(s)m(s)ds + \int_{b(t)}^{t} [1-y(s)]m(s)ds,\ G(t) = M(t)-R(t),$$

$$0 \le y(t) \le 1,\ 0 \le a(t) \le t,\ 0 \le b(t) \le t,\ f\^{}(t) \ge f(t) = m(t) + c(t) = k(f\^{}\^{})m(t),\ t \ge t\^{} \ge 0, \qquad (3)$$

where $t^\wedge$ is the starting time of modeling; $m(t)$ is the rate of creation of the new virus bio-mass of enzyme-substrate complex (another interpretation is the number of active or reactive centers) at the time instant t that provides internal needs or internal functions of virus as CAO; $y(s)m(s)$ is a share of $m(s)$ for the fulfillment of the internal functions in the virus subsystem A for recreation of $m(t)$; $\alpha(t,s)$ is the *efficiency index* of functioning of the *subsystem A*, i.e., the number of units of $m(t)$ to be created for the unit of the time starting from the instant t per one unit of $y(s)m(s)$; $a(t)$ is the temporal threshold for liquidation of the obsolete products in the subsystem A; $c(t)$ is the rate of creation of the new remaining virus components or products that provide its external functions (in particular, the initial and final interaction between the virus and the host cell); $[1-y(s)]m(s)$ is a share of $m(s)$ for the fulfillment of the external functions in the virus subsystem B for creation of $c(t)$; $\beta(t,s)$ is the *efficiency index* for the functioning of the *subsystem B* that is similar to $\alpha(t,s)$; $b(t)$ is the similar temporal threshold to $a(t)$, but for the subsystem B; u and v are certain control factors; $R(t)$ is the total virus bio-mass of enzyme-substrate complex, functioning at the instant t; $G(t)$ is the total out-of-date virus bio-mass of enzyme-substrate complex at the instant t; $f^\wedge(t)$ is the rate of the inflow of the external substrate accessible to the virus; $k(f^{\wedge\wedge})$ is the coefficient, $0 < k(f^{\wedge\wedge}) \le 1$, characterizing a share m in f and depending on qualitative composition of $f^\wedge$, named $f^{\wedge\wedge}$; $[0, t^\wedge]$ is the period of the *prehistory of the virus* on which all the functions are considered as given.

Thus, the MM (3) is a nonlinear system (with the prehistory), which consists from 8 equalities and 9 inequalities, connecting 14 different values. The relations (3) are faithful by definition (and here α and β can depend on all the elements of the model) provided that out-of-date biomass in bio-systems A and B is never used, but the new biomass created after $a(t)$ and $b(t)$ is used entirely. Assuming that the exogenous factors α, u, β, v, f, $f^\wedge$, $k(f^{\wedge\wedge})$ for the virus and that t are being assigned and regarding that partitioning y, which can be usually obtained from extreme principles, is being also assigned, we conclude that the system of the equalities in (3) is the determined system over the desired values m, a, b, c, M, C, R, and G.

Note the difference in MM (3) for enveloped usual RNA viruses and retroviruses according to well-known facts. For the usual virus, we can obtain the main elements of its model α and β from the viral RNA. In contrast, for the retroviruses, the functions α and β-types can also be obtained from the respective part of the host infected cell's DNA.

4.2. MM OF HIV

We accept MM (3) as the *base MM of HIV*. In contrast to the general case, $m(t)$ is the rate of creation of the new HIV bio-mass of enzyme-substrate complex (pro, RT, and IG) at the time instant t that provides internal needs of HIV as CAO; $c(t)$ is the rate of creation of the remaining HIV proteins including new thorns (proteins gp120 and gp41) which provide their external functions (in particular, the initial and final interaction between

HIV and the target Th cells); $P(t)$ is the total HIV bio-mass of enzyme-substrate complex functioning at the instant t; $G(t)$ is the total out-of-date HIV bio-mass of enzyme-substrate complex at the instant t; $f^{\wedge}(t)$ is the rate of the inflow of the external substrate accessible to HIV; $[0, t^{\wedge}]$ is the prehistory of HIV functioning.

Assuming that the factors α, u, β, v, f, $f^{\wedge}$, $k(f^{\wedge\wedge})$ for HIV and t are being assigned and regarding that partitioning y, which can be usually obtained from some extreme principles, is being also assigned, we conclude that the system of the equalities in (3) is the determined system over the desired values m, a, b, c, M, C, P, and G.

Since HIV is a retrovirus, the factors α and β-types can be obtained, in principle, from both the respective part of the infected host cell's DNA and from the viral RNA of HIV itself. Other ways of determination of α and β can be found in Ch. 5, s. 5, and in s. 4.3.

4.3. QUALITATIVE INVESTIGATION OF MM OF HIV

On the strength of (3), we easily find

$$m(t) = k(f^{\wedge\wedge})f(t),\ c(t) = [1-k(f^{\wedge\wedge})]f(t) \leq [1-k(f^{\wedge\wedge})]f^{\wedge}(t),$$

$$R(t) = \int_{a(t)}^{t} y(s)k(f^{\wedge\wedge})f(s)ds + \int_{b(t)}^{t} [1-y(s)]k(f^{\wedge\wedge})f(s)ds,\ G(t) = M(t)-R(t),\ t \geq t^{\wedge}, \qquad (4)$$

where $a(t)$ and $b(t)$ can be found from

$$\int_{a(t)} \alpha(t,s)u(s)y(s)k(f^{\wedge\wedge})f(s)ds = k(f^{\wedge\wedge})f(t),$$

$$\int_{b(t)} \beta(t,s)v(s)[1-y(s)]k(f^{\wedge\wedge})f(s)ds = [1-k(f^{\wedge\wedge})]f(t). \qquad (5)$$

It follows from (5) that under $f = f^{\wedge}$ the value $c(t)$, aggressive factor of HIV, depends only on composition and quantity of $f^{\wedge}(t)$. Under "favorable conditions" for HIV when $k(f^{\wedge\wedge}) \approx 1$, the value $c(t) \approx 0$. If $k(f^{\wedge\wedge})$ is decreasing for some reason or other, $c(t)$ will be able to increase, which can result in the invasion and killing of the host cells. Under the fixed quantity of $f^{\wedge}(t)$, only $k(f^{\wedge\wedge})$ affects the result.

From here, it follows, first, that under the favorable conditions HIV can exist *in vivo* a long time without noticeable change for an organism and, secondly, for the support of the above mentioned conditions "feeding" $f^{\wedge}(t)$ must mainly contain the component $m(t)$. Thirdly, when additional needs of proteins of $m(t)$-type in the cells infected with HIV appear, the contents of those proteins in $f^{\wedge}(t)$ can decrease, which can induce a growth of $c(t)$. And finally, with regard to the properties of Ag specificity, for HIV's sparing behavior it will be necessary, perhaps, to support the required quantity of Ig against $c(t)$ and tolerance condition to $m(t)$. On the contrary, for resolving elimination of HIV we need to destroy $m(t)$. The most weak or sensitive point for destroying any dynamic system is its structure (i.e., the components of α, β, a, b, y-type). If we do not know how to destroy the structure by special means, we may count on "desperate cure". One of these sorts is the stressing effect on HIV. To do so, we need to create the conditions under which HIV forces to keep most of $y \approx 1$.

Let $f^{\wedge}(t)$ permit significant variation of the solutions of (3). Then assuming R, G, and f are to be given and

$$y(t) = 1,\ t^{\wedge} < t < t^*,\ y(t) = 0,\ t > t^*, \tag{6}$$

where t^* is the instant of switching from internal to external functioning (which is also given or has to be found during the model identification), we can find

$$m(t) = [R(t)+G(t)]', \tag{7}$$

and $a(t)$, $b(t)$ from the relations

$$R(t) = \int_{a(t)}^{t} m(s)ds,\ G(t) = \int_{0}^{a(t)} m(s)ds,\ t^{\wedge} < t < t^*,\ R(t) = \int_{a(t)}^{t^*} m(s)ds + \int_{b(t)}^{t} m(s)ds,$$

$$G(t) = \int_{0}^{a(t)} m(s)ds + \int_{t^*}^{b(t)} m(s)ds,\ t^* < t < t',\ a(t') = t^*,\ a(t) \le t^* \le b(t);\ R(t) = \int_{b(t)}^{t'} m(s)ds,$$

$$G(t) = \int_{0}^{b(t)} m(s)ds,\ t' < t < t'',\ b(t'') = t';\ R(t) = 0,\ G(t) = \int_{0}^{t'} m(s)ds,\ t > t''; \tag{8}$$

and

$$c(t) = f(t) - m(t). \tag{9}$$

After this we can find αu and βv presupposing that our object has the piecewise stationary dynamics, that is, the stationary dynamics on each interval $(t^{\wedge}+(j-1)dT,\ t^{\wedge}+jdT)$, $j = 1, 2, \ldots$, and selecting the temporal intervals of the standard behavior of HIV, on which u and v can be considered as constants. Indeed, we have

$$m(t) = u\int_{a(t)}^{t} \alpha_j(t-s)m(s)ds = u\int_{0}^{t-a(t)} \alpha_j(s)m(t-s)ds,\ t \in (t^{\wedge}+(j-1)dT,\ t^{\wedge}+jdT),\ j = 1, \ldots, j^*,$$

$$m(t) = u\int_{0}^{t^*-a(t)} \alpha_j(s)m(t-s)ds,\ t \in (t^{\wedge}+(j-1)dT,\ t^{\wedge}+jdT),\ j = j^*, \ldots, j',\ j'dT = t',\ j^*dT = t^*,$$

$$c(t) = v\int_{b(t)}^{t} \beta_j(t-s)m(s)ds = v\int_{0}^{t-b(t)} \beta_j(s)m(t-s)ds,\ t \in (t^{\wedge}+(j-1)dT,\ t^{\wedge}+jdT),\ j = j^*+1, \ldots, j',$$

$$c(t) = v\int_{b(t)}^{t'} \beta_j(t-s)m(s)ds = v\int_{0}^{t'-b(t)} \beta_j(s)m(t-s)ds,\ t \in (t^{\wedge}+(j-1)dT,\ t^{\wedge}+jdT),$$

$$j = j'+1, \ldots, j'',\ j''dT = t'', \tag{10}$$

and hence

$$u\alpha_j(t-a(t))m(a(t))(1-a'(t)) = m'(t) - u\int_{0}^{t-a(t)} \alpha_j(s)m'(t-s)ds,$$

$$t \in (t^{\wedge}+(j-1)dT,\ t^{\wedge}+jdT),\ j = 1, \ldots, j^*,$$

$$-u\alpha_j(t^*-a(t))m(t-t^*+a(t))a'(t) = m'(t)-u\int_0^{t^*-a(t)} \alpha_j(s)m'(t-s)ds,$$

$$t \in (t\wedge+(j-1)dT,\ t\wedge+jdT),\ j = j^*, \ldots, j',$$

$$v\beta_j(t-b(t))m(b(t))(1-b'(t)) = c'(t) - v\int_0^{t-b(t)} \beta_j(s)m'(t-s)ds,$$

$$t \in (t\wedge+(j-1)dT,\ t\wedge+jdT),\ j = j^*+1, \ldots, j',$$

$$-v\beta_j(t'-b(t))m(t-t'+b(t))b'(t) = c'(t)-v\int_0^{t'-b(t)} \beta_j(s)m'(t-s)ds,$$

$$t \in (t\wedge+(j-1)dT,\ t\wedge+jdT),\ j = j'+1, \ldots, j''. \qquad (11)$$

Thus, the problem has been reduced to the solution of Volterra-type equations of the second kind. After the solution of the equations in (11), we can find $u(t)$ and $v(t)$ on the temporal intervals with nonstandard behavior of HIV. By the same token, we can detect the control mechanism of *nef* and *rev* into an infected cell including negative feedback, and that of the *tat* including a positive one, probably with the help of $u(t)$, $y(t)$, and $v(t)$, respectively.

It should be noted that in connections with the properties of HIV (see above and J. Coffin [35]) we likely have to accept the alternating step $(dT)_j$ instead of dT. In addition, we likely need to accept the system of the relations (6)-(11) for the viruses for which $t\wedge$ is the instant of injection into the host cells, and to use these relations until new virus particles bud from the surfaces of the host cells. The parts of HIV that are binding the host cells at the instant $t\wedge$ can be determined with the help of the truncated normal law F_7 for the decimal logarithm of number of all the viruses or a certain new law parameters, which have to be found on the stage of the model's testing. On the whole, we can result in *MM of HIV* by combining the mentioned relations for different values $t\wedge$. The obtained α and β can be used for their prediction in the future and, hence, for estimating the more exact values of m and c.

The relations (6)-(11) are the experimental-theoretical methods for achieving the model. In a similar way, we can achieve the model under the more general than above intensity of utilization of m-type biomass (see Ch. 5 and 14).

One of the possible optimization principles (for HIV) consists of

$$C(T) = \int_{t\wedge}^{T} c(t)dt = \max\,(uy) \qquad (12)$$

by means of the factors of strengthening or weakening of efficiency of functioning and the distribution between the two subsystems. The problem (12) has been investigated in detail in Ch. 15. Repeating the arguments of Ch. 15, we obtain the following important principle I.

For the solution of the problem (12) *on the "small" segment* $[t\wedge, T]$, *the values of* uy *are minimally possible, but for "large"* T-t *the desired* uy *may differ from the minimally possible values on the larger initial part of the segment* $[t\wedge, T]$, *and only on the smaller final part of* $[t\wedge, T]$ *the desired* uy *is minimally possible.*

The notions "small" and "large" depend on the values of functions α and β. Namely, the greater the functions in question, the nearer to $t^\wedge$ is the boundary between "small" and "large" segments. The obtained result was in essence an implication of the fact that for "large" $T\text{-}t^\wedge$ a growth of $\alpha(t,s)u(s)y(s)m(s)$ overtakes, on the average, a diminution of $\beta(t,s)v(s)[1\text{-}y(s)]m(s)$. We would like to emphasize here that the result has received, in the sequel, the important qualitative very general interpretation:

The record of external function for any CAO can be obtained only under the conditions of its sufficiently comfortable guarantee, that is, under the very significant fraction of resources sent to the internal needs of CAO.

It follows from the principle I, at least, the next probable strategy for HIV suppression: creation of sufficiently prolonged conditions of minimization of HIV m-type components for fulfillment of HIV active centers' restoration. One of the possible realizations of this strategy is the creation of prolonged strong external effects (stresses) localized in HIV.

4.4. MORE COMPLICATED MM OF HIV AND ITS QUALITATIVE INVESTIGATION

Possible complications and refinements of HIV MM can be obtained by analogy with Ch. 5. In particular, we believe that new "technologies" α, u and β, v are also the "products" of HIV product of m-type. So, admitting for simplicity that

$$\alpha(t,s)u(s) = \beta(t,s)v(s) = \gamma(s),\ a(t) = b(t), \tag{13}$$

we have instead of (3) the following *MM of HIV*:

$$m(t) = \int_{a(t)}^{t} \gamma(s)y(s)m(s)ds,\ \gamma'(t) = \int_{a(t)}^{t} \gamma(s)x(s)m(s)ds,$$

$$c(t) = \int_{a(t)}^{t} \gamma(s)[1-x(s)-y(s)]m(s)ds,\ R(t) = \int_{a(t)}^{t} m(s)ds,$$

$$G(t) = M(t) - R(t),\ 0 \le a(t) \le t,\ 0 \le x,\ y,\ 1-x-y \le 1,\ T \ge t \ge t^\wedge \ge 0 = a(t^\wedge). \tag{14}$$

For MM (14), R or G is usually given (and all the elements on the prehistory $[0, t^\wedge]$); x and y are found from the solution of some optimization problems, and m, a, γ, c, M, and G (or R) are found as the solution of the system (14). Now the problem (12) takes the form

$$C(T) = \int_{t^\wedge}^{T} c(t)dt = \max\ (x, y). \tag{15}$$

Detailed investigation (see Ch. 15) of the problem (15) has led to the principle II.

For "large" T-t^ the optimal distributions $x^*(t)$ and $y^*(t)$ possess the characteristics

$$x^*(t) = 1, t \in [t^\wedge, \theta], \; y^*(t) = 1, t \in [\theta, \Theta];$$

$$x^*(t) = y^*(t) = 0, t \in [\Theta, T];$$

$$\theta - t^\wedge \sim T - t^\wedge, \Theta - \theta \sim (T - t^\wedge)^{1/2}, T - \Theta \sim 1, \tag{16}$$

where ~ means the asymptotic equality in order when $T - t^\wedge$ tends to ∞.

This result qualitatively means that

under the optimal distributions of the given resources among different subsystems of HIV, the main priority probably belongs to the subsystem of "G"-type creating the new "more effective" technology (genes); on the second place by priority comes the subsystem A' creating the new active centers (products of m-type);
and finally comes the subsystem B of production of the external products (c-type).

From the principle II follows, at least, the next probable strategy for HIV suppression: creation of sufficiently prolonged conditions for minimization, mainly, of HIV *m*-type components' share for creation of $\gamma(t)$ that corresponds to HIV *genes*, then of HIV *m*-type component share for restoration of $m(t)$ that corresponds to HIV enzyme-substrate complex, and at last of HIV *m*-type component share for creation of $c(t)$ corresponding to HIV aggressive factors.

In the general case, assuming the optimal strategy of HIV as that of maximization of $C(T)$, we have the following *mini-max problem*

$$C(T) = \int_{t^\wedge}^{T} c(t)dt = \text{min max } (x, y, u, v) \; (w \in W), \tag{17}$$

where W is the set of various factors, which resist HIV. Among the factors are the immunological means against HIV (D. Leturcq et al. [119]).

4.5. MM OF HIV POPULATION

4.5.1. *MM of HIV population inside of the host cell*

According to the description above, the base MM of HIV population inside of the host cell may have the form

$$\alpha(t,s) = \alpha(s), \alpha'(t) = \int_{a(t)}^{t} \alpha(s)x(s)m(s)ds, m(t) = \int_{a(t)}^{t} \alpha(s)u(s)y(s)m(s)ds,$$

$$c_j(t) = \int_{a(t)}^{t} \alpha(s)z_j(s)m(s)ds, j = 1, 2, 3, 4, \beta(t) = \int_{a(t)}^{t} \alpha(s)v(s)z_5(s)m(s)ds,$$

$$c(t) = \int^{t}_{a(t)} \beta(s)v(s)w(s)[1-x(s)-y(s)-z_1(s)-\ldots-z_5(s)]m(s)ds,$$

$$R(t) = \int^{t}_{a(t)} m(s)ds,\ G(t) = \int_0^{a(t)} m(s)ds,$$

$$f(t) = \alpha'(t)+m(t)+c_1(t)+c_2(t)+c_3(t)+c_4(t)+\beta(t)+c(t),$$

$$0 \le x, y, z_j, 1-x(s)-y(s)-z_1(s)-\ldots-z_5(s) \le 1, 0 < t\hat{} \le t, \tag{18}$$

where α', m, c_1, c_2, c_3, c_4, β, c mean the rate of creation of, respectively, new provirus DNA, substrate-enzyme complex of HIV population, gene *tat*, gene *rev*, gene *nef*, gene *vif*, producing *c* gene, remaining bio-mass of HIV population. The first control factor *u* depends on the concentration of cellular receptor CD4 protein, $(cd)_4$, and gene *nef*. The second control factor *v* depends on genes *rev* and *nef*. The third control factor depends on genes *tat* and *vif*. The remaining elements of the model (18) have the usual definition as above.

As examples of *MM of the control factors of HIV* population inside an infected cell,

$$u = k_1c_3(cd)_4,\ v = k_2c_2/(1+c_3),\ w = k_3c_1c_4, \tag{19}$$

where k_i are coefficients subjected to identification.

4.5.2. *MM of HIV free population*

Now by analogy with s.1.2-1.3, we consider the theoretical course to achieve this model. We suppose that stimulated and infected Th cells are the same, and the share of stimulated cells is determined by the truncated normal law F_7 for the decimal logarithm of $V(t)$ of all alive viruses at instant *t*. In contrast to the above, the number of the proliferating infected Th cells

$$M_1{}^P(t) = \int_0^{t-12} M_1{}^P(t,r)dr,\ t > 12,\ g(t) = \lg V(t),\ M_1{}^P(t,r) =$$

$$f(t,r+12)M_1{}^P(r+12,r)\exp\{\int^{t}_{r+12} [0.08\text{-}0.75/24 - 0.02/24 - u(s) - c_1{}^P(s)]ds\}[1-F_8(t-r-12)],$$

$$M_1{}^P(r+12,r) = f_1(r+12,r)dM_1{}^s(r)\exp[-\int_r^{r+12} u(s)]ds],$$

$$dM_1{}^s(r) = (M_{1,0} + c^*(r))F_7(g(r))\exp[-\int_0^r u(s)ds],\ r > 0, \tag{20}$$

where the notations are the same as above. It follows from (20) that *MM of HIV population*

$$V(t) = 1000{\cdot}0.75/24\mathrm{w}(t)\int^{t}_0 M_1{}^P(s)ds\exp[-\int^{t}_0 v(s)ds], \tag{21}$$

where $w(t)$ is a share of infected $M_1{}^P$ cells producing free HIV and $v(s)$ is the death rate of free HIV. Certain information about w, v can be found in M. Nowak's Shanks lecture and the abstract [144]. In particular, the half-life of free HIV is about 6 hours, and 99% of plasma viruses are produced from the cells that have a half-life less than 2 days. Most HIV infected cells contain a defective pro-virus and have a life longer than 100 days. There are now blood tests that can actually measure the number of copies of HIV or the so-called viral load.

To determine $M_1{}^P(t)$ and hence $V(t)$, we have to add the other equations of IN from above that are similar to (20), but without the number 0.75/24. It seems unreliable that IS cannot eliminate $V(t)$ and the infected Th cells. Probably, MM above do not consider all the necessary factors. To be certain, we can either add new assumptions to the list in s.2, for example, the natural assumption about suppression of Ts and Tef by a large doze of HIV, or go to another interpretation of HIV as Ag.

Choosing the law F_7, we considered HIV as corpuscular Ag similar to cells. However, we can consider HIV as a very large collection of different types of molecular-dispersed Ag such that for their representatives MM of IN were constructed in this Ch. Let us assume the following guess regarding N of miscellaneous Ag (c f G. Marchuk [127], pp. 86-103):

1) If there exists one so-called immune-dominant molecular-dispersed *Ag* determinant, we can reduce the situation to the considered case.

2) With N "equivalent" *Ag*, we should take

$$z_i = k_i x_i y_i = 1/3F_7(\lg Y_i)/(NY_i)x_i y_i,\ i = 1, \ldots, N, \tag{22}$$

where $3k_i$ is a probability of contact of two cells x- and y-types that are ICC with respect to Ag with number i and Y_i are all control cells for x_i.

3) If there exist N "nonequivalent" *Ag*, we should take

$$z_i = 1/3q_iF_7(\lg Y_i)/(NY_i)x_i y_i,\ \sum_{i=1}^{N} q_i = 1,\ q_i \geq 0,\ \ i = 1, \ldots, N, \tag{23}$$

where q_i are appropriate weights.

4) Let $M_i(t)$ be the number of ICC of all levels to Ag with number i at the instant t. Then

$$\sum_{i=1}^{N} M_i(t) \leq M^+, \tag{24}$$

where M^+ is the maximum admissible number of IN cells.

5) Reduction for cell proliferation takes place in a way that preserves the relation (24).

Instead of (20) we now have

$$M_1{}^P(t,i) = \int_0^{t-12} M_1{}^P(t,u,i)du,\ t > 12,\ M_1{}^P(t,u,i) = M_1{}^P(u+12,u,i)f_1(t,u+12,i)\exp\left(\int_{u+12}^{t} [0.08-\right.$$

$$\left.0.75/24\text{-}0.02/24\text{-}u_i(s) - c_{1,i}{}^P]ds\right)[1-F_8(t-u-12)],$$

$$M_1{}^P(u+12,u,i) = dM_1{}^s(u,i)f_1(u+12,u,i)\exp\left[-\int_u^{u+12} u_i(s)]ds\right],\ u > 0,$$

$$dM_1{}^s(u,i) = [M_{1,i,0} + c^*(u,i)]\{(f_{1,2}(u,0,i)\exp[-\int^u_0 u_i(s)ds])' + (f_{1,1}(u,0,i)\exp[-\int^u_0 (u_i(s)ds)'\},$$

$$f_1(t,s,i) = \lim \prod \{[1-F_1(y_{0,i}(k\Delta t))][1- F_5(y_{0,i}(k\Delta t)][1-F_6(y_{5,i}(k\Delta t)]\},$$

$$f_{1,1}(t,s,i) = \lim \prod \{[1-F_1(y_{0,i}(k\Delta t))][1-F_2(y_{0,i}(k\Delta t))][1-F_5(y_{0,i}(k\Delta t)][1-F_6(y_{5,i}(k\Delta t)]\},$$

$$f_{1,2}(t,s,i) = \lim \prod \{[1-F_1(y_{0,i}(k\Delta t))]F_2(y_{0,i}(k\Delta t))[1-F_5(y_{0,i}(k\Delta t)][1-F_6(y_{5,i}(k\Delta t)]\},$$

$$(k\Delta t = s, \ldots, t), \Delta t \to 0, \tag{25}$$

where F_s is one of the given functions; the subscript i means that the respective formula has to be applied for the ith Ag. We now have the following *MM of HIV free population:*

$$V(t) = 1000{\cdot}0.75/24\sum w_i(t)\int^t_0 M_I{}^P(s,i)ds\exp[-\int^t_0 v(s)ds]\ (i = 1, \ldots, N). \tag{26}$$

5. MM of AIDS

It seems that for *MM of AIDS* we should at least add MM of a certain organ of IS (which suffers mostly from the helplessness of IN) to the relations of MM of HIV, HIV population inside of an infected cell, and HIV free population. Such organ is most likely the thymus and spleen (see Ch. 9). MM of thymus and spleen suffering can be introduced by the methods based on MM of thymus and spleen as CAO.

It has been shown, theoretically and by the respective experiments with the thymuses of mice, that rather long and intensive output of the external function of the thymus and spleen ($c^*(t)$) results in destruction of this organ. If the latter happened, it would mean that an organism becomes defenseless from attacks of great numbers of dangerous *Ag*. Apparently, HIV "knows" how to fight with IN and IS as a whole with the help of stress effects on them. At the same time HIV knows "too much" because it usually dies with the HIV patient.

Thus, *MM of AIDS* or MM of the HIV, HIV populations, and the thymus and spleen

$$m^\wedge(t) = \int^t_{a^\wedge(t)} \alpha^\wedge(t,s)u^\wedge(s)y^\wedge(s)m^\wedge(s)ds,\ M^\wedge(t) = \int^t_0 m^\wedge(s)ds,$$

$$c^\wedge(t) = \int^t_{b^\wedge(t)} \beta^\wedge(t,s)v^\wedge(s)[1-y^\wedge(s)]m^\wedge(s)ds,\ C^\wedge(t) = \int^t_{t^\wedge} c^\wedge(s)ds,\ 0 \le y^\wedge(t) \le 1,$$

$$R^\wedge(t) = \int^t_{a^\wedge(t)} y^\wedge(s)m^\wedge(s)ds + \int^t_{b^\wedge(t)} [1-y^\wedge(s)]m^\wedge(s)ds,\ G^\wedge(t) = M^\wedge(t)-R^\wedge(t),$$

$$0 \le a^\wedge(t), b^\wedge(t) \le t, f^\wedge(t) = m^\wedge(t) + c^\wedge(t) = k(f^{\wedge\wedge})m^\wedge(t),\ t \ge t^\wedge \ge 0; \tag{27}$$

$$\alpha(t,s) = \alpha(s),\ \alpha'(t) = \int^t_{a(t)} \alpha(s)x(s)m(s)ds,$$

$$m(t) = \int_{a(t)}^{t} \alpha(s)u(s)y(s)m(s)ds,\ c_j(t) = \int_{a(t)}^{t} \alpha(s)z_j(s)m(s)ds, j = 1, 2, 3, 4,$$

$$\beta(t) = \int_{a(t)}^{t} \alpha(s)v(s)z_5(s)m(s)ds,\ c(t) = \int_{a(t)}^{t} \beta(s)w(s)[1-x(s)-y(s)-z_1(s)-\ldots-z_5(s)]m(s)ds,$$

$$f(t) = \alpha'(t)+m(t)+c_1(t)+c_2(t)+c_3(t)+c_4(t)+\beta(t)+c(t),$$

$$R(t) = \int_{a(t)}^{t} m(s)ds,\ G(t) = \int_{0}^{a(t)} m(s)ds,\ F(t) = \int_0^t f(s)ds,$$

$$0 \le x, y, z, 1-x(s)-y(s)-z_1(s)-\ldots-z_5(s) \le 1,\ 0 < t\hat{} \le t,$$

$$u = k_1(cd)_4,\ v = k_2c_2/(1+c_3),\ w = k_3c_1c_4; \tag{28}$$

$$M_1^{\ p}(t) = \int_0^t M_1^{\ p}(t,r)dr + M_1^{\ p}(0),\ t \ge 0,$$

$$M_1^{\ p}(t,r) = M_1^{\ p}(r,r)f(t,r)\exp\{\int_r^t [0.05-h(s)-c_1^{\ p}(s)]ds\}[1-F_8(t-r-12)],$$

$$M_1^{\ p}(r,r) = m_1^{\ s}(r-12)f(r,r-12)\exp(-\int_{r-12}^{r} h(s)ds),\ M_1^{\ s}(0) = M_{1,0}F_2[g(0))][1-F_5[g(0)]],$$

$$m_1^{\ s} = -hM_1^{\ s}+dM_1^{\ s}-m_1^{\ s}(t-12)f(t,t-12)\exp[-\int_{t-12}^{t} h(s)ds],$$

$$dM_1^{\ s} = c^*(t)F_2[g(t)]+M_1^{\ s}(t)(F_2[g(t)])'/F_2[g(t)] - kf(t),$$

$$m_1^{\ s,s} = -hM_1^{\ s,s}+dM_1^{\ s,s}-m_1^{\ s,s}(t-6)f(t,t-6)\exp[-\int_{t-6}^{t} h(s)ds],$$

$$dM_1^{\ s,s} = M_1^{\ p}F'_8(t-12)/[1-F_8(t-12)]F_2(g),\ dM_1^{\ s,s}(0) = 0,\ m_1^{\ m}(t) =$$

$$-(h+0.01+c_1^{\ m})M_1^{\ m}+M_1^{\ p}(t-6)F'_8(t-18)/[1-F_8(t-12)]F_2(g(t-6))f(t,t-6)\exp[-\int_{t-6}^{t} h(s)ds],$$

$$M_1^{\ m}(0) = 0,\ m_1^{\ r}(t) = -h\hat{}M_1^{\ r}+M_1^{\ p}F'_8(t-12)/[1-F_8(t-12)][1-F_2(g)],$$

$$M_1^{\ r}(0) = M_{1,0}[1-F_2(g(0))],\ M_1 = M_1^{\ s}+M_1^{\ p}+M_1^{\ s,s}+M_1^{\ m}+M_1^{\ r},$$

$$f(t,s) = \lim \prod[1-F_6(y_5(k\Delta t))]\ (k\Delta t = s, \ldots, t),\ \Delta t \to 0,$$

$$c_1^{\ p} = 1/3k\hat{}_1^{\ p}(1-k_1^{\ m})[U\Sigma M_i^{\ j}(i = 4,2; j = s, s,s)+V\Sigma M_i^{\ j}(i = 9,2; j = s, s,s)],$$

$$c_1^{\ m} = 1/8k\hat{}_1^{\ m}[U\Sigma M_i^{\ j}(i = 4,2; j = s, s,s) + V\Sigma M_i^{\ j}(i = 9,2; j = s, s,s)]e^{-0.01},$$

$$U = [1-F_5(g)][1-F_6(y_{1,0})][1-k_{8,7,3,2}^{\ m,p}],\ V = [1-F_5(g)][1-F_6(y_{1,0})][1-k_{13,12,8,7,3,2}^{\ m,p}],$$

$$U^\wedge = [1-F_5(g)][1-F_6(y_{1,0})][1-k_{8,7}{}^{m,p}],\ k_i^j = 1/3F_7(\lg M_i^j),$$

$$h = \sum k_i^j\ (i = 2, 3, 7, 8; j = m, p),\ h^\wedge = \sum k_i^j\ (i = 7, 8; j = m, p),\ g(t) = \lg\,[W(t)c^\wedge(t)],$$

$$W(t) = 750/24d(t)\int{}^t{}_0 M_1{}^P(s)ds\exp\,[-\int{}^t{}_0\, q(s)ds] + W_0,\ t \geq r \geq 0; \tag{29}$$

$$m^*(t) = \int_{a^*(t)}^{t} \alpha^*(t,s)y^*(s)m^*(s)ds + \Delta m^*(t),\ M^*(t) = \int_{a^*(t6)}^{t} m^*(s)ds,$$

$$c^*(t) = \int_{a^*(t)}^{t} \beta^*(t,s)z^*(s)m^*(s)ds,\ C^*(t) = \int_{t^\wedge}^{t} c^*(s)ds,\ d(t) = \int_{a^*(t)}^{t} \gamma^*(t,s)[1\text{-}y^*(s)\text{-}$$

$$z^*(s)]m^*(s)ds,\ D^*(t) = \int_{t^\wedge}^{t} d^*(s)ds,\ R^*(t) = \int_{a^*(t)}^{t} m^*(s)ds,\ G^*(t) = \int_{a^*(t^\wedge)}^{a^*(t)} m^*(s)ds,$$

$$F^*(t) = \int_{t^\wedge}^{t} f^*(s)ds,\ F^*(t) = M^*(t) + C^*(t),\ 0 \leq y^*, z^*, 1-y^*-z^* \leq 1, 0 < t^\wedge \leq t, \tag{30}$$

where $t^\wedge$ is the starting time of modeling; $m^\wedge(t)$ is the rate of creation of the new enzyme-substrate complex at the time instant t that provides internal needs or internal functions of virus as CAO, $y^\wedge(s)m^\wedge(s)$ is a share of $m^\wedge(s)$ for the fulfillment of the internal functions in the virus subsystem A for recreation of $m^\wedge(t)$; $\alpha^\wedge(t,s)$ is the efficiency index of functioning of the subsystem A, i.e., the number of units of $m^\wedge(t)$ to be created for the unit of time starting from the instant t per one unit of $y^\wedge(s)m^\wedge(s)$, $a^\wedge(t)$ is the temporal threshold for liquidation of the obsolete products in the subsystem A, $c^\wedge(t)$ is the rate of creation of the new external virus components or products that provide its external functions of interaction between the virus and the host cell, $[1-y^\wedge(s)]m^\wedge(s)$ is a share of $m^\wedge(s)$ for the fulfillment of the external functions in the virus subsystem B for creation of $c^\wedge(t)$, $\beta^\wedge(t,s)$ is the efficiency index for the functioning of the subsystem B that is similar to $\alpha^\wedge(t,s)$, $b^\wedge(t)$ is the similar temporal threshold to $a^\wedge(t)$, but for the subsystem B, $R^\wedge(t)$ is the total number of molecules of virus enzyme-substrate complex functioning at the instant t, $G^\wedge(t)$ is the total number of molecules of out-of-date virus enzyme-substrate complex at the instant t, $f^\wedge(t)$ is the rate of the inflow of the external substrate accessible to the virus, $k(f^{\wedge\wedge})$ is the coefficient, $0 < k(f^{\wedge\wedge}) \leq 1$, characterizing a share $m^\wedge$ in $f^\wedge$ and depending on qualitative composition of $f^\wedge$, named $f^{\wedge\wedge}$; α', m, c_1, c_2, c_3, c_4, β, c, $(cd)_4$ mean the rate of creation respectively new pro-virus DNA, substrate-enzyme complex of HIV population inside of an infected cell, gene *tat*, gene *rev*, gene *nef*, gene *vif*, producing c gene, remaining bio-mass of the HIV population, the concentration of cellular receptor CD4 protein; u, v, w, x, y, z_i mean control factors of HIV population inside an infected cell, k_i are coefficients (subjected to identification), and $f(t)$ is the rate of the inflow of the substrate of an infected cell accessible to the viruses;

$M_1{}^P(t)$ is the number of Th proliferating cells; $M_1{}^s(t)$ is the number of Th cells stimulated by the free HIV Ag, $m_1{}^s(t) = (M_1{}^s(t))'$, $M_1(t)$ is the overall number of Th cells, W is the number of free HIV, $d(t)$ is a share of infected $M_1{}^P$ cells producing free HIV, $q(s)$ is the death rate of free HIV, $q(s) = 0.05$, k is a coefficient (subjected to identification);

m^* is the number of the new thymus and spleen cells in unit of time remaining in these organs, Δm^* is the cells' inflow from outside (e.g. from bone marrow), c^* is the number of the new thymus and spleen cells in the unit of time leaving these organs, d^* is the number of the new destroyed cells in the unit of time, α^* is the efficiency index for the cortex cells proliferation, β^* is the efficiency index for the cerebral substance cells differentiation, γ^* is the efficiency index for the cortex cells selection and destruction, $[0, t\hat{}]$ is the prehistory for which all the functions are considered as given; the remaining magnitudes have the usual definition as above and their notations are the same as above.

It should be noted that many notation symbols are changed here in comparison with Ch. 6. Otherwise, we would have confusing symbols.

We suppose that stimulated and infected Th cells are the same and the share of stimulated cells are determined by the truncated normal law F_2, with the parameters (13.5, 0.5^2), for the decimal logarithm of the number $W(t)c\hat{}(t)$ of molecules of all live free HIV Ag at the instant t, $t \geq 0$. Thus, we have here a different situation in comparison with Ch. 6, s.1.2. For simplicity, we assume that the function $W(t)c\hat{}(t)$ is not decreasing.

Choosing instead of F_2 the law F_7 and respectively instead of $W(t)c\hat{}(t)$ the function $W(t)$, we can consider HIV as corpuscular Ag similar to cells. Other, more complicated MM of HIV free population can be obtained from the relations of Ch. 5, s. 6.2.

Since the period of the time under consideration is much longer now, we also introduce the function c^* that means the number of additional Th cells from the thymus and spleen.

Let us explain the new relations in (29) with respect to $M_1{}^s(t)$. It is easy to see that the respective discrete MM have the form

$$M_1{}^s(t+1) = [M_1{}^s(t) + dM_1{}^s(t)]U(t) - dM_1{}^s(t-12)U(t-12)U(t-11). \dots U(t),$$

$$dM_1{}^s(t) = c^*(t)F_2[g(t)] + M_1{}^s(t)(F_2[g(t)] - F_2[g(t-1)])/F_2[g(t-1)]. \quad (31)$$

So,

$$M_1{}^s(t+\Delta t) - M_1{}^s(t) = M_1{}^s(t)[U_\Delta(t)-1] + \Delta M_1{}^s(t)U_\Delta(t) - \Delta M_1{}^s(t-12)U(t-12)\dots U(t),$$

$$\Delta M_1{}^s(t) = c^*(t)F_2[g(t)]\Delta t + M_1{}^s(t)(F_2[g(t)] - F_2[g(t-\Delta t)])/F_2[g(t-\Delta t)],$$

$$U_\Delta(t) = [1-F_6(y_5)\Delta t][1-k_{8,7,3,2}{}^{m,p}\Delta t]. \quad (32)$$

and the desired result follows from (32) under $\Delta t \to 0$.

If the functions $\alpha\hat{}$, $\beta\hat{}$, $P\hat{}$ (or $f\hat{}$), α^*, β^*, γ^*, P^* (or F^*), coefficients k_i and k are given and the control factors $y\hat{}$, $u\hat{}$, $v\hat{}$, y^*, z^* are determined from a certain optimization problem, we will have the determined system (27)-(30) with respect to c, $c\hat{}$, c^*, and $M_1{}^s$, $M_1{}^P$, and W. These last 3 functions can be found, of course, with regard to all the equations of IN from Ch. 6 that are similar to (29), but without the number 0.75/24 (≈ 0.03).

The combined MM of HIV, HIV populations, and some organs IS are probably capable of contributing to clinicians to fight with HIV. But, as one can see, those MM are very complicated.

Qualitative and numerical investigations of the models require cross-participation of interdisciplinary scientists, especially in the field of software, numerical methods, dynamic systems, immunology, and medicine.

We dwell here only on qualitative investigation of the models (27)-(30). The whole joint operation should be sponsored.

5.1. QUALITATIVE INVESTIGATION OF MM OF AIDS

First of all, let us consider different particular situations and respective MM.

If HIV populations inside the infected Th cells are so aggressive that

$$kf(\mathrm{t}) > c^*(t)F_2[g(t)] + M_1^{\mathrm{s}}(t)(F_2[g(t)])'/F_2[g(t)], \tag{33}$$

then $dM_1{}^s < 0$ and hence $m_1{}^s < 0$. Therefore, M_1 may be about 0. Most damage to $M_1{}^s$ may give the maximal possible $F(t)$.

Therefore, let us consider the problem

$$F(T) = \int_0^T [\alpha'(s)+m(s)+c_1(s)+c_2(s)+c_3(s)+c_4(s)+\beta(s)+c(s)]ds = \max\ (x, y, z_i, u, v, w) \tag{34}$$

under the natural conditions and

$$0 < u < u^*, 0 < v < v^*, 0 < w < w^*, \tag{35}$$

where u^*, v^*, w^* are certain restrictions on the control factors u, v, w.

Since $c(s)$ contributes mostly to $f(s)$ (see Ch. 5, 6), we have

$$F(T) \approx C(T), \tag{36}$$

and the matter reduces to the well-investigated problem (see Ch. 15) of maximization of quantity of the external product of CAO.

Based on the respective results for sufficient large T, assuming for simplicity constant values of v, w, and z_i, $i = 1, 2, 4, 5$, as well as non-decreasing of the function β, we have an approximate solution of the problem (35), (36) under

$$x(t) = 1, 0 < t < T - T^{1/2} - dT,\ dT \ll T^{1/2},$$

$$y(t) = 1, T - T^{1/2} - dT < t < T - dT,\ u = u^*,$$

$$x(t) = y(t) = z_3(t) = 0, z_1(t) = z_2(t) = z_4(t) = z_5(t) = 1/5,$$

$$v(t) = v^*, w(t) = w^*, T - dT < t < T. \tag{37}$$

On the other hand, for minimization of $F(T)$ we have to maintain x and y and/or u, v, and w near 0 as long as possible.

It is easy to see that the case $m_1{}^s < 0$ is also possible under the more general condition

$$dM_1{}^s < -hM_1{}^s + M_1{}^s(t-12)f(t,t-12)\exp\left[-\int_{t-12}^{t} h(s)ds\right],$$

$$dM_1{}^s = c^*(t)F_2[g(t)] + M_1{}^s(t)(F_2[g(t)])'/F_2[g(t)] - kf(t), \tag{38}$$

from where we have the following sufficient condition

$$kf(t) > c^*(t)F_2[g(t)] + M_1{}^s(t)\{h + [(F_2[g(t)])'/F_2[g(t)] - f(t,t-12)\exp\left[-\int_{t-12}^{t} h(s)ds\right]\}, \tag{39}$$

In particular, if

$$h + (F_2[g(t)])'/F_2[g(t)] < f(t,t-12)\exp\left[-\int_{t-12}^{t} h(s)ds\right], \tag{40}$$

a reason of which may be a small value of h and $(F_2[g(t)])'$ or relatively large value of $F_2[g(t)]f(t,t\text{-}12)$ (the last is possible when we have a large, but not too large, number of HIV Ag molecules and a small number of Ig of the second level at the same time), then we have the following sufficient condition

$$kf(t) > c^*(t)F_2[g(t)]. \tag{41}$$

This condition obviously means that, in particular, the number of Th cells will be decreased if the thymus and spleen supply is exhausted. To escape this situation, we have to maintain the sufficiently fair condition for the thymus and spleen functioning according to the principle of adaptable redistribution and regenerative accumulation of the thymus cells (see Ch. 9).

Decrease in the number $M_1{}^P(t)$ infected proliferating Th cells and, therefore, overall Th cells may also happen due to the condition

$$0.05 - h(s) - c_1{}^P(s) < 0. \tag{42}$$

Having in mind the weakened proliferating potential of infected Th cells (0.05 instead of 0.08 for the normal cells) and also the structure of the functions h and $c_1{}^P$ (see Ch. 8), it is easy to conclude that, for example, h may be more than 0.05 under the number of Ts cells of about 10^4, that is, about 2 times less in comparison with the norm.

One more reason for the deficiency of Th cells may be due to such a large number of molecules of HIV Ag ($g(t) \approx 16$, $F_5[g(t)] \approx 1$) that not only all Th cells, but also all Ts and Tef cells will be inhibited by Ag.

Under a rather small number of Ag ($g(t) \approx 11.5$, $F_1[g(t)] \approx 1$), we also have the case of complete suppression of Th cells (see Ch. 8). The role of this situation in HIV infection is still not well investigated.

5.2. ON AIDS IMMUNE THERAPY

On the whole, the AIDS immune therapy should consist in overcoming the tolerance induced by HIV to numerous Ag.

First, we should know ISO, including the main indices of the organs of IS (especially the thymus and spleen), of the patient infected with HIV. The most important factor of their restoration is probably the restoration of the organs of IS. It has been shown not once, especially in Ch. 6, 13, and 15, that for restoration of any CAO of the patient, the conditions for consuming an essential part of all resources to the internal needs of the CAO should be created. In the case of the thymus and spleen it means that in spite of necessity of giving the source of new Th cells, the conditions should be created for temporal decreasing of this source in order to restore the "health" of these organs.

Review of the strategies for combating HIV and AIDS can be found in A. Levin [120]. This includes gene-type and immune-type drug therapies.

We dwell only on the problem of possible applications of the models in order to understand certain ways of AIDS immune therapy. One of them has been to synthesize a large amount of the CD4 protein receptors. This synthesis has been done by placing the human CD4 gene into bacteria and stimulating the bacteria to synthesize large amounts of a soluble, non-membrane form of the CD4 protein that binds to the viral gp120 protein. Placed in the blood of AIDS patients, this CD4 protein blocks or competes with the viruses' attachment to cells. As a result, the viral infection is reduced. "Unfortunately, the amount of CD4 needed to block all viral attachment to cells is great, and the body degrades it too rapidly" [120], p.150.

In other words, the respective response of IS is too strong and too fast.

Using MM of IN, let us consider some possibilities for inhibition of the immune response to different doses of Ag.

The first possibility is to insert such a large amount of Ag that all sub-populations of T and B cells will be inhibited and the immune reaction will not start until the natural degradation of Ag decreases its amount by about 10 times and until the thymus enlarges the amount of unsuppressed cells. Under the rate of degradation of about 0.008 per hour, decrease of 10 times the amount requires about $0.008x = 2.3$, i.e., $x \approx 400$ hours or about 16 days. Approximately the same time is required to restore the normal quantity of the initial value of T and B cells by the thymus.

The second possibility is to insert not such a large amount of Ag, so only the respective Th cells will be inhibited (see Ch. 8, s.7.2). Without Th cells the immune reaction will include creation of Ig only by B_1 cells (10% of all B cells), and overall reaction will be weak, with the predominance of suppression factors. Under the right large amount of Ag, the normal reaction to Ag will be strong and rather fast, about a week. To inhibit this reaction, it is sufficient to insert the same Ag and/or to insert additional amount of resting Ts cells. Then we again have the immune reaction with a small number of Ig and a predominance of Ts cells.

In general, with the help of MM of IN and the *hybridoma technology* (G. Rifcin [160)]), we can find and realize such an initial value of the cells of IN that the immune reaction will be pre-assigned, in particular, tolerant to the given Ag.

For this, we need software and computers.

We can also determine the role of the second level of IN. Indeed, let us suppose that Ag (Ag-determinants of HIV) suppress most of M_1 (T helpers of the first level) and that Ig are created only by $M_{4,2}{}^m$ (plasma B_2 cells of the first level) production. Of course, $M_{4,2}{}^s$ and, the more so, $M_{4,2}{}^{s,s}$ can not be stimulated by M_1-cells, but, in principle, they can be stimulated by ample $M_6{}^{m,p}$ (mature and proliferating T-helpers of the second level). In turn, for $M_6{}^{m,p}$ we need the normal $M_{6,0}$ that can be stimulated by ample $M_{4,1}{}^p$ (proliferating B_1 cells) for which, in turn, we need ample $M_{4,1,0}$. Thus, the problem is to determine $M_{4,1\,0}$, $M_{4,2,0}$, and $M_{6,0}$ as well as the instants of their insertions in order to have an immune response to Ag in question. The exact quantities of all these values can be found with the help of these MM of IN as CAO and the respective software and computers.

The usual way we have conquered viruses in the past is through safe and effective *vaccines* to prevent viral infections, by maintaining high level of immunity in populations. Let us dwell on possible applications of MM of IN to the solution of this problem.

One of the vaccines can be based on a purified gp120 protein as Ag. However, there are at least four theoretical objections to a subunit gp120 vaccine [120].

The portion of the gp120 protein that binds to the CD4 receptor appears to form a pocket that may be poorly immunogenic.

A second problem is the high mutation rate of gp120, with up to 30% variability between gp120 proteins from different isolates of HIV, while these wide differences do not occur in the gp120 pocket.

Third, there may be a problem with stimulating high levels of anti-gp120 Ig in human beings. Under certain laboratory conditions, low levels of these Ig actually enhance entry of HIV into monocytes, which can replicate HIV and produce more viruses.

Finally, transmission of HIV by cell fusion might not be blocked by Ig because HIV does not have to enter the bloodstream where Ig are located.

One can see that the first and the third objections reduce the matter to the problem of stimulating advanced levels of anti-gp120 Ig. Translating into the language of MM of IN as CAO, this means that we should create a condition for advanced levels of the respective resting cells, numbers of $M_1{}^r$. Since the resting cells (or memory cells) are not reactive to the other cells of the same level (see Ch. 8, assumption 29), this can be done by inhibition of the respective proliferating $M_4{}^p$ cells and/or their process of proliferation while the vaccine is being created. Indeed, a large number of $M_4{}^p$ of B cells is the reason for stimulating the cells of the second level (assumptions 4), 17), and 28), Ch. 8). Logically, it contradicts the fact that for the strong immune response to HIV we are in a need of a large number of B cells. It means that the vaccine may be useful once at a time.

Another way is estimating the initial values of the cells of IN that give us the pre-assigned number of the memory cells and maintaining this number with the help of hybridoma technology. The obvious case is maintaining the lowered number of the initial values of cells of the second level. Of course, more exact estimates can be obtained with the help of MM of IN as CAO, the respective software, and computers.

The second problem is connected with modeling of variability of free HIV that was considered in s. 4.3. As to the fourth problem, it seems reliable to use another drug, for example, the soluble CD4 protein with a toxin [120] that kills infected cells and most probably the large infected "cell fusion."

In general, two or more independent drugs may be very useful because the chance of

HIV resistance to several drugs is far smaller than the chance of resistance to either drug alone. Another attractive approach is the so-called recombinant virus vaccine, which recombines the gene for gp120 into the carrier-virus chromosome. An example of such kind of the vaccine is vaccinia, the vaccine strain for smallpox. Indeed, vaccinia viruses, expressing several HIV proteins, have been made and are being tested in animals [120].

We believe that using MM of IN and IS allows us substantially to diminish the required number of animal models.

We have considered here a possible application of MM of IN and the latest biotechnology for elimination of hypothetical Ag similar to HIV Ag determinants of the molecular-disperse nature. It should be noted that for realization of this idea we should, in addition, consider the guess 1)-5), Ch. 8, so that the problem is getting much more complicated.

We should also consider HIV Ag as corpuscular similar to cells (see Ch. 8).

Of course, further investigation and perfection of MM of HIV, HIV populations, IN, and IS as a whole are necessary in order to consider the very fast possible variations of HIV Ag-determinants. Further investigation is also needed to have an accurate prediction for these determinants and determination of the necessary actions with immune correctors of IS.

As to the autoimmune aspects of HIV infection (G. Hoffman, M. Gran [81]), the role of MM of IN and IS as CAO can only grow.

6. Conclusion

In conclusion, we would like to emphasize that this Ch. is mainly devoted not to the means, but to the methods of AIDS treatment.

At present, there have been some immunological means of AIDS treatment (see, for example, [5]; D. Leturcq et al. [119]; and [208], some of which are associated with the IN concept), but the question about the best condition for the intake of the known agents is not trivial with regard to the peculiarities of the given patient and the complexity of AIDS.

It can be solved by means of mathematical simulation faster and more qualitatively than without them on the basis of application of the necessary MM and the appropriate software and computers.

Chapter 11

CANCERS

1. On Study of a Cancer at the Molecular Level

"*Our chance to understand a disease depends greatly upon whether it is based upon abnormalities in molecules with which we are already familiar.*" (J. Watson[206], p. 589).

The term *cancer*(*s*) (**C**) encompasses a large variety of diseases, all characterized by the property that cells grow and proliferate when they should not. The difference between a cancer cell and normal cell is often not an all-or-none matter, but is a question of the frequency of division. The normal cell requires the stimulus of a specific hormone, the cancer cell may not need this stimulus or may require less hormone. Thus, the only useful distinction is that the cancer cell is less subject to the normal control devices that tell a cell not to divide. Contiguous masses of cancer cells are called tumors.

Almost all types of differentiated cells can be transformed into cancer cells. Normal cells have a great affinity for solid surfaces and prefer to stick to others of their own kind (liver cells to liver cells and kidney cells to kidney cells). A loss of normal cellular affinity is most likely the reason why many malignant cancer cells invade a variety of normal organs. There is one known biochemical difference for cancer cells: many types of them excrete much larger quantities of lactic acid than do their normal counterparts. It arises from glucose via the glycolytic pathway. This suggests the loss of a normal control device, which regulates the rate at which glucose is taken into a cell. Tumor cells lack one or more specific surface proteins that control both glucose uptake and the "sticky" quality that give the cell the correct cellular affinity.

Alternatively, there is the hypothesis that many **C** are caused by viruses. In a sense, this is not really a hypothesis since there is already convincing evidence that a number of **C** in animals, ranging from fish to mammals, are virus-induced. Scientists strongly believe in the involvement of RNA viruses, proposing that most of these **C** are caused by defective RNA viruses that have become virtually a permanent part of a human chromosome.

Under this hypothesis, it is the functioning of the *oncogenes* which results in the production of the specific proteins that give their host cells their cancerous condition. Only when such tumor cells are simultaneously infected with helper viruses can their oncogenes be packaged into mature virus particles, thereby revealing the viral causation of the cancerous state. A variant of this theory proposes that many *oncogenes* exist in a quiescent form and generally are only induced to work by the action of external carcinogens like ionizing radiation, cigarette smoke, or conceivably the presence of another virus.

Much more information on **C** can be seen, for example, in [1]; G. Cooper [36]; C. Haskell [77]. Some stages of tumor development can be seen in Fig. 20.

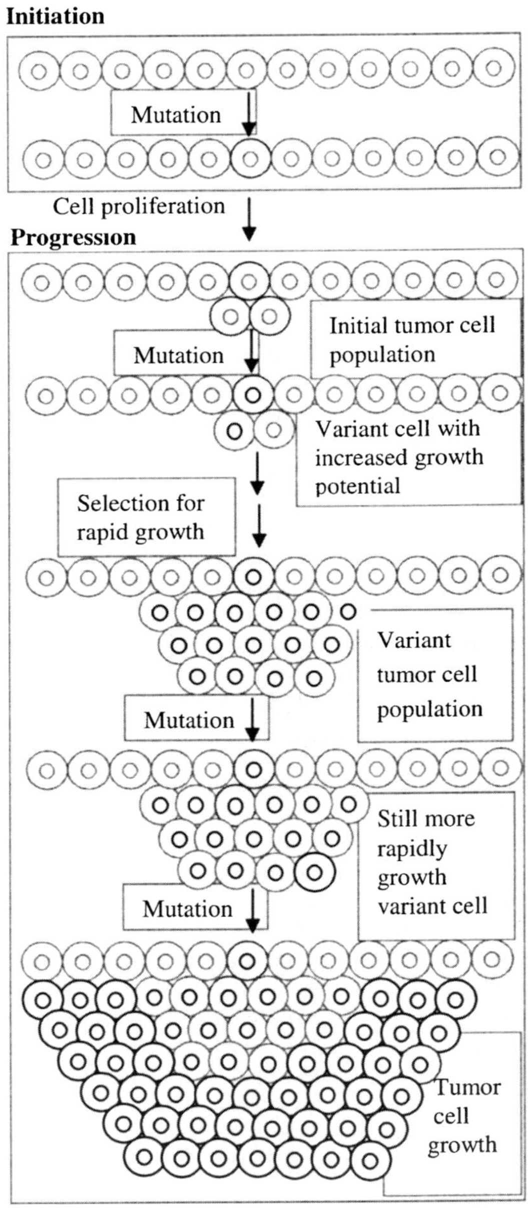

Figure 20. Stages of cancer development

2. MM of a Cancer Cell

2.1. BASE MM OF A CELL AT THE START OF PROLIFERATION

The Base, simplest MM in question has the form

$$\boldsymbol{m}(t) = \int_{a(t)}^{t} \alpha(t,s)\boldsymbol{y}(s)\boldsymbol{m}(s)ds = (m^1(t), m^2(t), m^3(t)),\ \boldsymbol{M}(t) = \int_0^t \boldsymbol{m}(s)ds,$$

$$\boldsymbol{c}(t) = \int_{a(t)}^{t} \beta(t,s)[\boldsymbol{I}-\boldsymbol{y}(s)]\boldsymbol{m}(s)ds = (c^1(t), \ldots, c^p(t)),\ \boldsymbol{C}(t) = \int_{t^\wedge}^{t} \boldsymbol{c}(s)ds,$$

$$\boldsymbol{R}(t) = \int_{a(t)}^{t} \boldsymbol{m}(s)ds,\ \boldsymbol{G}(t) = \boldsymbol{M}(t) - \boldsymbol{R}(t) > 0,\ \boldsymbol{f}(t) = \boldsymbol{m}(t) + \boldsymbol{c}(t),\ \boldsymbol{F}(t) = \int_0^t \boldsymbol{f}(s)ds,$$

$$\boldsymbol{y}(t) = (y^1(t), y^2(t), y^3(t)),\ \boldsymbol{a}(t) = (a^1(t), a^2(t), a^3(t)),\ m^1(t)+m^2(t)+m^3(t) = m(t),$$

$$c^1(t) + \ldots + c^p(t) = c(t),\ 0 \le \boldsymbol{y}(t) \le 1,\ 0 \le \boldsymbol{a}(t) \le t,\ t > t^\wedge > 0; \tag{1}$$

$$\alpha(t,s)\boldsymbol{y}(s) = \alpha^\wedge \exp\,[\boldsymbol{u}(t) - \boldsymbol{in}(t) + \ln \boldsymbol{p}(t)],\ \boldsymbol{u}(t) = (u^1(t), u^2(t), u^3(t)),$$

$$\beta(t,s)[\boldsymbol{I}-\boldsymbol{y}(s)] = \beta^\wedge \boldsymbol{v}(t)],\ \boldsymbol{v}(t) = (v^1(t), v^2(t), v^3(t)),\ \alpha^\wedge,\ \beta^\wedge = \text{constant},$$

$$\boldsymbol{m}(t) = \alpha^\wedge \exp\{\boldsymbol{u}(t) - \boldsymbol{in}(t) + \ln\,[\boldsymbol{p}(t)]\}\boldsymbol{R}(t),\ \boldsymbol{c}(t) = \beta^\wedge \boldsymbol{v}(t)\boldsymbol{R}(t),$$

$$\boldsymbol{R}'(t) = \boldsymbol{m}(t) - \boldsymbol{m}(\boldsymbol{a}(t))\boldsymbol{a}'(t) = 0,\ 0 \le \boldsymbol{a}(t) \le t,\ t > t^\wedge > 0; \tag{2}$$

$$\boldsymbol{u} = \boldsymbol{V'}\mathbf{s}/(\boldsymbol{K_m} + \mathrm{s}),\ \boldsymbol{V'} = \boldsymbol{V}/\boldsymbol{K_+},\ \boldsymbol{K_m} = \boldsymbol{K_S}/\boldsymbol{K_+},\ \boldsymbol{V} = \boldsymbol{k_2e_0},\ \boldsymbol{K_+} = (1+\mathbf{h}^+/\boldsymbol{K}_{1\mathrm{ES}}+\boldsymbol{K}_{2\mathrm{ES}}/\mathbf{h}^+),$$

$$\boldsymbol{K}_\mathrm{S} = (\mathbf{eh})\mathbf{s}/\mathbf{ehs},\ \boldsymbol{K}_{1\mathrm{ES}} = (\mathbf{es}^-)\mathbf{h}^+/\mathbf{ehs},\ \boldsymbol{K}_{2\mathrm{ES}} = (\mathbf{ehs})\mathbf{h}^+/[(\mathbf{eh}_2)\mathbf{s}^+]; \tag{3}$$

$$\alpha = \alpha^\wedge \exp\,[-\Delta H^0_\alpha/\mathrm{R}\delta T],\ \beta = \beta^\wedge \exp\,[-\Delta H^\circ_\beta/\mathrm{R}\delta T],\ \boldsymbol{p}(t) = \boldsymbol{p}^\wedge(t)\exp\,[-\Delta H^0_p(\mathrm{t})/\mathrm{R}\delta T],$$

$$K_S = K^\wedge{}_S \exp\,[\Delta H^0_s/\mathrm{R}\delta T],\ K_{i\mathrm{ES}} = K^\wedge{}_{i\mathrm{ES}} \exp\,[\Delta H^0_i/\mathrm{R}\delta T],\ i = 1, 2,\ \delta T = 1/T - 1/T^\wedge; \tag{4}$$

$$\alpha(t,s) = \alpha(t-s),\ \alpha(t) = \int_{a(t)}^{t} \alpha(t-s)\boldsymbol{x}(s)\boldsymbol{m}(s)ds,\ 0 \le \boldsymbol{x} \le 1,$$

$$\beta(t,s) = \beta(t-s),\ \beta(t) = \int_{a(t)}^{t} \alpha(t-s)\boldsymbol{z}(s)\boldsymbol{m}(s)ds,\ 0 \le \boldsymbol{z} \le 1; \tag{5}$$

where α is the matrix, corresponding to the regulatory, protein, and utilization genes that are ready to act; $\boldsymbol{m}$ corresponds to the regulatory enzyme-substrate factors (collection of activators, enhancers, transcription factors, promoters, suppressors, silencers, etc.), RNA polymerase, and utilization enzymes; β corresponds to production of external product genes, and $\boldsymbol{c}$ to the external products of the cell; ***in*** corresponds to inhibition factors (switch off), $\boldsymbol{u}$ and $\boldsymbol{v}$ to the protein factors, releasing the repressors (switch on), $\boldsymbol{p}$ to the factor, resulting from oncogene; R is the universal gas constant; T is the absolute temperature by Kelvin inside the cell; ΔH^0_α, ΔH^0_β, ΔH^0_p, ΔH^0_s, ΔH^0_1, and ΔH^0_2 are the standard enthalpy changes for the reactions with the equilibrium constants respectively α, β, p, K_S^{-1}, K_{1ES}^{-1}, and K_{2ES}^{-1} (their values under the temperature T^ have sign ^); **s** corresponds to the signal molecules (hormones) from the signaling cells; $\boldsymbol{u}$ to the resulting proteins; $\boldsymbol{e}$ corresponds to the respective necessary enzymes and **eh**, **es**, and **ehs** to enzyme-substrate complexes, where **h** means the vector of hydrogens.

Bold letters in the various functions above mean that we have to repeat these functions with the respective different non-bold letters, inequality for vector means the respective inequalities for their components, vector product means product column matrix on row matrix. It is not difficult to see that MM (1) -- (4) are a collection of MM in Ch. 5 and Ch. 7. A principle difference of MM of a **C** cell is increased value of α, what is mapped by the relation (1) under increased value of $\boldsymbol{u} - \boldsymbol{in} + \ln \boldsymbol{p}$.

We remind that there are two types of signal-transduction pathways leading to the synthesis of proteins. The first type occurs when the initial signals **s** are growth factors, and the target cell's ultimate response is the production of proteins $\boldsymbol{u}$ that stimulates the cell to divide. By contrast, the second type occurs when we have an inhibitory pathway, in which growth-inhibitory factors cause the target cell to make proteins ***in*** that inhibit cell division. In both cases, the newly made proteins function by interacting with components of the cell-cycle system launching the enzyme-substrate complexes $\boldsymbol{m}$ and DNA instructions α and β.

An oncogene protein can cause increased cell division even in the absence of growth factors. Mutant tumor-suppressor proteins can also affect cell division. The genes for the proteins that inhibit cell division can be remained turned off, and inappropriate cell division may occur. Examples of actual such *genes* are *proto-oncogene ras* and a *tumor-suppressor gene p*53 (for protein of 53,000 molecular weight).

2.2. MORE COMPLICATED MM

MM in question, regarding the spatial structure, have the form

$$\boldsymbol{m}(t,X) = \Sigma \int^{t}_{a(t)} \int_{V,j} \alpha(t,X,s,Y)\boldsymbol{m}(^{j})(s,Y)dYds\ (j = 1, \ldots, 5),\ X \in V,1;\ V,2;\ V,3;\ V,6;$$

$$\boldsymbol{m}(t,X) = \Sigma \int^{t}_{a(t)} \int_{V,j} \alpha(t,X,s,Y)\boldsymbol{m}(^{j})(s,Y)dYds\ (j = 1, 3, 5, 6),\ X \in V,4;$$

$$\boldsymbol{m}(t,X) = \Sigma \int^{t}_{a(t)} \int_{V,j} \alpha(t,X,s,Y)\boldsymbol{m}(^{j})(s,Y)dYds\ (j = 1, 3, 4, 7),\ X \in V,5;$$

$$\boldsymbol{m}(t,X) = \sum \int_{a(t)}^{t} \int_{V,j} \boldsymbol{\alpha}(t,X,s,Y)\boldsymbol{m}(^1)(s,Y)dYds \ (j = 1, \ldots, 6),\ X \in V,7;$$

$$\boldsymbol{m}(t,X) = \sum \int_{a(t)}^{t} \int_{V,j} \boldsymbol{\alpha}(t,X,s,Y)\boldsymbol{m}(^1)(s,Y)dYds \ (j = 1, \ldots, 7),\ X \in V,8;$$

$$\boldsymbol{m}(t,X) = \sum \int_{a(t)}^{t} \int_{V,j} \boldsymbol{\alpha}(t,X,s,Y)\boldsymbol{m}(^1)(s,Y)dYds \ (j = 1, \ldots, 8),\ X \in V,9;$$

$$\boldsymbol{c}(t,Z) = \int_{b(t)}^{t} \int_{V} \boldsymbol{\beta}(t,Z,s,Y)\boldsymbol{m}(^2)(s,Y)dYds,\ Z \in V\hat{},\ V\hat{} = R^3\backslash V,$$

$$\boldsymbol{c}(t) = \int_{V\hat{}} \boldsymbol{c}(t,Z)dZ,\ \boldsymbol{R}(t) = \int_{a(t)}^{t} \int_{V} \boldsymbol{m}(^1)(s,Y)dYds + \int_{b(t)}^{t} \int_{V} \boldsymbol{m}(^2)(s,Y)dYds,$$

$$\boldsymbol{m}(^1) + \boldsymbol{m}(^2) = \boldsymbol{m},\ \boldsymbol{f}(t) = \boldsymbol{m}(t) + \boldsymbol{c}(t),\ V = V(t) = \cup\, V,j\ (j = 1, \ldots, 9), \qquad (6)$$

where V is the domain that a cell occupies in R^3: $X = (x1, x2, x3)$, $V,1$ is the nucleolus, $V,2$ is the nucleus, $V,3$ is the mitochondria, $V,4$ is the ribosome, $V,5$ is the ER, $V,6$ is the Golgi apparatus, $V,7$ is the plasma membrane, $V,8$ is the lysosome, and $V,9$ is the cisterna (we mean that $V,j = V,j\,(t)$ are the domains of the respective parts of a cell's location); $\boldsymbol{m}(t,X)$ are the new cell's substrate-enzyme complex quantities at the instant t in their point X per unit of time (this means that X is simultaneously a notation for the products $\boldsymbol{m}(t,X)$ of X-type and by the same token the various products of a cell are distributed in the space); $\boldsymbol{\alpha}(t,X,s,Y)$ are the efficiency indices of a cell functioning along the channel $m(^1)(s,Y)$ -- $m(t,X)$, the remaining elements of the model have the same definition as above, the matrices $\boldsymbol{\alpha}$ and $\boldsymbol{\beta}$ may depend on all the other elements. If $\boldsymbol{\alpha}$, $\boldsymbol{\beta}$, and $\boldsymbol{c}$, $\boldsymbol{R}$ or $\boldsymbol{c}'_t = \boldsymbol{R}'_t = 0$ are given, then we have the determined system (6) with respect to the remained elements, all elements are known on the prehistory $[0, t\hat{}]$. Instead of $\boldsymbol{c}$, the function $\boldsymbol{f}$ or the condition $\boldsymbol{f}'_t = 0$ may be given.

We divide all matrices $\boldsymbol{\alpha}$ and $\boldsymbol{\beta}$ on three classes: the first that corresponds to the given procedures by the genome, the second that corresponds to the known laws of the nature, and the third that corresponds to combination of two former classes. Among the known laws is the mass action law that is used frequently under modeling of cells and CAO. In MM (6), for $X \in V,1$, $V,2$, the matrices $\boldsymbol{\alpha}$ and $\boldsymbol{\beta}$ probably relate to the first class, for $X \in V,3$, $V,4$, $V,5$, $V,6$, $V,7$ to the third class, and for $X \in V,8$, $V,9$ to the second class.

In the case of the first class, the respective control may come from the control of gene expression, that is, from gene therapy. In the case of the third and especially the second class, the respective control may come from the control of the known laws at the expense of their certain variables and parameters. In particular, for the mass action law, we have the relations upon the reaction rates of cell's pH (or $\mathbf{h}^+$) and temperature T similar to those in (3) and (4).

2.3. MM OF DEVELOPING DAUGHTER CELL

The MM in question have the form

$$\boldsymbol{m}'(t,X) = \int_{a(t,X)}^{t} \int_{O}^{X} \alpha'(t,s,Y)\boldsymbol{m}'(^{1})(s,Y)dYds,$$

$$\boldsymbol{M}'(t,X) = \int_{0}^{t} \boldsymbol{m}'(s,X)ds,\ \boldsymbol{M}'(t) = \int_{V'} \boldsymbol{M}'(t,X)dX,$$

$$\boldsymbol{c}'(t,X,Z) = \int_{a(t,X)}^{t} \int_{O}^{X} \beta'(t,Z,s,Y)\boldsymbol{m}'(^{2})(s,Y)dYds,$$

$$\boldsymbol{C}'(t,X,Z) = \int_{t\wedge}^{t} \boldsymbol{c}'(s,X,Z)ds,\ \boldsymbol{C}'(t,X) = \int_{R,3\backslash V'} \boldsymbol{C}'(t,X,Z)dZ,$$

$$\boldsymbol{C}'(t) = \int_{R,3\backslash V'} \boldsymbol{C}'(t,X,Z)dZ,\ \boldsymbol{R}'(t,X) = \int_{a(t,X)}^{t} \int_{O}^{X} \boldsymbol{m}'(s,Y)dYds,$$

$$\boldsymbol{R}'(t) = \int_{V'} \boldsymbol{R}'(t,X)dX,\ \boldsymbol{m}'(^{1})(t,X) + \boldsymbol{m}'(^{2})(t,X) = \boldsymbol{m}'(t,X),$$

$$\boldsymbol{G}'(t,X) = \boldsymbol{M}'(t,X) - \boldsymbol{R}'(t,X) > 0,\ \boldsymbol{G}'(t) = \boldsymbol{M}'(t) - \boldsymbol{R}'(t),$$

$$\boldsymbol{f}'(t,X,Z) = \boldsymbol{m}'(t,X) + \boldsymbol{c}'(t,X,Z) = \boldsymbol{k}'(\boldsymbol{f}',X,Z)\,\boldsymbol{m}'(t,X),$$

$$\boldsymbol{f}'(t,X) = \int_{R,3\backslash V'} \boldsymbol{f}'(t,X,Z)dZ,\ \boldsymbol{f}'(\mathrm{t}) = \int_{V'} \boldsymbol{f}'(t,X)dX,\ \boldsymbol{F}'(t) = \int_{0}^{t} \boldsymbol{f}'(s)ds,$$

$$V'(t) = \int_{b(t,X)}^{t} \int_{V} \alpha_{V}(t,X,s,Y)\boldsymbol{m}_{V}(s,Y)dYds,$$

$$0 \le a', b', c' \le t,\ t > t'\wedge > 0,\ O, X \in V'(t),\ Z \in R^{3}; \tag{7}$$

$$\alpha'(t,s,X) = \alpha'(t,X) = \int_{a(t,X)}^{t} \int_{V} \alpha_{\alpha'}(t,X,s,Y)\boldsymbol{m}_{\alpha'}(s,Y)dYds + \Delta\alpha',$$

$$\boldsymbol{f}'(t,X,Z) = \boldsymbol{m}'(t,X) + \boldsymbol{c}'(t,X,Z) + \alpha'(t,X),\ t > t'\wedge > 0,\ X \in V' = V'(t), \tag{8}$$

where the sense of the all functions is the same as in (1)-(6), excluding X and $\alpha'(t,s,X)$ development regularities, which may be given by DNA of the mother cell; the domain V' is not exactly the same as the domain V.

After the domain V' has become settled, the developed daughter cell separates from the mother cell.

2.4. MM OF A CANCER CELL BIO-FIELD

Let $V = V(t)$ be the domain in R^3 for an isolated cell, and $R^3\backslash V$ be the complement of V to R^3. We accept that the external bio-field with respect to a cell will be decided if the vector function

$$c(t,Z) = \{c_1(t,Z),\ c_2(t,Z),\ c_3(t,Z)\},\ Z \in R^3\backslash V, \tag{9}$$

is given. Here t is the time, (z_1, z_2, z_3) are Cartesian coordinates of the point Z; c_1 is the rate of creation along t,Z (i.e., along t, z_1, z_2, and z_3) of the (generalized) new material product of t,Z-type in the point Z at the instant t; c_2 is the rate of creation along t,Z of the (generalized) new energy product of t,Z-type in the point Z at the instant t; c_3 is the rate of creation along t,Z of the information product of t,Z-type in the point Z at the instant t.

As the base *MM of a cancer cell bio-field*, we can naturally assume the following system:

$$\boldsymbol{m}(t,X) = \int_{a(t,X)}^{t} \int_V \boldsymbol{\alpha}(t,X,s,Y)\boldsymbol{m}^{(1)}(s,Y)dYds,\ X \in V,$$

$$\boldsymbol{c}(t,Z) = \int_{a(t,X)}^{t} \int_V \boldsymbol{\beta}(t,X,s,Y)\boldsymbol{m}^{(2)}(s,Y)dYds,\ Z \in R^3,$$

$$\boldsymbol{R}(t,X) = \int_{a(t,X)}^{t} \int_V \boldsymbol{m}(s,Y)dYds,\ \boldsymbol{f} = \boldsymbol{m}(t,X) + \boldsymbol{c}(t,Z),\ \boldsymbol{m}^{(1)} + \boldsymbol{m}^{(2)} = \boldsymbol{m},$$

$$\boldsymbol{m}(t,X) = \{m_1(t,X),\ m_2(t,X),\ m_3(t,X)\},\ X \in V,\ t \geq t^\wedge > 0, \tag{10}$$

where $m_i(t,X)$ are the rates of creation along t,X of the (generalized) internal products of t,X-type respectively bio-mass ($i = 1$), energy ($i = 2$), and information ($i = 3$) in the X at the instant t; $\boldsymbol{\alpha}$ and $\boldsymbol{\beta}$ are matrices whose components are indices of efficiencies of a cell functioning along the corresponding channels.

The definitions of all the other values are similar to the previous definitions. It is natural to call V with (10) as the cancer cell internal bio-field and $R^3\backslash V$ with (9) and (10) as the cancer cell external bio-field.

The conditions under which the significant values of $c(t,Z)$ can arise for the values Z outside V and rather far removed from the boundary of V can be seen in V. Glushkov [61], where, in particular, it has been specified that the notion "bio-field" means physical fields of bio-systems.

3. Qualitative Investigation of MM of a Cancer Cell

3.1. INVESTIGATION OF MM (1)-(5)

It is obvious that under

$$q(t) = u(t) - in(t) + \ln [p(t)] > 0 \tag{11}$$

we have the case of increasing values of all the remained elements of the models. Let us consider in more detail the character of the solution of the respective equations and after this consider the problem of overcoming the condition (11).

From (1), it follows that

$$m' = mq',\ m(t) = m(t^{\wedge})\exp [q(t)],\ \ R = m(t^{\wedge})/\alpha^{\wedge},$$

$$c(t) = \beta^{\wedge}v(t)m(t^{\wedge})/\alpha^{\wedge},\ m(a(t))a'(t) = m(t),$$

$$\alpha(t,s) = \alpha^{\wedge}/y(\mathrm{s})m(t)/m(t^{\wedge}),\ \beta(t,s) = \beta^{\wedge}v(t)/[I-y(\mathrm{s})],\ t > t^{\wedge} > 0. \tag{12}$$

So, the functions q and v determine the solution of the equation (1).

From (1), (2), and (5), it follows that in the case $\alpha(t,s) = \alpha(t-s)$, $\beta(t,s) = \beta(t-s)$, $v(\mathrm{t}) = \exp [q(t)]$, and x, y, z are constant

$$\alpha(t) = x/ym(\mathrm{t}),\ \beta(t) = \mathrm{z}/\mathrm{y}m(t). \tag{13}$$

So, only $q(t)$ may determine the solutions of the equation (1) and (5).

In order to overcome the condition (11) (without gene changes), let us consider in more detail models (3) and (4). We should minimize q at the expense of pH (or $\mathbf{h}^+$) and T factors:

$$q(t) = u(t) - in(t) + \ln [p(t)] = \min (\mathbf{h}^+, T) \tag{14}$$

under the condition

$$u = V'\mathrm{s}/(K_m + \mathrm{s}),\ V' = V/K_+,\ K_m = K_S/K_+,\ V = k_2e_0,$$

$$K_+ = (1+\mathrm{h}^+/K_{1,\mathrm{ES}}+K_{2,\mathrm{ES}}/\mathrm{h}^+),\ K_S\ = K^{\wedge}_S\exp[\Delta H^0_s/\mathrm{R}\delta T],$$

$$K_{i\mathrm{ES}} = K^{\wedge}_{I,\mathrm{ES}}\exp[\Delta H^0_i/\mathrm{R}\delta T],\ \ i = 1, 2, p = p^{\wedge}\exp [-\Delta H^0_p/\mathrm{R}\delta T],\ \Delta H^0_p < 0. \tag{15}$$

By analogy with the results of Ch. 2, s. 5, it is not difficult to conclude that $\boldsymbol{q}$ has a maximum under certain $\mathbf{h}^{+}*$ and $T*$,

$$\mathbf{h}^{+}* = (\mathbf{K}_{1,\mathbf{ES}}\mathbf{K}_{2,\mathbf{ES}})^{1/2},\ T* = 0, \tag{16}$$

and

$$\boldsymbol{q}_{\min} = -\,\boldsymbol{in}(\mathrm{t}),\ \mathbf{h}^{+}*_{\min} = 0,\ \infty;\ T_{\min} = \infty,\ \Delta \boldsymbol{H}^{0}{}_{p} < 0. \tag{17}$$

It means that, in principle, there always exist values of $\mathbf{h}^{+}$ and $\Delta \boldsymbol{H}^{0}\delta T$ for which the condition (14) will be fulfilled. And what is more, the desired values are probably better on the left side from $\mathbf{h}^{+}*$, i.e. lower $\mathbf{h}^{+}$ (higher pH).

Of course, above we have only a guess about possible pH and T (and $\Delta \boldsymbol{G}^{0}$) influences on a **C** cell progression, and this problem requires more attention. Probably, we are in a need of, first, experimental determination of the equilibrium constants. In contrast to the methods of general chemistry ([26]; R. Caret et al. [32]), we now have more complicated case interaction $\mathbf{h}^{+}$ with the external signal of growth factor-type and, in addition, the protein-to-protein interaction. Then, we are also in a need of determination of ways for localization of pH and T influence, namely on cancer cells. It is well known [32] that our bodies use a combination of compounds called a *buffer system* to keep the pH within a narrow range. However, we can always prepare a buffer solution of a given pH [32], p. 729. And at last, but not least, we are in a need of scientific works that examine this problem in more detail and some animal models that can be used as models for eukaryotes.

In principle, we can overcome the condition (11) at the expense of only the value of $-\Delta \boldsymbol{H}^{0}{}_{p}(\mathrm{t})/\mathrm{R}\delta T$). For this, it is usually sufficient that

$$-\Delta \boldsymbol{H}^{0}{}_{p}(\mathrm{t})/\mathrm{R}\delta T < \boldsymbol{in}(\mathrm{t}), \tag{18}$$

since the factor $\boldsymbol{u}(\mathrm{t})$ usually degrades to zero rather quickly.

Besides, there is the more convenient formula for estimation of ΔH^{0} [32]:

$$\Delta H^{0} = \sum \Delta H^{0}{}_{i} = \sum n\Delta H^{0}{}_{f\,\mathrm{products}} - \sum n\Delta H^{0}{}_{f\,\mathrm{reactants}}, \tag{19}$$

where i refers to balanced equations that can be summed to give the equation for the desired reaction, $\Delta H^{0}{}_{f}$ is the standard molar enthalpy of formations, each multiplied by its coefficient, n, in the balanced equation. The relations (19) indicate a way for the products and reactants with the pre-assigned ΔH^{0}.

There is also another relationship of Gibbs free energy ΔG^{0} and K to the standard cell potential, $E^{0}{}_{\mathrm{cell}}$ (electrochemical cell), for a redox reaction [32]:

$$\Delta G^{0} = nFE^{0}{}_{\mathrm{cell}} = -RT\ln K_{T},\ F = 96{,}487\ \mathrm{J/Vmol}\ e^{-}. \tag{20}$$

where n is the number of moles of electrons involved in the overall process.

3.2. INVESTIGATION OF MM (6)-(8)

If our guess is true that the matrices α and β in (6) relate for $X \in V,8$, $V,9$ to the second class (i.e. they correspond to the known laws of nature, in particular, the mass action law for the lysosome and the mass-action law with saturation by the product for the cisternia), then we can repeat discussion of the previous s. about influence of pH and temperature T on the lysosome and cisternia. If these digestive and storage-type organelles of cancer cells were out of normal functioning, cancer cells would probably die.

Let us dwell on the case of the cisternia and put in (6)

$$\alpha(t,X,s,Y) = \alpha^{\wedge} \boldsymbol{M}(t,X)[\boldsymbol{I} - \boldsymbol{M}(t,X)/\boldsymbol{M}^*],$$

$$\boldsymbol{M}(t) = \int_{V,9} \boldsymbol{M}(t,X)dX,\ \boldsymbol{m}(t) = \int_{V,9} \boldsymbol{m}(t,X)dX,$$

$$\boldsymbol{R}(t) = \int_{a(t)}^{t} \boldsymbol{m}(s)ds,\ \boldsymbol{a}(t) = \boldsymbol{b}(t),\ \boldsymbol{m}(^{1}) = \boldsymbol{ym},\ X \in V,9, \tag{21}$$

where $\boldsymbol{M}^*$ is the desired limiting value of $\boldsymbol{M}$(t) and $\boldsymbol{M}^*, \boldsymbol{y}, \alpha^{\wedge}, \boldsymbol{R}$ are constants. Then we find

$$\boldsymbol{m}(t) = \boldsymbol{M}'(t) = \alpha^{\wedge} \boldsymbol{yRM}(t)[\boldsymbol{I} - \boldsymbol{M}(t)/\boldsymbol{M}^*],\ \boldsymbol{m}(\mathrm{t}^{\wedge}) = \boldsymbol{m}^{\wedge}, \tag{22}$$

and hence

$$\boldsymbol{M}(t) = \boldsymbol{M}^*/\{\mathbf{1}+\boldsymbol{C}\exp[-\alpha^{\wedge} \boldsymbol{yR}(t-t^{\wedge})]\},\ \boldsymbol{m}^{\wedge} = \boldsymbol{C}\alpha^{\wedge} \boldsymbol{yRM}^*/(\mathbf{1}+\boldsymbol{C})^2. \tag{23}$$

Assuming that $\boldsymbol{m}^{\wedge}$ is the initial rate of creation of the enzyme-substrate product with regard to influence of pH and $\Delta G^0/T$, besides (21) and (22) we have

$$\boldsymbol{m}^{\wedge} = \boldsymbol{V}'\mathrm{s}/(\boldsymbol{K}_m + \mathrm{s}),\ \boldsymbol{V}' = \boldsymbol{V}/\boldsymbol{K}_+,\ \boldsymbol{K}_m = \boldsymbol{K}_S/\boldsymbol{K}_+,\ \boldsymbol{V} = \boldsymbol{k}_2\boldsymbol{e}_0,$$

$$\boldsymbol{K}_+ = (\mathbf{1}+\mathbf{h}^+/\boldsymbol{K}_{1ES}+\boldsymbol{K}_{2ES}/\mathbf{h}^+),\ \boldsymbol{K}_S = \boldsymbol{K}^{\wedge}_S\exp[\Delta H^0_s/\mathrm{R}\delta T],$$

$$\boldsymbol{K}_{iES} = \boldsymbol{K}^{\wedge}_{iES}\exp[\Delta H^0_i/\mathrm{R}\delta T],\ i = 1, 2, \tag{24}$$

Repeating the discussion of s. 3.1., we can obtain a rather small value of $\boldsymbol{m}^{\wedge}$ under small $\mathbf{h}^+$ and T. It means that in the relation of (23) the constant $\boldsymbol{C}$ is rather large and hence the desired value $\boldsymbol{M}^*$ cannot be practically achieved. On the other hand, under possible large T and $\mathbf{h}^+_{max}$ value $m^{\wedge}$ may be rather large and the constant C is rather small. Hence, the limiting value M^* can be achieved too fast, which may be the reason for cisternia "storage diseases." Therefore, the normal regime of the cisternia can be broken in both cases.

The models (7) and (8) are devoted to the development of the daughter cell, and probably all regularities of its development are given by DNA of the mother cell.

However, both internal and external conditions influence the daughter cell development. The major factor of this development is the replication of DNA that corresponds to the equation (7).

Let us consider the particular simple case of this equation:

$$\alpha(t) = x\int_0^t \alpha(s)m(t-s)ds + \Delta\alpha,\ \alpha(s) = 0,\ s < 0,$$

$$f(t) = m(t) + \alpha(t),\ \ t > t\hat{} > 0, \tag{25}$$

from which

$$\alpha(t) = x\int_0^t \alpha(s)f(t-s)ds - x\int_0^t \alpha(s)\alpha(t-s)ds + \Delta\alpha. \tag{26}$$

On the strength of the well-known results for Laplace transform

$$L(p) = \int_0^\infty e^{-pt}l(t)dt, \tag{27}$$

we have

$$\mathbf{A} = x\mathbf{A}F - x\mathbf{A}^2 + \Delta\mathbf{A}, \tag{28}$$

from where

$$\mathbf{A} = (F-1/x)/2 \pm \{[(F-1/x)/2]^2 + \Delta\mathbf{A}\}^{1/2}. \tag{29}$$

We have, in particular, the double-solution of the equation (28) and hence (25)

$$\mathbf{A}_{1,2} = (F-1/x)/2,\ \alpha_{1,2}(t) = f(t)/2-\delta(t)/x, \tag{30}$$

where δ is Dirac delta-function (i.e., the case of the simple MM for DNA replication) under the condition

$$\Delta\mathbf{A} = [(1/x - F)/2]^2. \tag{31}$$

3.3. INVESTIGATION OF MM (9), (10)

If the external bio-fields

$$\boldsymbol{c}(t,Z) = \{c_1(t,Z), c_2(t,Z), c_3(t,Z)\}, Z \in R^3\backslash V, t \geq t\hat{} > 0, \tag{32}$$

of a **C** cell were identified, then we could find certain effects of material, energy, or information nature that inhibit **C**. Since

$$\boldsymbol{c}(t,Z) = \int_{a(t,X)}^{t} \int_V \beta(t,Z,s,Y)\boldsymbol{m}(^2)(s,Y)dYds, Z \in V\hat{}, V\hat{} = R^3\backslash V, \tag{33}$$

and the typical dynamics behavior of

$$\boldsymbol{C}(t) = \int_{t\hat{}}^{t} \int_{V\hat{}} \boldsymbol{c}(t,Z)dZ, \tag{34}$$

is probably saturation of all of its components, we may assume the gene-type function

$$\beta(t,Z,s,Y) = \beta\hat{}\boldsymbol{C}(t,Z)[\boldsymbol{I} - \boldsymbol{C}(t,Z)/\boldsymbol{C}^*],$$

$$\boldsymbol{C}(t) = \int_{V\hat{}} \boldsymbol{C}(t,Z)dZ, \;\; \boldsymbol{m}(t) = \int_V \boldsymbol{m}(t,X)dX,$$

$$\boldsymbol{R}(t) = \int_{a(t,X)}^{t} \boldsymbol{m}(s)ds, \boldsymbol{a}(t) = \boldsymbol{b}(t), \;\; \boldsymbol{m}(^2) = [\mathbf{I}\text{-}\boldsymbol{y}]\boldsymbol{m}, Z \in V\hat{}. \tag{35}$$

Then similar to (23)

$$\boldsymbol{C}(t) = \boldsymbol{C}^*/\{\mathbf{1}+\boldsymbol{D}\exp[-\beta\hat{}[\boldsymbol{I}-\boldsymbol{y}]\boldsymbol{R}(t-t\hat{})]\}, \boldsymbol{c}\hat{} = \boldsymbol{D}\beta\hat{}[\boldsymbol{I}-\boldsymbol{y}]\boldsymbol{R}\boldsymbol{C}^*/(\mathbf{1}+\boldsymbol{D})^2, \tag{36}$$

where $\boldsymbol{C}^*$ is the limiting value for $\boldsymbol{C}$ and $\beta\hat{}, \boldsymbol{y}, \boldsymbol{R}, \boldsymbol{c}\hat{}$ are given constants. Then the material-type component of $\boldsymbol{C}$ may be inhibited by the influence of pH and $\Delta G^0/T$ on $\boldsymbol{c}\hat{}$ the same way as in the previous s. The energy-type component of $\boldsymbol{C}$ may be inhibited by the same way based on the formulae of (36)-type. As to the information-type component of $\boldsymbol{C}$, in particular, information with respect to the mentioned values $\boldsymbol{C}^*$ and $\beta\hat{}$, $\boldsymbol{R}$, $\boldsymbol{c}\hat{}$, it may be actually inhibited simultaneously with the material and energy components. Of course, we only guess about the possible ways that influence a **C** cell progression, but we may repeat the arguments at the end of s. 3.1.

4. MM of a Cancer

MM of a cancer may be obtained by joining the MM of the **C** bio-field, immune system, and endothelial cell population:

$$\boldsymbol{m}(t,X) = \int^{t}_{a(t,X)} \int_V \alpha(t,X,s,Y)\boldsymbol{m}(^1)(s,Y)dYds,\ X \in V,$$

$$\boldsymbol{c}(t,Z) = \int^{t}_{b(t,Z)} \int_V \beta(t,Z,s,Y)\boldsymbol{m}(^2)(s,Y)dYds,\ Z \in V^\wedge,$$

$$\boldsymbol{R}(t,X) = \int^{t}_{a(t,X)} \int_V \boldsymbol{m}(s,Y)dYds, \boldsymbol{f} = \boldsymbol{m}(t,X) + \boldsymbol{c}(t,Z), \boldsymbol{m}(^1) + \boldsymbol{m}(^2) = \boldsymbol{m},$$

$$\boldsymbol{m}(t,X) = \{m_1(t,X), m_2(t,X), m_3(t,X)\}, X \in V,$$

$$\boldsymbol{c}(t,Z) = \{c_1(t,Z), c_2(t,Z), c_3(t,Z)\}, Z \in V^\wedge, t \geq t^\wedge > 0; \quad (37)$$

$$\partial \boldsymbol{c}^\wedge/\partial t + \boldsymbol{J}\nabla = \boldsymbol{F}(\boldsymbol{c}^\wedge,\boldsymbol{c},t), \boldsymbol{c}^\wedge(t,Z) = \{c^\wedge_1(t,Z), c^\wedge_2(t,Z), c^\wedge_3(t,Z)\}, \boldsymbol{J} = \{\boldsymbol{J}_1, \boldsymbol{J}_2, \boldsymbol{J}_3\},$$

$$\nabla = \partial/\partial z_1\mathbf{i} + \partial/\partial z_2\mathbf{j} + \partial/\partial z_3\mathbf{k}, \boldsymbol{F} = \{F_1, F_2, F_3\}; \quad (38)$$

$$\partial \boldsymbol{c}^\wedge/\partial t = \boldsymbol{F}(\boldsymbol{c}^\wedge,\boldsymbol{c}) + \partial/\partial \boldsymbol{r}[\boldsymbol{D}(\boldsymbol{c}^\wedge,\boldsymbol{r})\partial \boldsymbol{c}^\wedge/\partial \boldsymbol{r}],$$

$$\boldsymbol{r} = (x^\wedge_1 - z_1)\mathbf{i} + (x^\wedge_2 - z_2)\mathbf{j} + (x^\wedge_3 - z_3)\mathbf{k}, X^\wedge \in \partial V; \quad (38')$$

$$\partial \boldsymbol{c}^{\wedge\prime}/\partial t = \boldsymbol{A}\, c^{\wedge\prime} + \boldsymbol{D}\partial^2 \boldsymbol{c}^{\wedge\prime}/\partial \boldsymbol{r}^2], \boldsymbol{A} = \{\partial Fi/\partial c^{\wedge\prime}_j)_0\}, \boldsymbol{c}^{\wedge\prime} = \boldsymbol{c}^\wedge - \boldsymbol{c}^\wedge_0; \quad (38'')$$

$$\nabla^2\psi(X) + [E - W(X)]\psi(X) = 0; \quad (39)$$

$$\{\mathbf{i}\hbar\partial/\partial \mathrm{t} + E - W + J\partial^2/\partial\xi^2 + G|\psi(\xi, t)|^2\}\psi(\xi, t) = 0,$$

$$W = D + 2J,\ J = d^2R^{-3}(3\cos^2\theta - 1); \quad (40)$$

$$M_i^P(t) = \int^{t}_{0} M_i^P(t,r)dr + M_i^P(0),$$

$$M_i^P(t,r) = M_i^P(r,r) f_i(t,r)\exp\{\int^{t}_{r} [0.08 - h(s) - c_i^P(s)]ds\}[1 - F_8(t - r - 12)],$$

$M_i^p(r,r) = m_i^s(r-12)f_i(r,r-12)\exp(-\int_{r-12}^{r} h(s)ds)$, $M_i^s(0) = M_{1,0}F_2[g(0))]$,

$m_i^s = -hM_i^s + dM_i^s - m_i^s(t-12)f_i(t,t-12)\exp[-\int_{t-12}^{t} h(s)ds]$,

$dM_i^s = c^*(t)F_2[g(t)] + M_i^s(t)(F_2[g(t)])'/F_2[g(t)] - (k_i, f(t))$, $m_i^m(t) = -(h+0.01+c_i^m)M_i^m +$

$M_i^p(t\text{-}6)F'_8(t-18)F_2(g(t-6))f_i(t,t-6)\exp[-\int_{t-6}^{t} h(s)ds]$, $m_i^r(t) = -u\text{^}M_i^r + N_i^p F'_8(t-12)[1-$

$F_2(g)]$, $i = 1, 2, 3$; $M_{4,1}^p = M_{4,1}^p(18)f(t,18)\exp(\int_{18}^{t} [0.08 - h(s)]ds)[1-F_8(t-18)]$,

$m_{4,1}^m(t) = -0.01M_{4,1}^m + M_{4,1}^p(t-6)F'_8(t-24)f(t,18)F_3(g(t-6))\exp[-\int_{t-6}^{t} h(s)ds]$,

$m_{4,1}^r(t) = -u\text{^}M_{4,1}^r + M_{4,1}^p F'_8(t-18)[1-F_3(g)]$,

$M_{4,2}^p(\mathrm{t}) = \int_0^{t-12} \mathrm{M}_{4,2}^p(t,r)dr, M_{4,2}^p(r+12,r) = M_{4,2}^{s'}(r)f(r+12,r)\exp[-\int_r^{r+12} h(s)ds]$,

$M_{4,2}^p(t,r) = M_{4,2}^p(r+12,r)f(t,r+12)\exp(\int_{r+12}^{t} [0.08-h(s)]ds)[1-F_8(t-u-12)]$,

$m_{4,2}^m(t,r) = -0.01M_{4,2}^m(t,r)\, dM_{4,2}^{s,s'}(t-6)f(t,t\text{-}6)\exp[-\int_{t-6}^{t} u(s)ds]$, $m_{4,2}^m(t) = \int_0^t m_{4,2}^m(t,r)dr$,

$M_5(\mathrm{t}) = 10^7 M_4^p + 10^8 M_4^m$, $M_4^j = M_{4,1}^j + M_{4,2}^j$, $j = p, m$, $t > 0$; (41)

$m^*(t) = \int_{a^*(t)}^{t} \alpha^*(t,s)y^*(s)m^*(s)ds + \Delta m^*(t)$, $M^*(t) = \int_{a^*(t\text{^})}^{t} m^*(s)ds$,

$c^*(t) = \int_{a^*(t)}^{t} \beta^*(t,s)z^*(s)m^*(s)ds$, $C^*(t) = \int_{t\text{^}}^{t} c^*(s)ds, d^*(t) = \int_{a^*(t)}^{t} \gamma^*(t,s)[1-y^*(s)-$

$z^*(s)]m^*(s)ds$, $D^*(t) = \int_{t\text{^}}^{t} d^*(s)ds, R^*(t) = \int_{a^*(t)}^{t} m^*(s)ds$, $G^*(t) = \int_{a^*(t\text{^})}^{a^*(t)} m^*(s)ds$,

$F^*(t) = \int_{t\text{^}}^{t} f^*(s)ds, F^*(t) = M^*(t) + C^*(t)$, $0 \le y^*, z^*, 1\text{-}y^*\text{-}z^* \le 1$, $0 < t\text{^} \le t$; (42)

$m(t,Z) = \int_{a(t)}^{t} \int_{V\text{^}} \alpha(t,s,e(Z\cup Y))y(s,Y)m(s,Y)dYds$, $M(t) = \int_0^t \int_{V\text{^}} m(s,Y)dYds$,

$c(t,Z) = \int_{a(t)}^{t} \int_{V\text{^}} \beta(t,s.e(Z\cup Y))[1\text{-}y(s,Y)]m(s,Y)dYds$, $C(t) = \int_{t\text{^}}^{t} \int_{V\text{^}} c(sY)dYds$,

$$R(t) = \int_{a(t)}^{t} \int_{V\wedge} m(s,Y)dYds,\ G(t) = \int_{a(t\wedge)}^{a(t)} \int_{V\wedge} m(s,Y)dYds,$$

$$f(t,Z) = m(t,Z) + c(t,Z),\ 0 \le y \le 1,\ Z \in V\wedge,\ 0 < t\wedge \le t; \quad (43)$$

where V is the domain in R^3 for a **C**, ∂V is the boundary of V, and $V\wedge$ is the external environment of a **C**, c_i, i = 1, 2, 3, are the rate of creation along t,Z respectively the (generalized) new material, energy, information product of t,Z-type in the point Z at the instant t, m_i, i = 1, 2, 3, are the rates of creation along t, X the (generalized) new material, energy, and information internal product of t,X-type in the point X at the instant t, $\boldsymbol{\alpha}$ and $\boldsymbol{\beta}$ are matrices whose components are the indices of efficiencies of a **C** functioning along the corresponding channels;

the rate of change $\partial c\wedge/\partial t$ of the amount of the resource in $V\wedge$ is equal to the rate of flow of the (generalized) resource into $V\wedge$ plus the rate of the resource created in $V\wedge$, the vector $\boldsymbol{F}$ is the source of the resource, the matrix $\boldsymbol{J}$ is the flux of the recourse ($\boldsymbol{J}$ can be at the expense of not only diffusion, but also convection, galvanotaxis, chemotaxis, etc.), $\boldsymbol{D}$ is the matrix of the diffusion coefficients, $\boldsymbol{c}\wedge_0$ is the stationary state value of the initial system;

$|\psi(Z)|^2$ is a measure of the probability that the particle of energy E > 0 will be found at the location Z, given that the particular potential $W(X)$ is present;

$\hbar = 1.054 \cdot 10^{-27}$ erg, E, W, and $|\psi(\xi, t)|^2$ have the same definition as above, D is the strain potential of the molecular chain with distance R between adjacent molecules, interactions of which are determined by the potential $2J$ due to the electric dipole moment $\boldsymbol{d}$ directed under the angle θ with respect to the chain line, G is the parameter of non-linearity determining the part Δz of the chain that is energizing circuit, $\Delta z = \pi RJ/4G$;

M_i^j are the immune network cell of the first level numbers respectively Th (i = 1), Ts (i = 2), Tef (i = 3), B_1 (i = 4,1), B_2 (i = 4,2), proliferating cells ($j = p$), mature cells ($j = m$), resting cells ($j = r$), stimulated by Ag ($j = s$), stimulated by Ag and Th ($j = s'$), twice stimulated by Ag and Th ($j = s,s'$), M_5 is the number of Ig, (., .) means scalar production of the vectors ., ., $\boldsymbol{k}$ is the vector, subjected to identification, g is the logarithm of the number of cancer Ag molecules;

m^* is the number of the new thymus and spleen cells in the unit of time remaining into these organs, Δm^* is the cell's inflow from outside, c^* is the number of the new thymus and spleen immune system cells in the unit of time leaving these organs, d^* is the number of the new destroyed cells in the unit of time, α^* is the efficiency index for the cortex cell proliferation, β^* is the efficiency index for the cerebral cell differentiation, γ^* is the efficiency index for the cortex cell selection;

the indices of efficiencies of endothelial cell population functioning $\alpha = u\alpha\wedge$, $\beta = v\beta\wedge$, where, in turn, u and v depend on deviation of m and c from their average stationary values, depending on $\boldsymbol{c}\wedge$; [0, $t\wedge$] is the prehistory for which all the functions are considered as given;

the definition of all the other values are similar to the ones mentioned before.

It should be noted that many notation symbols are changed here in comparison with Ch. 5 and 7. Otherwise, we would have confusing symbols. It is natural to call V with (37) as a cancer internal bio-field and $V\wedge$ with (17)-(40) as a cancer external bio-field.

5. Base MM of Interaction of Cancer with Host Organism

On the strength of the theory above the minimal, *base MM of interaction of cancer and host organism* may have the form

$$m(t) = \int_{a(t)}^{t} \alpha(t,s)y(s)m(s)ds,\ M(t) = \int_{0}^{t} m(s)ds,\ c(t) = \int_{a(t)}^{t} \beta(t,s)[1-y(s)]m(s)ds,$$

$$R(t) = \int_{a(t)}^{t} m(s)ds,\ G(t) = M(t) - R(t),\ x(t)f(t) = m(t) + c(t),$$

$$n(t) = \int_{b(t)}^{t} \gamma(t,s)z(s)n(s)ds,\ N(t) = \int_{0}^{t} n(s)ds,\ d(t) = \int_{b(t)}^{t} \delta(t,s)[1-z(s)]n(s)ds,$$

$$Q(t) = \int_{b(t)}^{t} n(s)ds,\ H(t) = N(t) - Q(t),\ [1-x(t)]f(t) = n(t) + d(t),$$

$$0 \le x, y, z \le 1,\ 0 \le a(t), b(t) \le t,\ t \ge t^\wedge \ge 0, \qquad (44)$$

where m is a new resource of a "remained part" of an organism (RPO) providing its internal functions (of recreation of m); c is some new resource providing external functions of RPO; y is a distribution function for RPO; α and β are efficiency indices of functioning of RPO along the channel m--ym and the channel c--$(1-y)m$ respectively; $a(t)$ is a temporal bound: on the $[0, a(t)]$, resource m is considered as obsolete; R is the total amount of functioning resource of RPO; G is the total amount of the obsolete resource; f is the external resource; n, d, z, γ, δ, b, Q, and H are similar elements, but for a **C**; x is a relative share of f between RPO and **C**; $t^\wedge$ is the starting time for modeling; $[0, t^\wedge]$ is the prehistory on which all the elements of the model are considered as given.

Thus, the total number of different elements in (44) is 20, 4 of which, namely x, y, z, and f, can be considered as the control elements of the model. The total number of the equations in (44) is 12. If the exogenous character functions α, β, γ, and δ are given, we will have the determined system of the equations.

The *problem of C elimination* can be stated with the help of the model (44) as follows:

$$R'(t) = 0 \wedge Q'(t) < 0, \qquad (45)$$

that is, in RPO functioning mass is being constant, but in cancerous parts of the organism functioning mass is decreasing.

6. Qualitative Investigation of MM of a Cancer

6.1. INVESTIGATION OF MM (37)-(43)

This investigation is similar to s. 3.1 with the essential difference that instead of Th the sub-populations of IN Ts and Tef play the major role now.

First of all, let us consider a different particular situation and the respective MM. If **C** is so advanced that

$$(\boldsymbol{k}_3, f(\mathrm{t})) > c^*(t)F_2[g(t)] + M_3^s(t)(F_2[g(t)])'/F_2[g(t)], \tag{46}$$

then $dM_3^{\ s} < 0$ and hence $m_3^{\ s} < 0$ and not long after $M_3^{\ s}$ and M_3^P and hence M_3 may be about 0. Most damage to $M_3^{\ s}$ may give the maximum possible $\boldsymbol{F}(t)$.

Therefore, let us consider the problem

$$\boldsymbol{F}(T) = \int_0^T \int [\boldsymbol{m}(s,X)+\boldsymbol{c}(s,Y)]dXdY\,(X, Y \in R^3)ds = \max\,(\boldsymbol{y}, u, v), \boldsymbol{ym} = \boldsymbol{m}(^1), \tag{47}$$

under the natural conditions and

$$0 < u < u^*, 0 < v < v^*, \tag{48}$$

where u^*, v^* are certain restrictions on the control factors u, v.

Since $\boldsymbol{c}(s)$ contributes mostly to $f(s)$ (see Ch. 15), we have

$$\boldsymbol{F}(T) \approx \boldsymbol{C}(T), \tag{49}$$

and the matter reduces to the well-investigated problem of maximization of quantity of the external product of CAO (see Ch. 15). On the strength of the respective results for sufficient large T, assuming as well as non-decreasing function β-type, we have the approximate solution of the problem (46)-(48) under

$$\boldsymbol{y}(t) = \boldsymbol{1}, u = u^*, 0 < t < T - dT, 0 < dT \ll T; v = v^*, 0 < t < T. \tag{50}$$

On the other hand, for minimization of $\boldsymbol{F}(T)$, we have to maintain $\boldsymbol{y}$ and/or u and v near 0 as long as possible. Certain results about how to do that can be found in the next s. 6.2.

It is easy to see that the case $m_3^{\ s} < 0$ is also possible under the condition

$$dM_3^{\ s} < -hM_3^s + M_3^{\ s}(t-12)f(t,t-12)\exp\left[-\int_{t-12}^{t} h(s)ds\right],$$

$$dM_s^{\ s} = c^*(t)F_2[g(t)] + M_3^s(t)(F_2[g(t)])'/F_2[g(t)] - (\boldsymbol{k}_3, f(\mathrm{t})), \tag{51}$$

from where we have the following sufficient condition

$$(\boldsymbol{k}_3, \boldsymbol{f}(\mathrm{t})) > c^*(t)F_2[g(t)] + M_3^s(t)\{h + [(F_2[g(t)])'/F_2[g(t)] -$$

$$f(t,t-12)\exp\left[-\int_{t-12}^{t} h(s)ds\right]\}, \tag{52}$$

In particular, if

$$h + (F_2[g(t)])'/F_2[g(t)] < f(t,t-12)\exp\left[-\int_{t-12}^{t} h(s)ds\right], \tag{53}$$

a reason of which may be a small value of h and $(F_2[g(t)])'$ or relatively large value of $F_2[g(t)]f(t,t\text{-}12)$ (the last is possible when we have a large, but not too large number of **C** Ag molecules and a small number of Ig of the second level at the same time), then we have the following sufficient condition

$$(\boldsymbol{k}_3, \boldsymbol{f}(\mathrm{t})) > c^*(t)F_2[g(t)]. \tag{54}$$

This condition obviously means that, in particular, the number of Tef cells will be decreased if the thymus and spleen supply is exhausted. Not to be in this situation, we have to maintain a sufficiently fair condition for the thymus and spleen functioning according to the principle of adaptable redistribution and regenerative accumulation of the thymus cells (see Ch. 9).

Decreasing the number of $M_3{}^P(t)$ proliferating Tef cells and therefore the number of overall Tef cells may occur due to the condition

$$0.08 - h(s) - c_3{}^P(s) < 0. \tag{55}$$

Having in mind the structure of the functions h and $c_3{}^P$ (see Ch. 8), it is easy to conclude that, for example, h may be greater than 0.08 when the number of Ts cells is about $10^{4.5}$. Due to the "cannibalism" of Tef cells (see the assumption 17) in Ch. 8), we usually have substantially less Tef cells than Ts cells that may suppress Tef cells of the same level. One more reason for deficiency of Tef cells may be such a large number of molecules of **C** Ag ($g(t) \approx 16$, $F_5[g(t)] \approx 1$) that we have the case when not only all Tef cells, but also all Ts and Th cells will be inhibited by Ag. Under a rather small number of Ag ($g(t) \approx 11.5$, $F_1[g(t)] \approx 1$), we also have the case of complete suppression of Th cells (see Ch. 8) and hence a rather small number of Ig against **C**.

In general, we can find the pre-assigned behavior of IN with the help of its MM (see Ch. 8) and realize this behavior with the help of the hybridome and gene technology (J. Rifkin [160]). However, for this we are in a need of the respective means, including the software and computers.

The solutions of (38) and (38') are certain ones, according to the general results of mathematical physics. In particular, there are the solutions for the propagation of waves, heat, and potential outside of **C.** The solutions of (38") are linear combinations of the waves

$$\boldsymbol{c}\hat{}' = \boldsymbol{a}\exp(p_k t + \mathbf{i}\boldsymbol{k}\boldsymbol{r}), \tag{56}$$

where k is the wave number, determining the wave length $\lambda_k = 2\pi/k$, and $p_k = \delta_k + \mathbf{i}\omega_k$ are the solution of the equation of variance

$$\det\{a_{ij} - D_i k^2 - p_k\} = 0. \tag{57}$$

If the point $\boldsymbol{c}\hat{}_0$ is unstable, then there is at least one k for which $\delta_k > 0$. If the equation (57) for the wave length λ_k has an even number of roots with the $\delta_k > 0$, then unsteadiness is vibrating. In the case of odd roots we have *unsteadiness by Turing*, reducing to the stationary *dissipative* or damping *structures* (DS). We are interested in the stable condition for $\boldsymbol{c}\hat{}_0$, but in the case of open dynamic systems it is usually impossible. In that case, it would probably not be as bad if we could find the condition for the stationary DS. There are plenty results about values of $\boldsymbol{A}$ and $\boldsymbol{D}$, sufficient for DS (see Ch. 1).

In the particular case (39), $X = x$, $\nabla^2 = d^2/dx^2$, and $W(X) = W > E > 0$, $0 < x < x\hat{}$, $W(x) = 0$, otherwise, where $0 \in V$ and $x\hat{} \equiv \partial V$, denoting $k = (E)^{1/2}$ and $K = (W\text{-}E)^{1/2}$, the solution has the following form

$$\psi(x) = Ce^{-Kx} + De^{Kx}\ (0 < x < x\hat{}),\ BS(E)e^{\mathbf{i}k(x-x\hat{})}\ (x > x\hat{}), \tag{58}$$

where the constants B, C, and D can be found from the condition of continuity at $x = 0$ and $x = x\hat{}$. Besides, the transmission amplitude

$$S(E) = 2\mathbf{i}kK/\{2\mathbf{i}kK\text{ch}\ (Kx\hat{}) + (k^2 - K^2)\text{sh}^2\ (Kx\hat{})\}. \tag{59}$$

The quantity $AS(\mathrm{E}) \neq 0$, meaning that the wave or quantum particle has a non-zero probability of being found to the right of the $x\hat{}$ (i.e., outside of $\mathbf{C}$). This phenomenon called *tunneling* may be used by $\mathbf{C}$ for transmitting signal to the endothelial cells. A related quantity $|S(E)|^2$ can be regarded as the energy transmission factor:

$$|S(E)|^2 = 4E(W-E)/[4e(W-E) + W^2\text{sh}^2\,(Kx\hat{})]. \tag{60}$$

In the case of (40), the exact solution with the condition $\int_{-\infty}^{\infty} |\psi(\xi, t)|d\xi = 1$, has the following form

$$\psi(\xi, \mathrm{t}) = (\mu/2)^{1/2}\exp\{\mathbf{i}[\hbar v/(2J)(\xi-\xi_0) - Et/\hbar]\}/\text{ch}[\mu(\xi-\xi_0-vt)],\ \mu = G/(4J), \tag{61}$$

where vR is the rate of perturbation propagation along the chain, $vR < v^*$, v^* is the rate of sound propagation along the chain. In addition,

$$|\psi(\xi, t)|^2 = \mu/\{2\text{ch}^2[\mu[\xi-\xi_0-vt)]\} \tag{62}$$

is the probability of perturbation along the chain. The value ξ_0 is the argument of the maximal probability and deformation at the instant $t = 0$.

The perturbations described by the functions (61) and (62) (*solitons*) are capable of carrying over perturbation energy inside and outside of **C** very effectively.

One of the simple MM of the functions α, β, and control factors u, v, y for the endothelial cell population is u, v, y, which are constants, depending on the mean value of c^, and

$$\alpha(t,X,s,Y) = \alpha\text{^}u,\ \beta(t,Z,s,Y) = \beta\text{^}vC(t,Z)[1 - C(t,Z)/C^*(Z)], \tag{63}$$

where $C^*(Z)$ is the limiting value of $C(t,Z)$ and C^*, α^, β^, and R are also independent of t. Then, we find

$$C(t,Z) = C^*(Z)/\{1+C\exp[-\beta\text{^}v(1-y)R(t-t\text{^})]\},\ c\text{^}(Z) = C\,\beta\text{^}v(1-y)RC^*/(1+C)^2. \tag{64}$$

Assuming that c^ is the initial rate of creation of the enzyme-substrate product with regard to the influence of pH and $\Delta G^0/T$ we have

$$c\text{^} = V'\text{s}/(K_m + \text{s}),\ V' = V/K_+,\ K_m = K_S/K_+,\ V = k_2e_0,$$

$$K_+ = (1+\text{h}^+/K_{1\text{ES}}+K_{2\text{ES}}/\text{h}^+),\ K_S = \exp[\Delta G^0_s/(\text{R}T)],\ K\text{^}_{i\text{ES}} = \exp[\Delta G^0_i/(\text{R}T)],\ i = 1, 2. \tag{65}$$

Repeating the discussion of s. 3.1, we can obtain a rather small value of c^ under small v and/or h^+ and T. It means that in the relation of (64) the constant C is rather large and hence the desired value C^* cannot be practically achieved. On the other hand, under possible large v and/or T and h^+_{max} value c^ may be rather large and the constant C is rather small and hence the limiting value C^* can be achieved too fast. Thus, the normal regime of the endothelial cell population functioning can be broken into all of these cases. It should be noted that c^ may be considered as a gas component (for example NO, nitric oxide). Therefore, influence of pressure is also possible (see R. Caret et al. [32]).

In conclusion, we dwell briefly on comparison of our approach with the well-known approach, based on the kinetic cellular theory (see Ch. 3). The principal differences are in that our MM of cancer and certain environmental cells (in our case the endothelial cell population and the thymus and spleen cells) have different structure and sense of their components, the main characteristics of the model under consideration in this book. Among those are the self-organized structures of at least two-compartment models with the obsolete product and prehistory and also the gene and enzyme-substrate nature of the main components of the models. The respective MM of IN have more resembling components. However, they also have certain essential differences noted in Ch. 8.

6.2. INVESTIGATION OF MM (44), (45)

On the strength of (44) and (45), we have

$$R'(t) = m(t) - m(a)a'(t) = 0,\ G'(t) = m(a)a'(t),$$

$$Q'(t) = n(t) - n(b)b'(t) < 0,\ H'(t) = n(b)b'(t), \tag{66}$$

from which

$$m(t) = G'(t),\ n(t) < H'(t). \tag{67}$$

We have, in particular, if $c = d = 0$, then

$$G'(t) < x(t)/[1-x(t)]H'(t), \tag{68}$$

and if $x = 1/2$, then

$$G'(t) < H'(t). \tag{69}$$

The relations (66)-(69) obviously have the determined physiological interpretation and can serve as experimental testing of **C** elimination. Note that we always know $G' + H'$ as the total increment of a whole organism.

Let us consider the *problem of C elimination* in more detail. We consider the case when α, β, γ depend on s (or t-s) only. This assumption may be correct regarding a short period of the time for development of **C**. On the strength of (44), it is not difficult to find

$$R'(t) = m(t) - [(\alpha ym)(t) - m'(t)]/(\alpha y)(a) =$$

$$m(t) - [(\beta(1-y)m)(t) - c'(t)]/(\beta(1-y))(a) \tag{70}$$

and

$$Q'(t) = n(t) - [(\gamma zn)(t) - n'(t)]/(\gamma z)(b) =$$

$$n(t) - [(\delta(1-z)n)(t) - d'(t)]/(\delta(1-z))(b). \tag{71}$$

On the strength of (44), from (70) and (71) it follows

$$m(t) = [(\alpha y m)(t) - m'(t)]/(\alpha y)(a) = [(\beta(1-y)m)(t) - c'(t)]/(\beta(1-y))(a) \quad (72)$$

and

$$n(t) < [(\gamma z n)(t) - n'(t)]/(\gamma z)(b). \quad (73)$$

If $m' = c' = 0$, then from (72) it follows

$$y(t)/y(a) = \alpha(a)/\alpha(t),\ [1-y(t)]/[1-y(a)] = \beta(a)/\beta(t). \quad (74)$$

Let, for example,

$$\alpha(a) > \alpha(t) \wedge \beta(a) < \beta(t), \quad (75)$$

which seems to be natural. Then from (74) it follows

$$y(t) > y(a), \quad (76)$$

that is, a greater share of RPO resource should go to the realization of its internal functions.

If $n' = d' = 0$, then from (73) and the similar relation for δ and $1\text{-}z$ it follows

$$\gamma(t)z(t) > \gamma(b)z(b),\ \delta(t)(1-z(t)) > \delta(b)(1-z(b)). \quad (77)$$

If, for example,

$$\gamma(b) < \gamma(t) \wedge \delta(b) = \delta(t), \quad (78)$$

then

$$\gamma(b)z(b)/\gamma(t) < z(t) < z(b), \quad (79)$$

that is, a certain smaller share of **C** resource should go to fulfilling its internal functions. Similar estimates can also be done in the case of $t - s$ dependence.

Now let us consider the comparative conducts of xf and $(1-x)f$ depending on values $\alpha y + \beta(1-y)$ and $\gamma z + \delta(1-z)$. It is easy to see that

$$xf = \int_a^t [\alpha y + \beta(1-y)]mds,\ (1-x)f = \int_b^t [\gamma z + \delta(1-z)]nds. \tag{80}$$

So, if f, m, and n are positive constants and

$$\gamma z + \delta(1-z) \gg \alpha y + \beta(1-y),\ a = b, \tag{81}$$

then

$$xf \prec (1-x)f, \tag{82}$$

starting from a certain instant of time. It means that **C** suppresses RPO. We can prevent this suppression, in principle, with the help of control factors z. Indeed, under sufficiently small z

$$n(t) = \int_b^t \gamma z n(s)ds \tag{83}$$

becomes sufficiently small that in turn, due to (80), will result in sufficiently small $(1\text{-}x)f$. Unfortunately, we do not know, and will probably not learn (V. Glushkov [59]), **C**'s peculiarities in order to select in practice such a small z as above as well as x, and/or f such that $(1\text{-}x)f \ll xf$.

We dwell still on the control factor z. From $Q' < 0$ it follows that

$$n(t) = \int_b^t \gamma z n ds < n(b)b' = b'\int_{b(b)}^{b(t)} \gamma z n ds =$$

$$b'\int_b^t \gamma(b(t),b(s))z(b(s))n(b(s))b'(s)ds. \tag{84}$$

The condition (84) means that new $z(s)$ should be essentially less than previous ones on the average. As it will be shown in Ch. 15, a response of any CAO on an external significant increasing impulse (stress) (d) is just an essential decreasing impulse (z).

Thus, one of the desired therapies is to increase intensive impulse effects (stresses) localized on **C**. The frequency ratio of the effects is proportional to the values γ and δ, and can be determined more exactly on the basis of MM (44) if γ and δ are given. An example of such therapy can be an exposure of **C**, but not the way which kills the **C** cells and adjacent cells of RPO; it should probably spare the cells' life exposure, but be sufficiently strong to be a stress.

Of course, there are many other approaches to the **C** modeling and therapy. A survey of the traditional approaches and certain new results can be found in Ch. 1, 3 and in the next s.

7. On Immune Therapy of Cancers

This consideration may be similar to that in Ch. 10, s. 5.2. Probably on the whole, the **C** immune-therapy should consist in overcoming the tolerance induced by **C** to various **C** Ag.

First, we should know ISO of the patient with **C** including the main indices of the organs of IS, especially the thymus and spleen. The most important factor of their restoration is probably the restoration of the organs of IS. It will be shown not once, especially in Ch. 15, that for restoration of any CAO of the patient the conditions for consuming of essential parts of all resources to the internal needs of the CAO should be created. In our case it means that in spite of necessity of giving the source of new Tef cells, the conditions should be created for decreasing of this source to restore the "health" of the thymus and spleen.

A review of the strategies for combating **C** can be found in C. Haskell [77]. This includes gene-type and immune-type drug therapies. Here, we dwell only on the problem of possible applications of the models to understanding of certain ways of **C** immune therapy. One of them might be to synthesize a large amount of the **C** Ag. This might be done by placing the human gene of **C** Ag into bacteria and stimulating the bacteria to synthesize large amounts of a soluble **C** Ag that binds to the receptor of Tef cells and to Ig of the respective B cells. Placed into the blood of **C** patients, this **C** Ag might intensify stimulation of IN. As a result, the **C** progression might be reduced. Using MM of IN (Ch. 8) we can determine more precisely the doses of required **C** Ag. For this we are in a need of the respective software and computers. The important problem is connected to the modeling of strong variability of **C** Ag. For this we are also in a need of the respective software and computers.

Another approach is to synthesize a large amount of the Tef protein receptors. This might also be done by placing the human Tef gene into bacteria and stimulating the bacteria to synthesize large amounts of a soluble, non-membrane form of the Tef protein that binds to the **C** Ag. Placed into the blood of **C** patients, this Tef protein blocks or competes with the Ts cell attachment to Tef cells. As a result, the suppression of Tef cells by Ts cells might be reduced. The better approach is the creation of toxins for **C** and Ts that can carry Tef protein receptors.

Recently, an important new approach (see T. Wang et al. [202] and H. Yu, R. Jove [214]) to fight **C** was found, consisting partly of regulation of immune responses by Stat-3 signaling in tumor cells. "Interrupting Stat-3 signaling with selective inhibitors would not only have direct antitumor effects through growth suppression and apoptosis induction, but could also activate innate and adaptive antitumor immunity." [202], p. 53.

Most likely, the simultaneous application of the approaches is even better. In general, many independent approaches may be very useful because the chance of **C** resistant to many immune effects may be far smaller than to either of them.

It seems that one of the most attractive approaches is combating with the excess of the endothelial cells (the endothelial cell population) servicing **C**. For this, one of the possible ways is such "infection" of the population that influences a strong immune response.

It should be noted that for the realization of these ideas we should, in addition, consider the guess 1) - 5), Ch. 8, about complicated nature of **C** Ag and/or the endothelial cells Ag, so that the problem might be much more complicated. We should also consider **C** and/or the endothelial cells Ag as corpuscular, similar to the cells (see Ch. 8, s. 2.2).

8. On Gene Therapy of Cancers

This s. is based mainly on the paper by A. Zullo et al. [217] (also see C. Haskell [77]).

Indirect treatment strategies consist of gene immune therapy and tumor vaccination. There are three principal approaches to immune therapy using genetic manipulation: 1) *ex vivo* modification of autologous infiltrating T-lymphocytes, 2) *ex vivo* modification of autologous tumor cells, and 3) *in situ* transient expression of immuno-stimulatory genes. As a result of poor transduction efficiency and low cytokine expression, the first approach has proven difficult to validate.

Tumor cells *in vivo* can be removed and theoretically modified with a gene for one or any combination of molecules of such types as interleukin, IL-2, IL-4, IL-6, granulocyte-macrofage colony-stimulating factor, GM-CSF, and immune costimulatory factor, B7.1. The altered cells then can be expanded in culture and returned to the same patient to elicit a specific and robust immune response modulated by the expressed gene product (s). A related approach, making use of the immunomodulatory effects of cytokines, is the construction of antibody vaccines for B-cell activation. Tumor-specific idiotype DNA sequences can be fused to sequences encoding cytokines, which are then introduced as proteins directly back into the patient or transfected into autologous tumor cells. These cells would then function as a vaccine against the type of cancer specific to the idiotype sequence.

Animal studies in this direction have given certain positive results. The transient expression of the foreign histocompatibility antigen HLA-B7 on the cell surface of the tumor should induce an immune reaction against the altered tumor cells. Several patients have been treated using this approach.

Gene therapy has also been proposed as a means of protecting hematopoietic cells from the toxic effects of chemotherapy. This strategy makes use of the genes encoding the multi-drug resistance (MDR) proteins, which pump certain chemotherapy agents, leaving cells resistant to treatment with those agents. The manipulated stem cells would then be reinfused into the patient so as to repopulate the hematopoietic system. The gene-modified cells should then be more tolerant to the toxic agents that are used to kill the cancer cells. Animal models and human studies have been investigated to test the efficacy of this approach.

Direct cancer gene therapy capitalizes on the current understanding of crucial pathways and vital genes within the course of carcinogenesis. One strategy involves the direct genetic modification of tumor cells with the aim of correcting an activated proto-oncogene or replacing a missing or mutant tumor suppressor gene. Antisense oligonucleotides and ribozymes are being assessed as a means of correcting activated proto-oncogenes.

Introduction of genes, such as *p53* and *Rb*, are being used to correct mutations in tumor suppressor genes.

Another strategy is the direct alteration of tumor cells so as to provoke cell death. In this strategy, sensitivity to a non-toxic pro-drug, is conferred to the tumor by transfer of a so-called "suicide gene." Administration then of a pro-drug to a modified cell results in the generation of a cytotoxic anabolite and cell death. Gene therapy for malignant brain tumors using this approach has been tested in a clinical trial.

"Unfortunately, no gene transfer system exists which allows for specific targeting for metastatic cancer or which can guarantee gene transfer to all cells of a localized tumor." [217], p. 140. So, the best-combined application of different means and strategies is required to treat this devastating disease. A more complete understanding of the pathogenesis of **C**, in particular, with the help of MM, is also required.

Let us dwell here on the possible role of our MM in the process of gene therapy. MM of IN above, and in more details in Ch. 8, can help to determine better tactics of the strategies described in [217]. For example, of course, "presentation of tumor-associated antigens should lead to the attack of other neoplastic cells throughout the body" [217], p. 134. But to answer the question, 'how much of those Ag should be presented and what should be the dynamics of this presentation with regard to the given ISO in order that the attack would be successful?' we are in a need of application of MM of IN and the respective software and computers. The same kind of question should be answered in the case of the direct injection of an allogene encoding the histocompatability Ag HLA-B7.

As to the strategy of using genetically modified tumor-infiltrating T-lymphocytes (TILs), the significant adverse side-effects of the dose TILs required to evoke a suitable immune response is also likely due to not quite right tactics of its application.

We would like to emphasize here that the immune therapy as an indirect gene therapy strategy "is of particular interest because it could potentially eliminate every malignant cell including metastases within the body and respond to neoplastic recurrences." [217], p. 134.

In the opinion of the author of [217], the critical limitation to the direct gene therapy is the lack of sufficiently effective transfer system in order to deliver, for example, an expressing tumor suppressor gene or anti-oncogene to almost every tumor cell. So, more efficient transport system is required. Modeling of the retrovirus- or adenovirus-to-cell interaction based on MM of this book may help understand not only how to improve the transport system, but also how to improve the results of their interaction. Since retroviruses integrate randomly within the host genome, various undesirable events are possible. Thus, introducing a certain mechanism of regulation into this process may be useful.

Another direct approach, using suicide genes, is of particular interest because only a small percentage of transduced cells are required for complete regression of the tumor mass.

This is due to the so-called *bystander effect* [217], the phenomenon by which gene-modified and unmodified neighboring cells are killed almost at the same time. However, a tumor may be bystander sensitive or resistant, and probably some neighboring normal cells may also be affected. Additional means for protecting normal cells may be required.

In general, the best combination of traditional and experimental strategies is required. This systematic approach can hardly be done successfully without mathematical modeling based, in particular, on MM under consideration.

9. Conclusion

In conclusion, we would like to emphasize that this Ch. is mainly devoted not to the means, but the methods of **C** treatment based on qualitative investigation of MM.

At present, there are many new means of **C** treatment, including new angiogenesis inhibitors, see, for example, H. Yu, J love [214], but the question about the best condition for the intake of the known agents is not trivial with regard to the peculiarities of the given patient and complexity of **C** and IN. It can be solved by mathematical simulation faster and more qualitatively than without, on the basis of application of the necessary MM and the appropriate software and computer.

Note that, for example, the article by T. Byers at al. [29] devoted to the analysis of **C** rates decline in the U. S. by the year 2015. The conclusion is 25% of the rate decline (i.e., 60,000 less deaths from **C**).

Chapter 12

ON LIFE LONGEVITY PROBLEMS

1. Clock of Cellular Aging

There are two main *theories* of *cellular aging* (L. Hayflick [78], J. Medina [133]): *Error Accumulation*, explaining aging as an imperfect ability to repair all the breakdowns that normally occur and *Programmed Death*, explaining aging by much more deliberate breakdowns with the help of 'suicide genes" that turn on at specific times.

1.1. ERROR ACCUMULATION THEORY

Let us focus on the mitochondria of cells as an example. Like most manufacturing processes, this energy-production factory also produces toxic waste. The waste consists of excess electrons. Occasionally, those electrons do not crash into molecular oxygen and do make water. Instead, the electrons crash into other atoms creating molecules with unpaired electrons called *free radicals*. Free radicals are highly reactive and cause a lot of internal cellular damage. Oxygenated free radicals are also called ROS (Reactive Oxygen Species) molecules. Like a lot of industrial waste, ROS molecules can damage anything that they encounter in their way, including proteins, fats, RNA, and even DNA.

On the other hand, there are the molecules that destroy free radicals, including vitamin E. They are called *anti-oxidants*. There is evidence [78] that in spite of anti-oxidants and some products of genes that destroy free radicals, ROS molecules damage DNA as we age. Similar processes take place not only in mitochondria, but also in other parts of cells.

The similar processes of error accumulation take place not only in mitochondria, but also in other parts of cells.

1.2. PROGRAMMED DEATH THEORY

Cytogerontologists now can remove the nuclei from cultured normal cells and then insert nuclei from young cells into old cytoplasts and nuclei from old cells into young cytoplasts. The number of times that both types of hybrid cells divided led them to conclude that the age of the nucleus governed how long the hybrid cell divided in culture.

A cell clock might actually work due to *telomeres*, DNA sequences that are located on the tips of human chromosomes and that consist of repeating sub-units. With each cell division the repeated sequence in the telomeres is reduced. Because it happens at a fixed rate, the length of telomeres behaves like a clock. Immortalized tumor cells found a way to keep their telomers the same length. There is the enzyme called *telomerase* that can replicate telomeres.

In most non-tumor cells, the genes that encode telomerase are turned off and the cells subsequently age and die. Is this turn-off an error or the beginning of programmed death? It could be both.

When young cells are given a growth factor, it stimulates gene, called *fos*, expression. This gene expression makes the transcription factor Fos. Aging cells do not make the Fos protein. Actually, there are other genes that act in similar manner as the *fos* gene and control the cell cycle.

Among newly discovered genes is the gene *apo*-1. The respective protein Apo-1 is known as a receptor. The hands that bind to this molecular lever are called *ligands*. As soon as Apo-1 binds to its ligand, the cell begins to undergo apoptosis. Another death protein is *BAX*. If a cell has enough Bax the cell will die, otherwise the cell will survive. Thus, the idea of regulating the amount of deadly molecules is real.

Are there also life genes? The answer is yes and the data came from cancer research. One of these genes is *bcl*-2. If this gene is placed into a cell that normally has a death sentence the cell will not undergo apoptosis. The respective protein Bsl-2 inhibits formation of ROS molecules. As to BAX protein, Bsl-2 will bind to all the BAX, if there is enough Bsl-2. Thus, the decision to die or live depends on the ratio between these two gene products.

The world of death-dealing and life-saving genes is complex and ever growing. Various laboratories studied not only particular gene products (among these products is *senstatine protein* that can reverse the aging process in skin cells, in nerves within the brain, and in certain blood vessel cells), but also how multiple interactions between many genes can cause the cell to live or die (longevity assurance genes (LAG genes), miscellaneous genes including multiple aging genes (the final total may run into the thousands)).

1.3. WINDING BACK THE CLOCK

While we are beginning to understand how cells commit to growth or to suicide, "we have found two extraordinary properties: 1. By blocking the death genes, we extend not only the life of the cell where the block occurred, but also the life of the organism that contains the cell. 2. Genes that control these life-and-death decisions in humans work just as effectively in worms. This means that the process is conserved, and gives greater weight to the data we see in animals when we think of our own bodies." [133], p. 312.

Among many factors, the *human growth hormone*, *hGH*, has a great deal to do with human longevity. After the age of 60, this hormone begins to shut down. The process is called *hGH menopause*. The certain procedures of injecting this hormone have resulted in not only stopping normal march toward aging, but also reversing certain biological functions. In conclusion, we would like to quote from Dr. Michael Jazwinski, Louisiana State Univ. Medical Center.

"Possibly in 30 years we will have in hand the major genes that determine longevity, and will be in a position to double, triple, even quadruple our maximum life span of 120 years."

2. Modeling of Cell Longevity Problems

We consider modeling of cell longevity with regard to the *error accumulation theory* that explains aging as an imperfect ability to repair all the breakdowns that normally occur and the *programmed death theory* that explains aging with the help of 'suicide genes" turning on at specific times. The main prospective conventional way to overcome aging is handling the major genes that determine longevity.

Below, we also consider, in short, a possible application of our class MM to modeling of cell longevity without gene changes.

2.1. MODELING OF ERROR ACCUMULATION

Error accumulation factors (see s. 1.1) can be mapped in the models by decreasing values of the main functions of the type $\alpha(t,s)$ and $\beta(t,s)$ while t is increasing. The reason of that decrease is the influence of toxic waste, produced in most manufacturing processes.

On the abstract level of the function of $\alpha(t,s)$ and $\beta(t,s)$-types, a failure to accumulate results of all restoration efforts can be explained by necessity of maintaining and restoring, in turn, those restoration efforts themselves. This convergent recursive chain practically has to be stopped somewhere, and sometimes this occurs after the process of development had been launched.

The simplest *MM of degradation of* $\alpha(t,s)$ *and* $\beta(t,s)$ have the form

$$\alpha(t,s) = \exp(-c_\alpha t)\,\alpha^*(t-s),\ \beta(t,s) = \exp(-c_\beta t)\beta^*(t-s),$$

$$c_\alpha,\ c_\beta > 0, \tag{1}$$

where α^* and β^* likely correspond to DNA.

Let us consider the influence of decreased value of α on an example of the MM of cisternia (21)-(23), Ch. 11.

In this example we have the solution

$$\boldsymbol{M}(t) = \boldsymbol{M}^*/\{1+\boldsymbol{C}\exp[-\alpha\hat{}^{-}\boldsymbol{yR}(t-t\hat{})]\},\ \boldsymbol{m}\hat{} = \boldsymbol{C}\alpha\hat{}^{-}\boldsymbol{yRM}^*/(1+\boldsymbol{C})^2, \tag{2}$$

where $\boldsymbol{M}^*$ is the desired limiting value of the enzyme-substrate complex quantity $\boldsymbol{M}$(t), $\boldsymbol{m}\hat{}$ is the initial rate of creation of the enzyme-substrate complex, $\alpha\hat{}^{-}$ is the decreased initial value of α, and $\boldsymbol{M}^*$, y, $\alpha\hat{}^{-}$, $\boldsymbol{R}$ are constants.

Under small $\alpha\hat{}^{-}$ in the relation (2), the constant $\boldsymbol{C}$ has to be rather large and hence the desired value $\boldsymbol{M}^*$ practically cannot be achieved.

2.2. MODELING OF PROGRAMMED DEATH

We remind that there is the enzyme *telomerase*, which can replicate telomeres that function like a life clock. In most non-tumor cells, the genes that encode telomerase are turned off and the cells subsequently age and die.

When young cells are given a growth factor, it stimulates gene, called *fos*, expression. This gene makes the transcription factor Fos.

Aging cells do not make the Fos protein. Actually, there are other genes that act in similar manner as the *fos* gene and control the cell cycle.

Among newly discovered genes is the *gene apo*-1. The respective protein Apo-1 is known as a receptor. The hands that bind to this molecular lever are called *ligands*. As soon as Apo-1 binds to its ligands, the cell begins to undergo apoptosis.

Another death protein is BAX. If a cell has enough BAX the cell will die, otherwise the cell will survive.

Thus, the idea of regulating not only the presence of deadly molecules, but also the amount of deadly molecules is real.

On the other hand, there are also life genes. One of them is *gene bcl*-2. If this gene is placed into a cell that normally has a death sentence the cell will not undergo apoptosis. More than that, the respective protein Bsl-2 inhibits formation of ROS molecules.

As to BAX protein, the protein Bsl-2 will bind to all the BAX protein if there is enough Bsl-2.

Thus, the decision to die or live depends on the ratio of these gene products. So, the modeling problem is similar to that considered in the case of **C**.

As an example, let us introduce the functions similar to those functions in Ch. 11, s. 3 and s. 6,

$$\boldsymbol{\alpha}(t,s)\boldsymbol{y}(s) = \boldsymbol{\alpha}^*\exp\,[\boldsymbol{u}(t) - \boldsymbol{in}(t)],$$

$$\boldsymbol{m}(t) = \boldsymbol{\alpha}^*\exp\,[\boldsymbol{u}(t) - \boldsymbol{in}(t)]\boldsymbol{R}(t),\; t > t^\wedge > 0, \tag{3}$$

where now $\boldsymbol{u}$ is the cell life-saving factor, and $\boldsymbol{in}$ is the cell suicide factor.

It is quite clear that the modeling question for the cell to be or not to be depends on the behavior of the function

$$\boldsymbol{q}(\mathrm{t}) = \boldsymbol{u}(t) - \boldsymbol{in}(t). \tag{4}$$

If $\boldsymbol{q}(\mathrm{t})$ continue to be essentially less than 0, the cell will obviously die.

It was observed experimentally on tumor cells (C. Haskell [77]) that apoptosis happened more frequently in the cells that had obtained the suppressor gene *p*53.

2.3. MODELING OF WINDING BACK THE CLOCK

In the considered case of error accumulation modeling, with regard to the influence of pH and $\Delta G^0/T$, beside (2) we have

$$\boldsymbol{m}\hat{} = \boldsymbol{V'}\text{s}/(\boldsymbol{K}_m + \text{s}),\ \boldsymbol{V'} = \boldsymbol{V}/\boldsymbol{K}_+,\ \boldsymbol{K}_m = \boldsymbol{K}_S/\boldsymbol{K}_+,\ \boldsymbol{V} = \boldsymbol{k}_2\boldsymbol{e}_0,$$

$$\boldsymbol{K}_+ = (1+\text{h}^+/\boldsymbol{K}_{1\text{ES}}+\boldsymbol{K}_{2\text{ES}}/\text{h}^+),\ \boldsymbol{K}_S = \boldsymbol{K}\hat{}_S \exp\,[\Delta \boldsymbol{H}^0{}_s/\text{R}\delta T],$$

$$\boldsymbol{K}_{i\text{ES}} = \boldsymbol{K}\hat{}_{i\text{ES}}\exp\,[\Delta \boldsymbol{H}^0{}_i/\text{R}\delta T],\ \ i = 1, 2,$$

$$\alpha = \alpha\hat{}^{-}\exp\,[-\Delta \boldsymbol{H}_\alpha/\text{R}\delta T]. \tag{5}$$

Repeating the discussion of Ch. 11, s. 3 and s. 6, we can correct the values of $\boldsymbol{m}\hat{}$ and $\boldsymbol{\alpha}$ at the expense of Ph, T and ΔH^0. In particular, under $-\Delta \boldsymbol{H}_\alpha > 0$ we can change the small value $\boldsymbol{\alpha}\hat{}^{-}$ to the increased value $\boldsymbol{\alpha}$. It may mean that in the relation (2) the desired value $\boldsymbol{M}^*$ can be practically achieved. Thus, the normal regime of the cisternia functioning can be restored.

In the considered case of programmed death modeling, due to the influence of the same factors as above, we have

$$\boldsymbol{q} = \boldsymbol{V'}\text{s}/(\boldsymbol{K}_m + \text{s}),\ \boldsymbol{V'} = \boldsymbol{V}/\boldsymbol{K}_+,\ \boldsymbol{K}_m = \boldsymbol{K}_S/\boldsymbol{K}_+,\ \boldsymbol{V} = \boldsymbol{k}_2\boldsymbol{e}_0,$$

$$\boldsymbol{K}_S = (\text{eh})\text{s}/\text{ehs},\ \boldsymbol{K}_{1\text{ES}} = (\text{es}^-)\text{h}^+/\text{ehs},\ \boldsymbol{K}_{2\text{ES}} = (\text{ehs})\text{h}^+/[(\text{eh}_2)\text{s}^+],$$

$$\boldsymbol{K}_+ = (1+\text{h}^+/\boldsymbol{K}_{1\text{ES}}+\boldsymbol{K}_{2\text{ES}}/\text{h}^+),\ \boldsymbol{K}_S = \boldsymbol{K}\hat{}_S\exp\,[\Delta \boldsymbol{H}^0{}_s/\text{R}\delta T],$$

$$\boldsymbol{K}_{i\text{ES}} = \boldsymbol{K}\hat{}_{i\text{ES}}\exp\,[\Delta \boldsymbol{H}^0{}_i/\text{R}\delta T],\ i = 1, 2,$$

$$\alpha = \alpha\hat{}\exp\,[-\Delta \boldsymbol{H}_\alpha/\text{R}\delta T], \tag{6}$$

where all the control factors, of course, have other definitions corresponding to the definition of $\boldsymbol{q}$ (instead of $\boldsymbol{m}\hat{}$).

We can also introduce, similar to (6), the relations for each of the factors $\boldsymbol{u}$ and $\boldsymbol{v}$ in (3) and (4).

As one can see, we have in principle at least 6 control factors, h^+, $\Delta H^0{}_i$, $i = 1, 2$, s, α, and T, of which at least 5 may be independent. Therefore, in principle, we can select them such that temporarily they "wind back the clock."

3. On Modeling of Organism Longevity Problems

Let us accept the *base, simplest MM of an organism* as a whole in the form

$$m(t) = \int_{a(t)}^{t} \alpha(t,s)u(s)v(s)/v(t)y(s)m(s)ds,\ R(t) = \int_{a(t)}^{t} m(s)ds,\ G(t) = \int_{0}^{a(t)} m(s)ds,$$

$$c(t) = \int_{a(t)}^{t} \beta(t,s)[1-y(s)]m(s)ds, f(t) = m(t)+c(t) = k(t)c(t),\ 0 \le y \le 1,\ t \ge 0, \tag{7}$$

where m is the rate of recreation of the new biomass or generalized resource of an organism, $R(t)$ is the total acting resource, $G(t)$ is the total obsolete recourse, $c(t)$ is the rate of creating of the external product, u and v are the general control functions (see Ch.5, s.6.2.), y is the distribution function, α and β are the indices of efficiency of functioning, f is the resource from outside, k is a share of c with respect to f.

There are the following independent possibilities for $m(t)$ tendency to zero:

$\alpha(t,s) \to 0$, *ii*) $u(s) \to 0$, *iii*) $v(t) \to \infty$, *iv*) $y(s) \to 0$, *v*) $a(t) \to t$, *vi*) $k(t) \to 1$, *vii*) $f(t) \to 0$. (8)

We remind that $u(s)$ is usually a non-stationary factor connected with the interaction of OS, $v(s)$ is a factor connected with the interaction of an organism and external surroundings, $y(s)$ is connected with the interaction of OS and external surroundings, $a(t)$ decides the rate of formation of the obsolete recourse, $k(t)$ decides qualitative composition of the external resource $f(t)$.

There is extensive literature about the impact of these types of factors on aging. In particular, B. Bolotov [27] and H. Clark [34] contain nontraditional interesting data about the impact of the factors of $k(t)$ and $f(t)$-types. As to the factors u, v, y, and a, they have been investigated above.

We dwell on the case *i*) as the strongest one, connected probably with the genome's main program of aging. Let in the case *i*) $\alpha(t,s) \le \alpha^*/t$, $t > t^*$, where α^* and t^* are certain constants. Then from the first equation in (7), we have

$$m(t) \le \alpha^*/t \max\,[u(s)v(s)/v(t)y(s)]\ (s \in [a(t),t])R(t). \tag{9}$$

Thus, the natural condition

$$u(s)v(s)/v(t)y(s) \le c^*,\ R(t) \le R^*, \tag{10}$$

where c^* and R^* are positive constants, results in $m(t) \to 0$. But this does not mean that $R(t)$ and $c(t)$ also tend to 0. It has been shown in Ch. 5 that $R(t)$ can be almost a positive constant under the condition $m(t)$, $c(t) \to 0$. It is true, in this connection, $t - a(t)$ cannot be bounded.

One can see that for the longevity maximization we are in need of holding

$$R^* = \int_{a(t)}^{t} u(s)v(s)/v(t)y(s)m(s)ds, \ c(t) > c^- > 0, \tag{11}$$

where c^- is the desired minimal level of external activities, as large as possible.

4. Certain Perspectives

There are two extraordinary properties: 1. By blocking the death genes, we extend not only the life of the cell where the block occurred, but also the life of the organism that contains the cell. 2. Genes that control these life-and-death decisions in humans work just as effectively in animals. This means the process is conserved and gives greater weight to the data we see in animals when we think of our own bodies.

Among many factors, the *human growth hormone*, *hGH*, has a great deal to do with human longevity. After the age of 60, this hormone begins to shut down. The process is called *hGH menopause*. The certain procedures of injecting this hormone have resulted in not only stopping normal march toward aging, but also reversing certain biological functions.

In a stunning step toward producing donor tissues and organs for transplants on demand, two group scientists have grown human cells with the potential to become virtually any kind of body tissue. These embryonic stem cells are the parent cells from which all organs in the body are produced. The ability to grow them in large quantities in the laboratory opens the door to new treatments for a broad variety of diseases, including heart disease, cancers, diabetes, etc.

Scientists have demonstrated that stem cells can be converted in other types of cells, including bone, muscle, neural, and gut cells. These cells will be grown in the laboratory and then used to regenerate failing tissues. These cells will also provide new tools for screening potential drugs and for examining the early stages of embryo development.

However, to realize all the possibilities above researches have to learn how stem cells develop and especially how to control this development *in vivo* (in particular, how to ensure the recipient's immune system tolerance to the cells).

We believe that MM of the cells and CAO can contribute a great deal to solve these and many other problems.

Chapter 13

ON MM OF A DISEASE

1. Introduction

The reader can be acquainted with the base of the techniques under consideration in Ch. 1, 5-9. There are about 150 works of different authors devoted to the further development of the techniques with applications mainly in economics and biology. However, until now, there is no systematic presentation of these mathematical modeling techniques for various subsystems of an organism with applications to medicine and health care.

Meanwhile, as we have shown above and will show below, the new technique allows us to describe much more thoroughly the main structure and functions of an organism both in the norm and pathology. In addition, the description is essentially different from the traditional one (see Part I) in the form of ordinary or partial differential equations, which requires us to revise the whole statement and realization of technology of necessary biomedical experiments. We believe that new difficulties, which arise here, can be overlapped by new useful results.

For many applications the following *MM of organism* (generalization of MM (7), Ch. 12) should be considered:

$$m(t) = \int_{a(t)}^{t} \alpha(t,s)u(s)v(s)v^{-1}(t)y(s)m(s)ds,\ c(t) = \int_{a(t)}^{t} \beta(t,s)[1-y(s)]m(s)ds,$$

$$R(t) = \int_{a(t)}^{t} m(s)ds,\ G(t) = \int_{0}^{a(t)} m(s)ds,\ M(t) = R(\mathrm{t}) + G(\mathrm{t}),$$

$$f(t) = m(t)+c(t) = k(f^*,t)c(t),\ 0 \le y \le 1,\ t \ge 0, \tag{1}$$

where α, u, v, y, and β are matrices, 1 is the unit matrix (inequalities for matrices mean the respective inequalities for their components), a, m, c, R, G, M, and k are vectors, and the relations for f (f^* is the qualitative structure of f) should be replaced by

$$f(t) = \sum m_i(t)\ (i = 1, \ldots, r) + \sum c_j(t)\ (j = 1, \ldots, s) = \sum k_j(t)c_j(t)\ (j = 1, \ldots, s),\ r+s = n. \tag{2}$$

In addition, the *equations of homeostasis* should be included; that is,

$$R'_i(t) = m_i(t) - m_i(a_i(t))a_i'(t) = 0,\ i = 1, \ldots, r. \tag{3}$$

It should be emphasize that considering the given OS in detail, we should include in its

MM an aggregate MM of the rest of the whole organism.

Note yet that an example of application of this technique to the problem of modeling of blood sugar subsystem and diabetes can be found in V. Ivanov [92].

2. Conception of the Norm and Pathology

All the values in MM under consideration, (see, for example, (1-3)) have rather profound sense from the point of view of the structure and functions of the subsystem under consideration. So, if the vector of these values is denoted by $\boldsymbol{R} = (r_1,..., r_q)$, then it is natural to suppose that one of the *conceptions* of the *norm* consists in the validity of the relations

$$r^-_s \leq r_s \leq r^+_s, s = 1, \ldots, q, \quad (4)$$

where r^-_s and r^+_s are admissible bounds for variations of the introduced values from point of view of the norm. On the other hand, if at least one of the relations in (4) is violated, then it is natural to speak about *conception* of the *pathology* or the pathological state of the subsystem.

The pathology can be detailed by discrimination of two classes of (4) violations: A) those of the resource values, namely m_i, c_j, R_i, G_i, and k_j, and B) those of the structural values, namely a_i, α, β, u, v, y, and f^*. In turn, the indices of the class A) can be decomposed into two groups: A') the indices connected with the internal functions of resource recreation, namely m_i, R_i, and G_i, and A'') the indices connected with the external functions of active centers, namely c_j, and k_j. Of course, in the clinical practice there are various other indices of the norm and the pathology. Part of those indices and their bonds as well as the problems of introducing generalized indices are considered in detail in A. Asachenkov et al. [11].

The *treatment of a disease* based on the model (1) and (4) can be considered, in general, under three aspects:

1. prophylaxis, when we hold the relations (4)-type by using the permissible control factors;

2. diagnosis, when we determine the violations of at least one of the relations (4) by using the direct and/or indirect indices of OS;

3. therapy, when we reduce the improper values of the magnitudes in (4) to the norm by using the permissible factors.

3. Norm Restoration Problems

The most simple and expanded restoration is that the determined values of the group A") are injected into the organism directly or, on the contrary, are removed from the organism. The respective medicines should be injected into the organism to achieve validity of the necessary bounds in (4) relative to this group. However, under the condition that the indices of violations of the group A") are accompanied by those of the group A') and class B), such a simple method is usually the temporal measure and does not deduce restoration.

Let us consider, at first, the case of index violation of the group A', for example, m_i. Components of the distributive matrices y, u, and v are the main control magnitudes for MM (1). It depends on their value how the whole internal resource $m_i(s)$ of the subsystem is distributed among its various parts.

For increasing (decreasing) the values of $m_i(t)$, provided that the values of the other magnitudes are not to be broken, it is usually necessary to increase (decrease) the value components of the distributive matrices. As it appears, that for increasing (decreasing) the values $c_j(t)$ for the majority of values t from $[t^\wedge, T]$ under sufficiently large $T - t^\wedge$, the respective values y_{sj} do not usually have to be decreased (increased), but on the contrary increased (decreased). It follows from overtaking of $m_i(s)$ increase (decrease) in comparison with decrease (increase) of $1\text{-}y_{sj}(s)$ such that $(1\text{-}y_{sj}(s))m_i(s)$ still increases (decreases) for the majority of s from $[t^\wedge, T]$ (see the respective rigorous proof in Ch. 15).

Thus, the crucial condition for restoration is the increase of a share of all resources of the subsystems to its internal requirements.

The most serious and profound violations are connected to the structural shift of the subsystem, i.e., with the violation of (4) for the indices of the class B. Here only the interference into the structure of f^* and the genome apparatus can likely restore the norm. Although in principle, in this case, there can be developed a theory similar to the above theory.

Indeed, for creation, support, and variation of the structures of indices α and β- types, a certain share of the whole resources should be also used.

To complete MM (1) in the case, for example, of α_{ik}, we can assume

$$\alpha_{ik}(t,s) = \alpha_{ik}(s)\exp\,[d_{ik}(s-t)],\ d_{ik} \geq 0,\ s \leq t,$$

$$\alpha_{ik}(t) = \int_{a,i(t)}^{t} \Sigma\, \alpha_{jk}(t,s)x_{jk}(s)m_j(s)\ (j = 1, \ldots, r)ds,\ i, k = 1, \ldots, r,$$

$$0 \leq \Sigma\, x_{jk}(s)\ (k = 1, \ldots, r) + y_j(s) \leq 1, j = 1, \ldots, n,\ t \geq t^\wedge, \tag{5}$$

where d_{ik} are decrements of deterioration characterizing the rate of decreasing $\alpha_{ik}(t,s)$ by t; $x_{ik}(s)m_i(s)$ are a share of $m_i(s)$, incoming for the creation of the new technologies $\alpha_{ik}(t)$; the rest magnitudes are the same as in (1).

Investigation of MM similar to (5) (see Ch. 14, 15) shows that the priorities of x_{ik} are usually higher than those of y_i, which, in turn, as was underlined earlier, are higher than those of

$$1 - y_j - \sum x_{jk}(s) \; (k = 1, \dots, r). \qquad (6)$$

It means the following *principle of the norm of OS restoration*:

the crucial condition for restoration (under weakened functions of the classes A and B) is the increase of the share of all the resources to the needs of a comfortable state of OS structure, first, then its internal sphere, and after that its external sphere.

4. On Optimization Problems

Let us consider one of the natural *optimization problems*, namely that of optimal control by high speed, i.e., the problem of determination of those distributions x_i, y_i, and z^i_j that reduce the pathological values to the norm in the minimal possible time so that they will remain inside the norm limits later on.

Let for definiteness $r_1 < r^-_1$ and let us require to find

$$T = T^* = \min (x_{ik}, y_j \,|\, r^-_s \leq r_s \leq r^+_s, t > T^*). \qquad (7)$$

As it is shown in Ch. 15, the general regularity for the desired distribution x^*_{ik}, y_j behavior is the same as noted above: an essential share of resources must go to maintain the structure and internal sphere of the subsystem.

However, the requirement of optimization can result in the rather complicated character for the desired distributions. They can have a large, theoretically an infinite, number of switching times from their minimal to maximal possible values.

Many other optimization problems for OS can be considered in a similar way as above (see Ch. 5 and 15).

5. On the Problem of Individualization of MM

The high value of MM of (1) and (5)-types resides, particularly, in the following. Rather general qualitative information about initial data (usually about the functions α and β-types) is sufficient for investigation of these models by qualitative methods for the future period of the time and, by the same token, for anticipation of possible undesirable consequences. Of course, the numerical values of initial data are necessary in the cases of bringing the results to the numbers. These numerical initial data can be obtained with the help of the models themselves under the known prehistory of functioning of the system considered. However, here we should consider in detail new requirements to the biomedical technology for realization of the respective procedures and experiments about what we have already spoken above. All these are related to the so-called identification problems, solving of which are usually obtained at the expense of application of the mathematical methods based on the regularities of mass events.

Thus, we have the fact of a certain controversy here: identification is needed for the concrete object, but the method requires mass events. Therefore, the problem of identification for one given organism was recently called the *problem* of *individualization* or personification, and the new respective approach was created. We dwell on some results from I. Pogozhev [151].

The main parameter of personification - Q - characterizes the relative flow liquids (lymph, plasma of blood) velocity in the organism considered in comparison with the "base" or "basal" organism. In our notations

$$Q^2 = Q^2(t) = \sum_{i=1}^{n} \int_{t}^{t+t,2} m_i(s)ds/[\sum_{i=1}^{n} \int_{t}^{t+t,2} m^*_i(s)ds], \qquad (8)$$

where $t,2$ is the typical time ($t,2 \approx 1$ minute) for which sufficiently complete intermixing of separate micro-motions into liquid mediums of the organism takes place ($t,1$ is the typical time of a separate particle pulsation, $t,1 \approx 1$ second), $\sum_{i=1}^{n} m^*_i(s)$ is the internal generalized resource circulating into the subsystem of the "basal" organism or into the "basal" organism on the whole. It is supposed that for the given and "basal" organism, the properties of the liquid mediums as well as disposition and sizes of organs do not have essential differences.

The parameter Q^2 is called in [151] the (relative) "alive" temperature of the organism. The scale of Q has been selected such that $Q = 1$ coordinate with the "norm", the case $Q = 0$ means complete stopping of "alive" motion. It has been shown in [151] that Q is decreased with aging: from the age of 25 to 75 Q varies from 1 to 0.82. Q is considered as constant during the typical time $t3$ of keeping by the organism of a certain physiological state, $t,3 \approx 1$ hour.

Let the main characteristics of MM for the "basal" organism be α^*_i and β^*_j. Then the respective averaged characteristics of MM for the given organism is

$$\alpha_i = Q^2\alpha^*_i,\ \beta_j = Q^2\beta^*_j,\ i, j = 1, \dots, n. \tag{9}$$

The ground of the relations of (8)-types is given in [151] for linear MM in the form of ordinary differential equations.

For the time that is much more than *t3* instead of Q^2 we can use, in particular,

$$Q_\delta{}^2 = Q^2 + \zeta_Q(t/\delta),\ \delta = t{,}3/t{,}4, \tag{10}$$

where $t{,}4$ is the typical time for changing of the phase variables, $t{,}4 \approx 1$ day, ζ_Q is stationary stochastic time series with a finite variance and zero mean value. With the substitution of $Q_\delta{}^2$ from (9) for Q^2 in (8), we have the perturbed system of the equations, the solutions of which are converging to the solutions of the original systems, when $\delta \to 0$.

This approach has become especially important in connection with the completion of the Human Genome Project because in many cases we may consider the main characteristics of the "basal" organism α^*_i and β^*_j as given by "basal" genome, and find the main characteristics of the individual with the help of (7)-(9).

6. On MM of a Doctor's Business

The simplest, *base MM of a doctor's business*, according to the general theory of CAO and ES (also see V. Ivanov [92]), has the form

$$m(t) = \int^t_0 \alpha(t,u)\lambda(t,u)y(u)m(u)du,\ c(t) = \int^t_0 \beta(t,u)\mu(t,u)[1-y(u)]m(u)du,$$

$$R(t) = \int^t_0 \lambda(t,u)y(u)m(u)du + \int^t_0 \mu(t,u)[1-y(u)]m(u)du,\ G(t) = M(t) - R(t),$$

$$M(t) = \int^t_0 m(u)du, f(t) = m(t) + c(t) = kc(t),\ t \geq t^\wedge \geq 0, \tag{11}$$

where m is the new resource (per unit of time) of a doctor's work place, providing its internal functions ; c is some new resource providing external functions of a doctor, including the function of curing; λ, μ are the intensities of functioning along the channels ym--m, $[1\text{-}y]m$--c respectively; $R(t)$ is the total amount of functioning internal resource of a doctor's work place; y, α, β, G, k, f, and $t^\wedge$ have the same definitions as many times above.

The function of curing decides a doctor's business. The number of patients c_i who have been cured for ith disease in the unit of the time is actually decided by the same function. Therefore, this function should be well estimated. It should be estimated by the total number of the patients who have been successfully cured in a certain period of time.

The methods of curing or restoring to the norm depend on many factors. We would like to dwell only on some approaches to curing, following from the models. The most simple and extended restoration is when the determined values of the group A") are injected into the organism directly or, on the contrary, are removed from the organism and/or the respective medicines are injected into the organism so as to achieve validity of the necessary bounds (2) relative to this group. However, under the condition that the indices of violations of the group A") are accompanied by those of the group A') and class B), such a method is the temporal measure and does not deduce restoration.

To increase (decrease) the values of $m(t)$-type, if the values of the other magnitudes are not broken, it is usually necessary to increase (decrease) the values $y(s)$, $s < t$. Thus, the crucial condition for restoration (under weakened functions of the class A) is the increase of a share of all the resources of a doctor to his/her internal requirements: development of new, more effective technology.

One may hope that any modes of action, among them nontraditional ones (see, for example, B. Bolotov [27] and H. Clark [34]), including the various physical loading on a certain part of an organism and contributing to increasing of resource distribution in advantage to the internal sphere of the sick subsystem, will usually bring an essential positive effect. The most serious and profound violations are connected with the structural shift of the subsystem, i.e., with the violation of (4) for the indices of the class B. Here, only interference into the gene apparatus (gene therapy) can likely restore the norm.

In conclusion of this s. and this Ch., we would like to emphasize that application of MM and all the techniques under consideration to modeling of the cells and CAO may be expanded far beyond the examples already examined. The same class of MM may also be employed for modeling of other evolutionary systems (ES). The main reason for all of these is not "black-box", but profound "grey-box" dynamics described MM in question (see Introduction and Ch. 5, 14, and 15).

Appendix:

MATHEMATICS OF DEVELOPMENT

Chapter 14

INVESTIGATION OF EQUATIONS

Here are special simple cases of MM of CAO and examples of analytic solutions. The conditions for existence and uniqueness of solutions for the systems of equations and some optimization problems and also stability of solutions are established. Sequences of formulae for determination of gradients of specific functionals are given. Some cases of non-uniqueness of solutions are considered. Cases of oscillating and wave solutions of the systems are found. Asymptotic properties of solutions are investigated.

1. Examples of Analytical Solutions

Let in (1, Ch. 5) $\alpha\lambda = A(s) \geq 0$. Then the first equation in (1, Ch. 5) is a linear differential equation (or a system of equations) of the form

$$m'(t) = A(t)y(t)m(t) + \Delta m'(t),\ m(t^{\wedge}) = m^{\wedge}. \qquad (1)$$

If $\alpha\lambda = A(s)m(s)$, $A(s) \geq 0$, then we have a *bilinear differential equation* of the form

$$m'(t) = A(t)y(t)m^2(t) + \Delta m'(t),\ m(t^{\wedge}) = m^{\wedge}, \qquad (2)$$

or a system of such equations

$$m'(t) = A(t)y(t)m(t)m^T(t) + \Delta m'(t),\ m(t^{\wedge}) = m^{\wedge}, \qquad (2')$$

where $m(t)$ is a column vector and $m^T(t)$ is a row vector. The solutions of the equations (1)-(2') can be obtained by methods C. Baker [14]; G. Korn, T. Korn [115]; [176].

Let us consider now the solutions of the system of equations (5, 6, Ch. 5) under the conditions: α, β, y, and R are given and also α, $y^{\wedge}$, and R are constants, or f is given and α, β, and y are given constants, $\alpha y + \beta(1-y) \neq 0$.

It is easy to see that for $t \in [t^{\wedge}, a^{-1}(t^{\wedge})]$, $a(a^{-1}(t^{\wedge})) = t^{\wedge}$, we have

$$m(t) = \alpha y^{\wedge}\int_{a(t)}^{t^{\wedge}} m^{\wedge}(s)ds + \alpha\int_{t^{\wedge}}^{t} ym(s)ds,\ R(t) = \int_{a(t)}^{t^{\wedge}} m^{\wedge}(s)ds + \int_{t^{\wedge}}^{t} m(s)ds, \qquad (3)$$

from which

$$m(t) = m^{\wedge}\exp\,[\alpha(y-y^{\wedge})(t-t^{\wedge})],\ M(t) = M(t^{\wedge}) + m^{\wedge}\{\exp\,[\alpha(y-y^{\wedge})(t-t^{\wedge})] - 1\}/[\alpha(y-y^{\wedge})],$$

$$a(t) = M^{\wedge -1}[M(t) - M(t^{\wedge})],\ M^{\wedge}(t) = \int_0^t m^{\wedge}(s)ds,\ M(t^{\wedge}) = M^{\wedge}(t^{\wedge}),$$

$$c(t) = \beta^{\wedge}(1-y^{\wedge})R + \int_{t^{\wedge}}^{t} [\beta(1\text{-}y) - \beta^{\wedge}(1-y^{\wedge})]m(s)ds, \tag{4}$$

where $M^{\wedge -1}$ is the *reciprocal function* or *operator* for $M^{\wedge}$. If $f = m + c = [\alpha y + \beta(1\text{-}y)]R(t)$ is given, then

$$R(t) = f(t)/[\alpha y + \beta(1-y)],\ m(t) = \alpha y R(t),\ c(t) = \beta(1-y)R(t). \tag{5}$$

For the interaction of two CAO according to Fig. 3, Introduction, let $\alpha,2$, $y,2$, $y,2^{\wedge}$, $R,2$ be given constants, and let $m,2'' = qm,2$ (see (27, Ch. 5)), q be also constants. Then $m,2'(t) = [\alpha,2(y,2 - y,2^{\wedge}) - q]m,2$ and therefore

$$m,2(t) = m,2^{\wedge}\exp\,\{[\alpha,2(y,2 - y,2^{\wedge}) - q](t-t^{\wedge})\}. \tag{6}$$

From (6) it follows that $m,2(t)$ is decreasing under

$$q > \alpha,2(y,2 - y,2^{\wedge}). \tag{7}$$

The inequality (7) is called the *condition of irreproducible loss* for the 2nd CAO.

For the system (46, Ch. 5), we assume that x, y, and R are positive constants. Then in the case $\nu = \mu = 1$, $c_1 = 0$, we have $(\alpha_1 = \alpha)$

$$\alpha'(t) = x/ym(t);\ m'(t) = y[\alpha(t)m(t) - \alpha(a)m(a)a'(t)],\ m(t) - m(a)a'(t) = 0,\ a(t^{\wedge}) = 0, \tag{8}$$

from where

$$\alpha(t) - \alpha(a) = xR/y,\ m(t) = m^{\wedge}\exp\,[xR(t-t^{\wedge})],\ \alpha(t) = \alpha(t^{\wedge}) + m^{\wedge}/y\exp\,[xR(t-t^{\wedge})],$$

$$a(t) = t + 1/(xR)\ln\,\{1+[1-\exp\,(xRt^{\wedge})]\exp\,(-xRt)\}. \tag{9}$$

For arbitrary $R(t)$, $R(t) > 0$, and positive constants x, y, using the relations $m - m(a)a' = R'$, we reduce the case to the second order linear differential equation

$$M'' - xRM' - xR'M = (y\alpha(t^{\wedge}) - xR)R',\ M' = m, \tag{10}$$

the solutions and properties of which have been investigated in detail by many mathematicians (see, for example, J. Murray [142] and [176]).

2. Existence and Uniqueness of Solutions

Let us consider MM (1, Ch. 5).

Theorem 1. *If* $\alpha\lambda$, $\beta\mu$, y, *and* Δm *are given nonnegative integrable bound functions, then the system of equations in* (1, Ch. 5) *concerning* m, c, R, G, *and* f *has the unique nonnegative bound solution, and also* $m(t)$ *has the form*

$$m(t) = u(t) + \int_{t^\wedge}^{t} g(t,s)u(s)ds,\ t \geq t^\wedge,\ g(t,s) = \sum_{k=0}^{\infty} g_{k+1}(t,s),\ g_1(t,s) = \alpha(t,s)\lambda(t,s)y(s),$$

$$g_{k+1}(t,s) = \int_s^t g_1(t,z)g_k(z,s)dz,\ k = 1, 2, \ldots,\ u(t) = \int_0^{t^\wedge} g_1{}^\wedge(t,s)m^\wedge(s)ds + \Delta m(t), \tag{11}$$

and c, R, G, *and* f *are obtained by the respective replacement of* $m(t)$.

This result is a corollary of the well-known theory of Volterra linear integral equations (C. Baker [14]). The similar theorem takes place when $\alpha\lambda$, $\beta\mu$, y are given matrices.

Let us put

$$\boldsymbol{m}(f,[a, b]) = \min |f(t)|\ (a \leq t \leq b),\ \boldsymbol{M}(f,[a, b]) = \max |f(t)|\ (a \leq t \leq b) \tag{12}$$

for any function f and any segment $[a, b]$. And also

$$\boldsymbol{m}(f,[0, t]) = \boldsymbol{m}(f,t),\ \boldsymbol{M}(f,[0, t]) = \boldsymbol{M}(f,t);\ \boldsymbol{m}(f,[0, T]) = \boldsymbol{m}(f),\ \boldsymbol{M}(f,[0, T]) = \boldsymbol{M}(f). \tag{13}$$

Theorem 2. *Let in the systems* (5, 6, Ch. 5) *the elements* αy, $\beta(1\text{-}y)$, R, *and* $m^\wedge(t)$, $t \in [0, t^\wedge]$, *be given positive functions, and here*

$$m^\wedge \in C([0, t^\wedge]),\ \alpha y,\ \beta(1-y) \in C^l([0, T] \times [0, T]),\ R \in C^1([t^\wedge, T]), \tag{14}$$

where C *and* C^1 *are continuous and continuously differentiable spaces of functions, respectively. Then in order fort these systems to have the unique nonnegative solutions* m, a, c, $f \in C^1([t^\wedge, T])$, $0 \leq t^\wedge < T < \infty$, *it is sufficient that the following condition holds*:

$$0 \leq R(t) \leq M^\wedge(t^\wedge) + \int_{t^\wedge}^{t} \boldsymbol{m}(\alpha y,s)R(s)ds,\ 0 \leq t^\wedge \leq t \leq T < \infty. \tag{15}$$

In addition, the desired $m(t)$ *and* $a(t)$ *can be found by the iterative process*

$$m_{i+1}(t) = \int_{a(t)}^{t} \alpha(t,s)y(s)m_i(s)ds,\ R(t) = \int_{a(t)}^{t} m_i(s)ds,\ m_i(s) = m^\wedge(s),$$

$$s \in [0, t^\wedge],\ i = 1, 2, \ldots,\ m_1(t) = \boldsymbol{m}(\alpha y,t)R(t),\ t^\wedge \leq t \leq T, \tag{16}$$

which converges at the rate exceeding that of geometrical progressions.

<u>Proof.</u> On the strength of (15), (16) the equation

$$\int_{a,i(t)}^{t} m_i(s)ds = R(t) \tag{17}$$

has a solution. Indeed, if $\int_{a,i-1(t)}^{t} m_{i-1}(s)ds = R(t)$, then $m_i(t) \geq \boldsymbol{m}(\alpha y,t)R(t)$ and therefore

$$\int_0^t m_i(s)ds = M\hat{}(t\hat{}) + \int_{t\hat{}}^t m_i(s)ds \geq R(t), \int_t^t m_i(s)ds = 0 \leq R(t). \tag{18}$$

But $\int_{a,1(t)}^{t} m_1(s)ds = R(t)$, since $\int_0^t m_l(s)ds = M\hat{}(t\hat{}) + \int_{t\hat{}}^t \boldsymbol{m}(\alpha y,s)R(s)ds \geq R(t)$.
Further,

$$m_2(t) - m_1(t) = \int_{a,1\hat{}}^{t} [\alpha y - \boldsymbol{m}(\alpha y,t)]m_l ds \leq \boldsymbol{M}(\alpha y)R(t)(t - t\hat{}),\ a,1\hat{} = \max(t\hat{}, a,1). \tag{19}$$

And also for $i \geq 2$

$$|m_{i+1} - m_i| \leq \boldsymbol{M}(\alpha y)[\int_{a,i-1\hat{}}^{t} |m_i - m_{i-1}|ds + |\int_{a,i}^{a,i-1} m_i ds|],$$

$$\int_{a,i-1}^{t} m_{i-1}ds = \int_{a,i}^{t} m_i ds, \tag{20}$$

from which $|\int_{a,i-1}^{a,i} m_i ds| \leq \int_{a,i-1\hat{}}^{t} |m_i - m_{i-1}|ds$, $a,i-1\hat{} = \max(t\hat{}, a,i-1)$. So,

$$|m_i - m_{i+1}| \leq 2\boldsymbol{M}(\alpha y)\int_{a,i-1\hat{}}^{t} |m_{i-1} - m_i|ds \leq R(t)[2\boldsymbol{M}(\alpha y)(t - t\hat{})]^i/i!,$$

$$2\boldsymbol{M}(\alpha y)\boldsymbol{m}(\alpha y)\boldsymbol{m}(R)|a,i - a,i-1| \leq R(t)[2\boldsymbol{M}(\alpha y)(t - t\hat{})]^i/i!. \tag{21}$$

From (15), (20), and (21) follows the validity of Theorem 2.

Note that the condition (7, Ch. 5), which is $a'(t) \geq 0$, $t \geq 0$, holds iff $m(t) \geq R'(t)$. It follows from $R'(t) = m(t) - m(a)a'$ and $m(a) > 0$. For $a'(t) \geq 0$, $t \geq 0$, to hold, it is sufficient that $\boldsymbol{m}(\alpha y) \geq R'(t)$. If $\boldsymbol{m}(\alpha) \geq R'(t)$, then there exists y, $0 < y \leq 1$, for which $\boldsymbol{m}(\alpha y) \geq R'(t)$. If $R'(t) \leq 0$, then for any y, $0 \leq y \leq 1$, $m(t) > R'(t)$.

The initial results about the existence and uniqueness of the solutions of the equations under consideration have been related to the case $t\hat{} > 0$. The present, more general formulation of Theorem 2 is based on the paper S. Girlin [56], which considers originating evolutionary systems.

If $t\hat{} = 0$, we will obviously have $m(t) = a(t) = 0$. This will not be the case if we consider the system similar to (5, 6, Ch. 5) when Δm, $\Delta c \neq 0$ or we assume $\alpha y = \delta(t-s)$, where δ - Dirac delta-function. The proof itself has been different in the case $t\hat{} > 0$.

Here, briefly, is an outline of the main idea of that proof. Let us introduce the relations

$$m(t) = \Psi(m)(t) = \Phi\{M\hat{}^{-1}[M(t)-R(t)],t\} + \int_{t\hat{}}^{t} \alpha(t,s)y(s)m(s)ds,$$

$$M(t) = \int_0^t m(s)ds,\ a(t) = M\hat{}^{-1}[M(t)-R(t)],$$

$$\Phi(a,t) = \int_a^{t\hat{}} \alpha(t,s)y\hat{}(s)m\hat{}(s)ds,\ M\hat{}(t) = \int_0^t m\hat{}(s)ds. \tag{22}$$

Since Φ and $M\hat{}$ and hence $M\hat{}^{-1}$ are continuously differentiable monotone increasing functions, the *Volterra-type operator* Ψ realizes contracted mapping of $C([t\hat{},t^*])$ into itself, where $t^* > t\hat{}$, and has the property: $a(t) \le t\hat{}$, $t \in [t\hat{}, t^*]$. By virtue of (22) and $M(t\hat{}) - R(t\hat{}) = 0$ such t^* will always be found. It follows that the unique positive solution of the equation $m = \Psi(m)$ on the segment $[t\hat{}, t^*]$ can be found by the *method of simple iteration*

$$m_{i+1} = \Psi(m_i),\ i = 0, 1, \ldots;\ m_0 = 0;\ m = \Sigma_{i=0}^{\infty} (m_{i+1}-m_i). \tag{23}$$

On the strength of the Weierstrass' theorem, we have from (23) that m is a continuous function. But then it follows from (22) that $m, a, c, f \in C^1([t\hat{}, t^*])$. The first instant t^* for which $a(t^*) = t\hat{}$, is until now unknown and found in the process of solving. If $a(t) \le t\hat{}$ *for* $t\hat{} \le t \le T$, then it follows that the constructed solution is valid over the whole given segment and the theorem is proved.

In the similar way, it can be proven that for $T > t^*$ the problem has the unique positive solution over $[t^*, t_1]$, $a(t_1) = t^*$, then over $[t_1, t_2]$, $a(t_2) = t_1$, and so on until the solution over the whole segment $[t\hat{}, T]$ has been obtained. The latter are possible since by virtue of the conditions of the theorem it is easy to show that

$$t - a(t) \ge d > 0,\ t \in [t\hat{}, T], \tag{24}$$

where d is a certain constant.

Theorem 3. *Under the assumption of Theorem* 2 *and under the condition*

$$0 < f(t) \le \int_0^{t\hat{}} [\alpha y+\beta(1-y)]m\hat{}ds + \int_{t\hat{}}^{t} f(s)[\alpha y+\beta(1-y)][1+\boldsymbol{M}(\beta(1-y),s)/\boldsymbol{m}(\alpha y,s)]^{-1}ds \tag{25}$$

instead of (15), *where f is a given positive continuously differentiable function, the system of equations* (5, 6, Ch. 5) *has the unique nonnegative differentiable solution m, a, c, R for* $0 \le t\hat{} \le T < \infty$. *In addition,* $m(t)$ *and* $a(t)$ *can be found by the iterative process*

$$f(t) = m_{i+1}(t) + c_{i+1}(t),\ c_{i+1}(t) = \int_{a,i(t)}^{t} \beta(t,s)[1-y(s)]m_i(s)ds,$$

$$m_i(s) = m\hat{}(s),\ s \in [0, t\hat{}],\ i = 1, 2, \ldots,$$

$$m_I(t) = [1+\boldsymbol{M}(\beta(1-y),t)/\boldsymbol{m}(\alpha y,t)]^{-1},\ t \in [t\text{^}, T], \qquad (26)$$

m_{i+1} *is expressed in the terms of* m_i *the same way as in* (16), *which converges at the rate exceeding that of geometrical progressions.*

Proof. The equation (26) with respect to $a,i(t)$ has the unique solution. Indeed,

$$\int_0^t [\alpha y+\beta(1-y)]m_i ds = \int_0^{t\text{^}} [\alpha y+\beta(1-y)]m\text{^}ds + \int_{t\text{^}}^t [\alpha y+\beta(1-y)]m_i(s)ds$$

$$\geq \int_0^{t\text{^}} [\alpha y+\beta(1-y)]m\text{^}ds + \int_{t\text{^}}^t f(s)[\alpha y+\beta(1-y)][1+\boldsymbol{M}(\beta(1-y,s))/\boldsymbol{m}(\alpha y,s)]^{-1}ds \geq f, \quad (27)$$

since $c_i \leq m_i\boldsymbol{M}(\beta(1-y),t)/\boldsymbol{m}(\alpha y,t)$, $m_i + c_i = f$, and $\int^0{}_0 [\alpha y+\beta(1-y)]m_i ds = 0 \leq f$. Then on the strength of (25) and (26) we can receive the estimates similar to (21) by the similar way to the proof of Theorem 2.

In a similar way, we can prove the theorem when instead R or f the function c is given. If f is given, then the condition $a'(t) \geq 0$, $t \geq 0$, takes place iff

$$\{\alpha(t,t)y(t)+\beta(t,t)[1-y(t)]\}m(t) + \int_{a(t)}^t \{\alpha'_t(t,s)y(s)+\beta'_t(t,s)[1-y(s)]\}m(s)ds \geq f'(t). \quad (28)$$

It follows from

$$f(t) = \int_{a(t)}^t \{\alpha(t,s)y(s) + \beta(t,s)[1-y(s)]\}m(s)ds. \qquad (29)$$

For $a'(t) \geq 0$, $t \geq 0$, to hold, it is sufficient that

$$\alpha(t,t)y(t) + \beta(t,t)[1-y(t)] \geq$$

$$\{f'(t)/f(t)-\min\,[\boldsymbol{m}(\alpha'_t/\alpha),\boldsymbol{m}(\beta'_t/\beta)]\}[1+\boldsymbol{M}(\beta(1-y)/\boldsymbol{m}(\alpha y)]. \qquad (30)$$

If c is given, then the condition $a'(t) \geq 0$, $t \geq 0$, takes place iff

$$\beta(t,t)[1-y(t)]m(t) + \int_{a(t)}^t \beta'_t(t,s)[1-y(s)]m(s)ds \geq c'. \qquad (31)$$

And we can receive the sufficient condition that is similar to the one above.

Theorem 4. *Under the assumptions of Theorem* 2 *and under the condition*

$$0 < R_1(t) \leq \int_0^{t\text{^}} ym\text{^}ds + \boldsymbol{m}(\alpha)\int_{t\text{^}}^t yR_1(s)ds,$$

$$0 < R_2(t) \leq \int_0^{t\text{^}} (1-y)m\text{^}ds + \boldsymbol{m}(\alpha)\int_{t\text{^}}^t (1-y)R_1(s)ds, \qquad (32)$$

where R_1 and R_2 are given differentiable functions, the system of equations (9, Ch. 5) *has the unique nonnegative differentiable solution m, a, and b, which can be found by the iterative process*

$$R_1(t) = \int_{a,i(t)}^{t} y(s)m_i(s)ds,\ R_2(t) = \int_{b,i(t)}^{t} [1-y(s)]m_i(s)ds,\ m_i(s) = m\hat{}(s),\ s \in [0, t\hat{}],$$

$$i = 1, 2, ...;\ m_1(t) = \mathbf{m}(\alpha)R_1(t),\ t \in [t\hat{}, T], \tag{33}$$

$m_{i+1}(t)$ are expressed in terms of $m_i(t)$ the same way as in (16), *converging faster than a geometrical series.*

Proof. It is similar to the proof of Theorem 2.
Now, the condition $a'(t) \geq 0$, $b'(t) \geq 0$, $t \geq 0$, takes place iff

$$m(t)y(t) \geq R_1'(t),\ m(t)[1-y(t)] \geq R_2'(t).$$

For this to hold it is sufficient that

$$\mathbf{m}(\alpha)y(t)R_1(t) \geq R_1'(t),\ \mathbf{m}(\alpha)[1-y(t)R_1(t) \geq R_2'(t).$$

Theorem 5. *Under the assumption of Theorem* 2 *and under the condition*

$$R(t) < \min\left(\int_0^{t\hat{}} ym\hat{}ds, \int_0^{t\hat{}} (1-y)m\hat{}ds\right), f(t) < \min\left(\int_0^{t\hat{}} \alpha ym\hat{}ds, \int_0^{t\hat{}} \beta(1-y)m\hat{}ds\right),$$

$$[\mathbf{m}(\alpha,t) > f(t)/R(t) > \mathbf{M}(\beta,t)],\ [\mathbf{m}(\beta,t) > f(t)/R(t) > \mathbf{M}(\alpha,t)] \tag{34}$$

the system of equations (9, Ch. 5) *has the unique nonnegative differentiable solution m, a, b, c, which can be found by the iterative process*

$$\Phi(a,i,b,i) = m_{i+1}(t) + c_{i+1}(t) = f(t),\ \Psi(a,i,b,i) = \int_{a,i(t)}^{t} y(s)m_i(s)ds +$$

$$\int_{b,i(t)}^{t} [1-y(s)]m_i(s)ds = R(t),\ c_{i+1}(t) = \int_{b,i(t)}^{t} \beta(t,s)[1-y(s)]m_i(s)ds,$$

$$i = 1, 2, ...;\ m_l(t) = 0,\ t \in [t\hat{}, T], \tag{35}$$

m_{i+1} are expressed in terms of m_i in the same way as above, converging faster than geometrical progressions.

Proof. It is not difficult to determine that $\Phi(t,t) = \Psi(t,t) = 0 < f, R$; $\Phi(0,0) > f$, $\Psi(0,0) > R$. So, there exist values a,i and b,i for which $\Phi(a,i,b,i) = f$ and $\Psi(a,i,b,i) = R$..

These values will meet the conditions: $0 < a,i,\ b,i < t$. Indeed, we can rewrite the system

(35) in the form

$$x\hat{} = \int_{a,i}^{t} ym_i ds = (Rc_{i+1}/z\hat{} - f)/d,\ z\hat{} = \int_{b,i}^{t} (1-y)m_i ds = (f - Rm_{i+1}/x\hat{})/d,\ d = c_{i+1}/z\hat{} - m_{i+1}/x\hat{},$$

from which on the strength of (34), it follows rather easily that $d \neq 0$ and if $\beta > \alpha$, then

$$x\hat{} > R[\boldsymbol{m}(\beta,t) - f/R]/[\boldsymbol{M}(\beta) - \boldsymbol{m}(\alpha)] > 0,\ z\hat{} > R[f/R - \boldsymbol{M}(\alpha,t)]/[\boldsymbol{M}(\beta) - \boldsymbol{m}(\alpha)] > 0,$$

and if $\alpha > \beta$, then

$$x\hat{} > R[f/R - \boldsymbol{M}(\beta,t)]/[\boldsymbol{M}(\alpha) - \boldsymbol{m}(\beta)] > 0,\ z\hat{} > R[\boldsymbol{m}(\alpha) - f/R]/[\boldsymbol{M}(\alpha) - \boldsymbol{m}(\beta)] > 0.$$

These inequalities mean that $ai,\ bi < t$. If ai or bi is equal to 0, then respectively

$$\Phi(0,b,i) > \Phi(0,t) > f,\ \Psi(0,b,i) > \Psi(0,t) > R$$

or

$$\Phi(a,i,0) > \Phi(t,0) > f,\ \Psi(a,i,0) > \Psi(t,0) > R..$$

It means that $a,i,\ b,i > 0$. This solution is unique, since Jacobian of the system (35)

$$J = (ym)\,|_a((1-y)m)\,|_b(\beta(t,b) - \alpha(t,a)) \tag{36}$$

has a fixed sign. Further proof is similar to the one above.

The condition $a'(t),\ b'(t) \geq 0,\ t \geq 0$, takes place iff

$$z_1\beta(t,b) \geq z_2 \geq z_1\alpha(t,a)\ (J > 0)\ or\ z_1\beta(t,b) \leq z_2 \leq z_1\alpha(t,a)\ (J < 0), \tag{37}$$

where

$$z_1 = m(t) - R'(t),\ z_2 = \{\alpha(t,t)y(t) + \beta(t,t)[1-y(t)]\}m(t) - f'(t) +$$

$$\int_{a(t)}^{t} \alpha'(t,s)y(s)m(s)ds + \int_{b(t)}^{t} \beta'(t,s)[1-y(s)]m(s)ds. \tag{38}$$

We can receive the respective sufficient condition by the similar way to the one above.

Note that in the case of NES and CAO,

$$G(t) = \int_0^{a(t)} y(s)m(s)ds + \int_0^{b(t)} [1-y(s)]m(s)ds, \tag{39}$$

can be given instead of $R(t)$.

We can receive the result similar to Theorem 5 with the replacement of $R(t)$ by $G(t)$.

Theorem 6. *In the case of* (46, Ch. 5), *except when* $\mu = 1$, $\nu = 0$, *we have the system, which has the unique differentiable and positive solution* α_μ, m, a, β_κ, c, M, G, f *on any segment* $[t^\wedge, T]$, $0 < t^\wedge < T < \infty$, *under the assumptions similar to Theorem 2 and under*

$$0 < R(t) \leq \int_0^{t^\wedge} m^\wedge(s)ds,\ y(t) > 0,\ t \in [t^\wedge, T]. \tag{40}$$

In addition, the desired functions α_μ, m, *and* a *can be found by the iterative process*

$$\int_{a,i(t)}^{t} m_i(s)ds = R(t),\ \alpha_\mu{}^{\nu}{}_{i+1}(t) = \int_{a,i(t)}^{t} \alpha_{\mu i}(t,s)x(s)m_i(s)ds,\ m_{i+1}(t) = \int_{a,i(t)}^{t} \alpha_{\mu i}(t,s)y(s)m_i(s)ds,$$

$$m_i(s) = m^\wedge(s),\ \alpha_{\mu i}(s) = \alpha_\mu{}^\wedge(s),\ s \in [0, t^\wedge],\ i = 1, 2, \ldots;\ m_1(s) = \alpha_{\mu 1}(s) = 0,\ s \in [t^\wedge, t], \tag{41}$$

which converges faster than geometrical progressions.

Proof. Let $c_i = d_i = 0$, $i = 1, 2$, in (46, Ch. 5). Otherwise, the proof is similar. We eliminate the case $\mu = 1$, $\nu = 0$ since it seems senseless (see Ch. 5).

On the strength of (40), (41) the equation $\int_{a,i(t)}^{t} m_i(s)ds = R(t)$ always has the unique solution $a,i(t)$. The respective equations in (46, Ch. 5) have the unique a positive differentiable solution on the segment $[t^\wedge, T]$. Indeed, consider in detail the case $\nu = 1$, $\mu = 1$. Here we have

$$\alpha_1'(t) \leq M(x)\alpha_1(t)R(t),\ m(t) \leq M(y)\alpha_1(t)R(t), \tag{42}$$

from which

$$\alpha_1(t) \leq \alpha_1(t^\wedge)\exp\left[M(x)\int_{t^\wedge}^{t} R(s)ds\right],\ m(t) \leq M(y)R(t)\alpha_1(t^\wedge)\exp\left[M(x)\int_{t^\wedge}^{t} R(s)ds\right]. \tag{43}$$

It means that the functions m and α_1 are bounded on any segment $[t^\wedge, T]$. The desired result follows from the general theory of Volterra-type equations.

And the iterative process (41) converges faster than geometrical series on the strength of the estimates similar to (21).

In the case of $\nu = 0$, $\mu = 2$, we have

$$\alpha_2(t) \leq M(x/y)m(t),\ m(t) \leq m^\wedge + M(x/y)\int_{t^\wedge}^{t} m(t-s)m(s)ds, \tag{44}$$

from which consecutively

$$m(t) \leq m^\wedge + M(x/y)\int_{t^\wedge}^{t} m^\wedge(t-s)m^+(s)ds,\ t \in [t^\wedge, 2t^\wedge],$$

$$m(t) \leq m^{+\wedge} + M(x/y)\int_{2t^\wedge}^{t} m^{+\wedge}(t-s)m^+(s)ds,\ t \in [2t^\wedge, 4t^\wedge], \tag{45}$$

and so on. It means that $m(t)$ and hence $\alpha_2(t)$ are bound on any segment $[t^\wedge, T]$. Further proof is similar to the one above.

The case $\nu = 1$, $\mu = 2$ can be considered by analogy with that of $\nu = 1$, $\mu = 1$.

After determination of the functions α_μ, m, and a, we can find β_κ, c, M, G, and f on the basis of their respective equations in (46, Ch. 5).

It should be noted that instead of (40) we could accept the more general sufficient conditions, which are similar to (15). For example, in the case of $\nu = 1$, $\mu = 1$ we can accept

$$0 < R(t) \leq m^\wedge + \alpha_1(0)\boldsymbol{m}(y)\int_{t^\wedge}^{t} \exp(-c_1 s)R(s)ds \tag{46}$$

and in the case $\nu = 1$, $\mu = 2$, the condition

$$0 < R(t) \leq m^\wedge + \alpha_2(0)\boldsymbol{m}(y)\int_{t^\wedge}^{t} \exp(c^\wedge{}_2 s)R(s)ds,\ c^\wedge{}_2 = \min(c_2, 0). \tag{47}$$

The condition $a'(t) \geq 0$ takes place iff $m(t) \geq R'(t)$. For $\nu = 1$ it is sufficient that

$$\alpha_\mu(0)\boldsymbol{m}(y)R(t) \geq R'(t). \tag{48}$$

It follows from

$$m(t) \geq \alpha_\mu(0)\boldsymbol{m}(y)R(t). \tag{49}$$

The remaining cases can be considered in a similar way.

<u>Theorem 7</u>. *Under the conditions of Theorems* 2 *and* 6 *and under the condition*

$$0 < f(t) \leq \int_0^{t^\wedge} \{\alpha_\mu{}^\wedge(t,s)[x(s)+y(s)+z(s)] + \beta_\kappa{}^\wedge(t,s)[1-x(s)-y(s)-z(s)]\}m^\wedge(s)ds, \tag{50}$$

where f is given, the system of equations (46, Ch. 5) *has the unique differentiable solution with respect to m, a, α_μ, c, β_κ, R. In addition, the desired functions m, a, α_μ, c, and β_κ can be found by the iterative process*

$$f(t) = m_{i+1} + c_{i+1} + \alpha_\mu{}^\nu{}_{i+1} + \beta_{\kappa i+1},\ \beta_{\kappa i+1} = \int_{a,i(t)}^{t} \alpha_{\mu i}(t,s)z(s)m_i(s)ds,$$

$$c_{i+1} = \int_{a,i(t)}^{t} \beta_{\kappa i}(t,s)[1-x(s)-y(s)-z(s)]m_i(s)ds,\ i = 1, 2, \ldots; \tag{51}$$

m_{i+1}, $\alpha_\mu{}^\nu{}_{i+1}$ *are the same as in* (41), *converging faster than geometrical progressions.*

<u>Proof</u>. It is similar to the proof of Theorem 6.

Theorem 8. *In the case of MM* (49, Ch. 5), *except* $\mu = 1$, $\nu = 0$, *under the condition*

$$0 < R_j(t) \le \int_0^{t^\wedge} m_j^\wedge(s)ds, j = 1, 2, 3, 4, \quad (52)$$

where $R_j(t)$ *and* $m_j^\wedge(s)$, $0 \le s \le t^\wedge$, $t^\wedge \le t \le T$, *are given positive differentiable functions, the system of respective equations has the unique positive differentiable solution. In addition, the desired functions a,j, m, and* α_μ *can be found by the iterative process*

$$R_j(t) = \int_{a,ji}^{t} m_{j,i}(s)ds,\ \alpha_\mu{}^\nu{}_{i+1}(t) = \int_{a,1i(t)}^{t} \alpha_{\mu i}(t,s)m_{1i}(s)ds,\ m_{i+1}(t) = \int_{a,2i(t)}^{t} \alpha_{\mu i}(t,s)m_{2,i}(s)ds;$$

$$\alpha_{\mu i}(t,s) = \alpha_\mu{}^\wedge(t,s),\ m_{ij}(s) = m_j^\wedge(s),\ s \le t^\wedge,\ m_{j,i} = x_j m_i,\ x_1 = x,\ x_2 = y,\ x_3 = z,\ x_4 = (1\text{-}x\text{-}y\text{-}z), \quad (53)$$

where x, y, and z are given, converging faster than geometrical progressions.

Proof. It is similar to the proof of the Theorems above.
The condition $a,j'(t) \ge 0$, $t \ge 0$, takes place iff $m_j(t) \ge R_j'(t)$, $j = 1, 2, 3, 4$. If $\mu = 1$, $\nu = 1$,

$$x_j\alpha_1(0)R_2(t) \ge R_j'(t), j = 1, 2, 3, 4. \quad (54)$$

It follows from

$$m_j(t) \ge x_j\alpha_1(0)R_2(t), j = 1, 2, 3, 4. \quad (55)$$

Theorem 9. *In the case of MM* (50, Ch. 5), *unless* $\mu = 1$, $\nu = 0$, *if*

$$0 < R_j(t) \le \int_0^{t^\wedge} m_j^\wedge(s)ds, j = 1, ..., 6, \quad (56)$$

where $R_j(t)$ *and* $m_j^\wedge(s)$, $0 \le s \le t^\wedge$, $t^\wedge \le t \le T$, *are given positive differentiable functions, then the system of respective equations has the unique positive differentiable solution. In addition, the desired functions a,j, m, and* α_μ *can be found by the iterative process*

$$R_j(t) = \int_{a,ji}^{t} m_{j,i}(s)ds,\ \alpha_\mu{}^\nu{}_{i+1}(t) = \int_{a,1i(t)}^{t} \alpha_{\mu i}(t,s)m_{1i}(s)ds,\ m_{i+1}(t) = \int_{a,2i(t)}^{t} \alpha_{\mu i}(t,s)m_{2i}(s)ds,$$

$$i = 1, 2, \dots;\ \alpha_{\mu i}(t,s) = \alpha_\mu{}^\wedge(t,s),\ m_{j,i}(s) = m_j^\wedge(s),$$

$$0 \le s \le t^\wedge,\ m_{j,i} = x_j m_i,\ x_1 = x,\ x_2 = y,\ x_3 = z,\ x_4 = e,\ x_5 = g,\ x_6 = 1-x-y-z-e-g, \quad (57)$$

where *x, y, z, e, and g are given, converging faster than geometrical progressions. After determination of the functions a,j, m, and* α_μ, *we can find the functions* β_κ, *c*, γ_κ, *f, and* cost(*f*) *from their respective equations in* (50, Ch. 5).

Proof. It is similar to the one above.

Note that we can consider some other variants in the case MM (50, Ch. 5). For example, we can suppose cost(f) to be given and R_6 not to be given. Then we can find $a,6$ from the equation for f (under the condition that f is less than $\int_0^t \gamma_\kappa m ds$), after this we can find R_6.

Theorem 10. *For MM* (51, Ch. 5), *unless* $\mu = 1$, $\nu = 0$, *under the condition*

$$0 < R_j(t) \leq \int_0^{t\wedge} m_j\wedge(s)ds, j = 1, ..., 8, \tag{58}$$

where $R_j(t)$ *and* $m_j\wedge(s)$, $0 \leq s \leq t\wedge$, $t\wedge \leq t \leq T$, *are given positive differentiable functions, the system of respective equations has the unique positive differentiable solution. In addition, the desired functions* a,j, m, and α_μ *can be found by the iterative process*

$$R_j(t) = \int_{a,ji}^{t} m_{j,i}(s)ds,\ \alpha_\mu{}^\nu{}_{i+1}(t) = \int_{a,1i(t)}^{t} \alpha_{\mu i}(t,s)m_{1i}(s)ds,$$

$$m_{i+1}(t) = \int_{a,2i(t)}^{t} \alpha_{\mu i}(t,s)m_{2,i}(s)ds,\ i = 1, 2, \dots ;\ \alpha_{\mu i}(t,s) = \alpha_\mu\wedge(t,s),\ m_{j,i}(s) = m_j\wedge(s),$$

$$0 \leq s \leq t\wedge,\ m_{j,i} = x_j m_i,\ x_1 = x,\ x_2 = y,\ x_3 = z,\ x_4 = e,\ x_5 = g,\ x_6 = r,\ x_7 = w,$$

$$x_8 = (1-x-y-z-e-g-r-w), \tag{59}$$

where x, y, z, e, g, r, and w are given, converging faster than geometrical progressions. After determination of the functions a,j, m, and α_μ *we can find the functions* β_κ, c, γ_κ, f, δ_κ, *N, and* cost(f) *from their respective equations in* (51, Ch. 5).

Proof. It is similar to the one above.

For MM (51, Ch. 5) we can also consider various variants. If, for example, we need to put $N(t) = R(t)$, then R_8 cannot be given in general. We can find $a,8$ from the equation

$$R(t) - R_8(t) = \int_{a,8(t)}^{t} [\delta_\kappa(t,s)w(s)m(s) - m_8(s)]ds = R_1 + \dots + R_7, \tag{60}$$

and after this we can find R_8 and R.

The sufficient condition for solvability of (60) is

$$R_1 + \dots + R_7 \leq \int_0^{t\wedge} [\delta_\kappa(t,s)w(s)m(s) - m_8(s)]ds. \tag{61}$$

We can interpret the relation (61) as the restriction on the distributions x_j, $j = 1, ..., 8$, which are found from the solutions of certain optimization problems.

3. Cases of Non-Uniqueness of Solutions

Let us consider the system of the equations

$$f(t) = m(t) + c(t) = \int_{a(t)}^{t} \alpha(t,s)y(s)m(s)ds + \int_{b(t)}^{t} \beta(t,s)[1-y(s)]m(s)ds,$$

$$R(t) = R_1(t) + R_2(t) = \int_{a(t)}^{t} y(s)m(s)ds + \int_{b(t)}^{t} [1-y(s)]m(s)ds, \; t^\wedge < t, \qquad (62)$$

with respect to a and b, $0 < a, b < t$.

Let $\alpha = \beta = \alpha(t) > 0$. Then the system (62) has no solution under $f/\alpha \neq R$, and infinite number of solutions under $f/\alpha = R$, $R(t) < \int_0^t m(s)ds$.

We have found in Ch. 5 the cases when the system (62) has two solutions, in particular, coincident solutions or a *two-multiple solution*, which is related to a case of CAO reproduction. We also have the cases of *two-multiple repeated* solutions for the model

$$m(t) = \int_{a(t)}^{t} \alpha(t,s)y(s)m(s)ds, \; c(t) = \int_{b(t)}^{t} \beta(t,s)[1-y(s)]m(s)ds,$$

$$R(t) = \int_{a(t)}^{t} y(s)m(s)ds + \int_{b(t)}^{t} [1-y(s)]m(s)ds, \; 0 < t^\wedge \leq t. \qquad (63)$$

It is not difficult to verify that under the condition

$$m(t) = \text{constant}, \; y(t) = y^\wedge = 0, \; t \in [0, t^\wedge]; \; \int_{(2k-1)t^\wedge}^{t} \alpha(t,s)ds = 1, \; \beta(t,s) = 2[s-(t-t^\wedge)]/m^\wedge,$$

$$s \in [t\text{-}t^\wedge, (2k-1)t^\wedge], \; y(t) = 1, \; c(t) = (2kt^\wedge - t)^2, \; t \in [(2k-1)t^\wedge, 2kt^\wedge]; \; \int_{t-t^\wedge}^{2kt^\wedge} \alpha(t,s)ds = 1, \; \beta(t,s) =$$

$$2(t-s)]/m^\wedge, \; s \in [2kt^\wedge, t], \; y(t) = 0, \; c(t) = (2kt^\wedge - t)^2, \; t \in [2kt^\wedge, (2k+1)t^\wedge]; \; k = 1, 2, \ldots, \qquad (64)$$

we have *two-multiple repeated* solutions

$$b(t) = t - t^\wedge, \; m(t) = m^\wedge, \; R(t) = m^\wedge t^\wedge, \; t \in [(2k-1)t^\wedge, 2kt^\wedge],$$

$$a(t) = t - t^\wedge, \; m(t) = m^\wedge, \; P(t) = m^\wedge t^\wedge, \; t \in [2kt^\wedge, (2k+1)t^\wedge], \; k = 1, 2, \ldots. \qquad (65)$$

Note that the condition (65) is valid for

$$\alpha(t,s) = 2\delta[s-(2k-1)t^\wedge], \; s \in [(2k-1)t^\wedge, 2kt^\wedge]; \; \alpha(t,s) = 2\delta(2kt^\wedge - s), \; s \in [2kt^\wedge, (2k+1)t^\wedge], \qquad (66)$$

where δ is Dirac generalized function.

The relations (65) and (66) can be interpreted as imitations of unlimited reproduction of CAO, in particular, unlimited reproduction of cancer cells.

We can summarize the results of this section as the following

Theorem 11. *In the class of MM considered, under the solution of the determined system of equations may be encountering no solutions, a unique solution, two-multiple solutions, and unlimited number solutions.*

4. On Stability of Solutions

We show that solutions of equations of MM under consideration have the *technical stability*, that is, any infinitesimal variations of the input data result in infinitesimal variations of the solutions. Let δx be a variation for any x. For the equations (5, 6, Ch. 5) and under the condition of Theorem 2 we have for $t \in [t^\wedge, T]$, regarding the main terms,

$$\delta m = \int_{a(t)}^{t} [\alpha(t,s)y(s) - \alpha(t,a(t))y(a(t))]\delta m(s)ds + \int_{a(t)}^{t} \delta[\alpha(t,s)y(s)]m(s)ds + \alpha(t,a(t))y(a(t))\delta R,$$

$$\delta c = \int_{a(t)}^{t} \{\beta(t,s)[1-y(s)] - \beta(t,a(t))[1-y(a(t))\}\delta m(s)ds + \int_{a(t)}^{t} \delta\{\beta(t,s)[1-y(s)]\}m(s)ds +$$

$$\beta(t,a(t))[1-y(a(t))]\delta R(t),\ \delta a = [\int_{a(t)}^{t} \delta m(s)ds - \delta R(t)]/m(a(t)), \quad (67)$$

where m, a is the solution of the system in question, and $\delta m(t)$ is given for $0 \le t \le t^\wedge$. In a similar way, we can receive $\delta m(t)$, $\delta c(t)$ and $\delta a(t)$ under the conditions of Theorems 3 and 4.

For the equations (9, Ch. 5) and under the condition of Theorem 5, we have

$$\delta f = -\alpha y m\,|_{a(t)}\delta a(t) - \beta(1-y)m\,|_{b(t)}\delta b(t) + \int_{a(t)}^{t} \delta(\alpha y m)ds + \int_{b(t)}^{t} \delta[\beta(1-y)m]ds,$$

$$\delta R = -ym\,|_{a(t)}\delta a(t) - (1-y)m\,|_{b(t)}\delta b(t) + \int_{a(t)}^{t} \delta(ym)ds + \int_{b(t)}^{t} \delta[(1-y)m]ds,$$

$$\delta m = -\alpha y m\,|_{a(t)}\delta a(t) + \int_{a(t)}^{t} \delta(\alpha y m)ds,\ \delta c(t) = -\beta(1-y)m\,|_{b(t)}\delta b(t) + \int_{b(t)}^{t} \delta[\beta(1-y)m]ds. \quad (68)$$

The determinant of the system (68)

$$\Delta = ym\,|_{a(t)}(1-y)m\,|_{b(t)}\delta b(t)[\alpha(t,a(t)) - \beta(t,b(t)] \neq 0. \quad (69)$$

So, the system (68) has the unique solution with respect to $\delta a(t)$, $\delta b(t)$.

For the equations (46, Ch. 5, $\nu = 0$) and under the condition of Theorem 6, we have

$$\delta\alpha_\mu = \int_{a(t)}^{t} [\alpha_\mu(t,s)x(s) - \alpha_\mu(t,a(t))x(a(t))]\delta m(s)ds + \int_{a(t)}^{t} \delta[\alpha_\mu(t,s)x(s)]m(s)ds + \alpha_\mu x\,|_a^t\,\delta R(t),$$

$$\delta m = \int_{a(t)}^{t} [\alpha_\mu(t,s)y(s) - \alpha_\mu(t,a(t))y(a(t))]\delta m(s)ds + \int_{a(t)}^{t} \delta[\alpha_\mu(t,s)y(s)]m(s)ds + \alpha_\mu y\,|_a^t\,\delta R(t),$$

$$\delta\beta_\kappa = \int_{a(t)}^{t} [\alpha_\mu(t,s)z(s) - \alpha_\mu(t,a(t))z(a(t))]\delta m(s)ds + \int_{a(t)}^{t} \delta[\alpha_\mu(t,s)z(s)]m(s)ds + \alpha_\mu z \Big|_a^t \delta R(t), \quad (70)$$

and

$$\delta c = \int_{a(t)}^{t} \{\beta_\kappa(t,.)[1-x(.)-y(.)-z(.)]\Big|_a^s\}\delta m(s)ds + \int_{a(t)}^{t} \delta\{\beta_\kappa(t,s)[1-x(s)-y(s)-z(s)]\}m(s)ds +$$

$$\beta_\kappa(t,a(t))[1-x(a(t))-y(a(t))-z(a(t))]\delta R(t),\ \delta a = [\int_{a(t)}^{t} \delta m(s)ds - \delta R(t)]/m(a(t)). \quad (71)$$

So, we have the linear system of two equations for determination of $\delta\alpha_\mu(t)$ and $\delta m(t)$.

In a similar way, we can receive the estimates of $\delta\alpha_\mu(t)$, $\delta m(t)$, $\delta\beta_\kappa(t)$, $\delta c(t)$, $\delta\gamma_\kappa(t)$, $\delta f(t)$, $\delta\delta_\kappa(t)$, $\delta N(t)$, and $\delta aj(t)$, $j = 1, \ldots, 10$, under the conditions of Theorems 7-10.

However, usual stability in mathematics means that the above-mentioned estimates do not depend on T. Let us consider, for simplicity, the system

$$\delta m(t) = \int_{a(t)}^{t} [\alpha(t,s)y(s) - \alpha(t,a(t))y(a(t))]\delta m(s)ds + \int_{a(t)}^{t} \delta[\alpha(t,s)y(s)]m(s)ds +$$

$$\alpha(t,a(t))y(a(t))\delta R(t),\ \delta a(t) = [\int_{a(t)}^{t} \delta m(s)ds - \delta R(t)]/m(a(t)),\ t \in [t^\wedge, T], \quad (72)$$

where m, a is the solution of the system in question, and $\delta m(t)$ is given for $0 \le t \le t^\wedge$. If $t - a(t)$ remains bounded, then obviously the technical stability is the same as usual stability.

In general, we introduce the definition of *uniform stability*, following to Y. Kuang [116], assuming that $R(t) = \int_{a(t)}^{t} m(s)ds$ is always bounded. The *solution* of the system (5), Ch. 5 is said to be *uniformly stable* if, for any $\varepsilon > 0$, there exists $\delta = \delta(\varepsilon)$ such that the initial data variation in the vicinity of the sphere of the radius δ in order (in the space of C-type) implies the solution variation in the sphere of the radius ε in order independent of T.

One of the principle differences between our MM and DDE is the presence in our MM of the kernels depending on two variables. So, we cannot simply carry over the results from [116], in particular, the method of Liapunov functionals, to our case. Instead, we should most likely use the *Volterra-type method* or "Volterra typical" properties (M. Vath [193]) based, in particular, on the relations of Theorem 1, (11)-types. Applying these relations to the equations (72), we can conclude that the necessary condition for the independence of $\delta m(t)$ on T is non-fulfillment

$$\alpha(t,s)y(s) - \alpha(t,a(t))y(a(t)) > 0,\ s, t \in [0, T]\backslash T',\ T' \subset [0, T], \quad (73)$$

where T' is a restricted set. If the opposite to (73) takes place, then the sufficient condition is any test for uniform convergence of the alternating series in the relations of (11)-types obtained from (72) to the absolute and uniform integrable kernels. In particular (c f L. Smale [172], p. 472), an alternating series $a_1 - a_2 + a_3 - a_4 + \cdots$ is uniformly convergent if each term a_n is uniformly less than the preceding term a_{n-1}, and if the limit of the nth term is 0.

We can summarize the results of this section as the following:

Theorem 12. *For the class of MM considered we have the technical stability that can be estimated by the formulae* (67)--(72)-*types and in certain cases* (e.g. (73)) by *the uniform stability.*

5. Cases of Oscillating, Wave, and Some Other Solutions

Let α, β, and y be constants. Then for (5, 6, Ch. 5) we have

$$m(t) = \alpha y R(t),\ c(t) = \beta(1-y)R(t),\ m(t) + c(t) = f(t),\ \int_{a(t)}^{t} m(s)ds = R(t), \tag{74}$$

from which

$$R(t) = f(t)[\alpha y + \beta(1-y)],\ m(t) = \alpha y f(t)[\alpha y + \beta(1-y)],$$

$$c(t) = \beta(1-y)f(t)[\alpha y + \beta(1-y)],\ \int_{a(t)}^{t} f(s)ds = f(t)/[\alpha y]. \tag{75}$$

It means if f is a periodic function, then the functions R, m, c are periodic as well. Let

$$\alpha(t,s) = \alpha[1-\cos\omega(t-s)],\ a(t) = t-a,\ y = \text{constant}, \tag{76}$$

where α, ω, and a are also constants. Then we have

$$m(t) = \alpha y\int_{t-a}^{t} [1-\cos\omega(t-s)]m(s)ds,\ R(t) = \int_{t-a}^{t} m(s)ds, \tag{77}$$

from which

$$m''(t) + \omega^2 m(t) = \alpha y\omega^2 R(t) - \alpha y(1-\cos\omega a)m\hat{}'(t-a) - \alpha y\omega\sin\omega a m\hat{}(t-a),$$

$$t \in [a, 2a],\ m(a) = \alpha y\int_0^a [1-\cos\omega(a-s)]m\hat{}(s)ds,$$

$$m'(a) = -\alpha y(1-\cos\omega a)m\hat{}(0) + \alpha y\omega\int_0^a \sin\omega(a-s)m\hat{}(s)ds, \tag{78}$$

The characteristic equation for (78) and its roots are

$$k^2 + \omega^2 = 0,\ k_{1,2} = \mathbf{i}\omega,\ \mathbf{i} = (-1)^{1/2}. \tag{79}$$

It means that the solution $m(t)$ of the system (77) is rather general oscillating function depending on the right side of the equation and the initial data in (78).

Let us consider the system

$$m = \int_{t-a}^{a} \alpha y m\hat{}ds + \int_{a}^{t} \alpha y m ds,\ c = \int_{t-a}^{a} \beta(1-y)mds + \int_{a}^{t} \beta(1-y)mds,\ t \in [a, 2a]. \quad (80)$$

For $\alpha = \alpha(s)$, $\beta = \beta(s)$ it follows from (80) that

$$m' = (\alpha y m)(t) - (\alpha y m)(t-a),\ c' = (\beta(1-y))(t) - (\beta(1-y))(t-a). \quad (81)$$

It means that, in particular, when $\alpha y m$ and $\beta(1-y)$ are arbitrary constants for $t \in [0, a]$ and $\alpha y m$ and $\beta(1-y)m$ are arbitrary positive functions of m, c for $t \in [a, 2a]$, the system (81) is an arbitrary autonomous system of differential equations of the second order. One can apply to this system the well-known qualitative methods of investigation based on construction of the phase picture of the solutions.

Let, in particular,

$$\alpha m = A - A'm + A''c,\ \beta m = B - B'm + B''c,$$

$$Ay - \alpha y m(t-a) = d',\ B(1-y) - (\beta(1-y))(t-a) = d'', \quad (82)$$

where $A, A', A'', B, B', B'', d', d''$ are positive constants, $\Delta = A''B' - A'B'' \neq 0$. Then

$$m' = -d' + y(-A'm + a''c),\ c' = -d'' + (1-y)(-B'm + B''c). \quad (83)$$

The *singular point* of the system (83) is

$$m^* = [B''d'/y - A''d'/(1-y)]/\Delta,\ c^* = [B'd'/y - A'd''/(1-y)]/\Delta.. \quad (84)$$

The sufficient condition for positiveness of m^*, c^* is

$$B'/A' > B''/A'' > d''y/[d'(1-y)] > 0,\ \Delta > 0. \quad (85)$$

We can rewrite the system (83) in the form

$$d(m-m^*)/dt = y[-A'(m-m^*)+A''(c-c^*)],$$

$$d(c-c^*)/dt = (1-y)[-B'(m-m^*)+B''(c-c^*)]. \quad (86)$$

The *characteristic equation* for (86) is

$$k^2 - k[B''(1-y) - A'y] + \Delta y(1-y) = 0, \quad (87)$$

the roots of which

$$k_{1,2} = [B''(1-y) - A'y]/2\ \{[B''(1-y) - A'y]^2 - 4\Delta y(1-y)\}^{1/2}/2. \quad (88)$$

We consider the phase picture's variation while y is changed from 0 to 1. We need to find y_1, y_2 for which the discriminant of the equation (87)

$$D(y) = [B''(1-y)-A'y]^2 - 4\Delta y(1-y) =$$

$$y^2[(A'-B'')+4A''B'] - 2[2A''B'-B''(A'-B'')]y + B''^2 = 0. \tag{89}$$

We have

$$y_{1,2} = [A''B' + B''^2 + \Delta \ \ 2(A''B'\Delta)^{1/2}]/(2A''B' + 2\Delta + A'^2 + B'^2). \tag{90}$$

We also have y critical,

$$y_{cr} = B''/(A' + B''), \tag{91}$$

for which $B''(1-y) - A'y = 0$.

It is not difficult to verify that $0 < y_1 < y_{cr} < y_2 < 1$.

Thus, we have the following result.

Theorem 13. *The singular point* (m^*, c^*) (84) *of the system* (83) *is*
i) *the unsteady node for* $0 < y \le y_1$;
ii) *the unsteady focus (increasing by amplitude oscillations) for* $y_1 < y < y_{cr}$;
iii) *the center (undamped oscillations) for* $y = y_{cr}$;
iv) *the steady focus (decreasing by amplitudes oscillations) for* $y_{cr} < y < y_2$;
v) *the steady node for* $y_2 \le y < 1$.

If we put

$$m' = -d' + \alpha y m = c, \ c' = -d'' + \beta(1-y)m = -m + \delta(1-m)c, \tag{92}$$

where δ is a parameter, then we have the classic system of *Van-der-Pol's equations* which describes steady auto-oscillations of variable m or c. On the plane (m, c) this system has the phase picture in the form of the steady *limiting cycle*.

Setting

$$m' = (m-m\hat{})^2 + c - c\hat{}, \ c' = \beta(1-y)m \ (\alpha y m = (m-m\hat{})^2 + c - c\hat{}), \tag{93}$$

we obtain the case of the so-called *catastrophe of folding* in the vicinity of the point $(m\hat{}, c\hat{})$. Setting in (17, Ch. 5) $n = 3$, $r = 1$, $s = 2$,

$$m' = (m-m\hat{})^3 + (c_1 - c_1\hat{})(m-m\hat{}) + (c_2 - c_2\hat{}), \ c_i' = \beta_i z_i m, \ i = 1, 2, \tag{94}$$

we obtain the case of the *catastrophe of assembling* in the vicinity of the point (m^, c_1^, c_2^), and so on.

If we put

$$\alpha m = A + 4m'(t-d)[1-m'(t-d)], Ay-(ym\alpha)(t-a) = 0, d \ll a, t \in [a, 2a]. \quad (95)$$

then we have

$$m'(t) = 4ym'(t-d)[1-m'(t-d)], t \in [a, 2a]. \quad (96)$$

It is well known (see, for example, C. Marzec, E. Spiegel [128]) that under increasing y from 0 to 1 the solution of the equation (96) has the limiting cycles (on the plane ($m'(t-d)$, $m'(t)$)) and the so-called *strange attract* (with the random oscillations). Note that in spite of a rather artificial reduction (82) and (93)-(96), the reduction makes sense in the theory of MM under consideration.

Indeed, input data of the models are usually the functions of α, β-types that can be arbitrary, but nonnegative. So, it is rather important to consider some special cases of these functions and by the same token to understand better that their (and m and c) non-negativity does not mean some kind of restriction on the possibility of applications of the models.

Various particular classes of the models can be also obtained by a slightly different way. Let $\alpha = \alpha(M(t), C(t))$, $\beta = \beta(M(t), C(t))$, $m(t) = M'(t)$, $c(t) = C''(t)$. Then as a particular case of MM (5), (6), Ch. 5, where y is constant, we have

$$M'(t) = \alpha(M(t), C(t))yR(t), \; C'(t) = \beta(M(t), C(t))yR(t). \quad (97)$$

Thus, by this way we actually have the general system of ODY of the second order. We can obtain the phase portrait on the plane (M, C) of N-shaped (*hysteresis*) or Z-shaped as in the cases of the cell cycle or electrical activity of neurons and pancreatic β-cells, respectively (see Ch. 2). In the case of a continuous MM of CAO (for example, MM (19, Ch. 5)), we can obtain in a similar way the equations that have different *wave solutions*.

The usual method is the supplement of the so-called diffusive terms to our models:

$$\partial M(r,t)/\partial t = m(r,t) = \int^t_0 \int^r_0 \alpha(t,r,s,v)\lambda(t,r,s,v)y(s,v)m(s,v)dvds +$$

$$\partial/\partial l[D_1(M,C)\partial M/\partial l] + A, \partial C(r,t)/\partial t = c(r,t) =$$

$$\int^t_0 \int^r_0 \beta(t,r,s,v)\mu(t,r,s,v)z(s,v)m(s,v)dvds + \partial/\partial l[D_2(M,C)\partial C/\partial l] + B,$$

$$r = (x_1, x_2, x_3), v = (v_1, v_2, v_3), l = (l_1, l_2, l_3), \quad (98)$$

where D_1 and D_2 are coefficients of diffusion along the direction l from the point r; A and B are constants guaranteeing non-negativity of m and c.

Note that various cases of the MM (98) were investigated (see V. Ivanov [92]). In particular, the cases corresponding to elliptic, hyperbolic and parabolic-type equations were found and examined.

6. Asymptotic Properties of Solutions

In the case of CAO and MM (5, 6, Ch. 5), we assume that the input data fulfil the condition of Theorem 2 and have the following *asymptotic properties*

$$\alpha(t,s)y(s) \cong \alpha y,\ \beta(t,s)[1-y(s)] \cong \beta(1-y),\ R(t) \cong R, \tag{99}$$

where αy, $\beta(1-y)$, and R are positive constants. Here for any x and y the expression $x \cong y$ is identical to $|x - y| = O(|y|/t)$ or, what is the same, there exist (J. Murray [142]) positive constants C and K such that $|x - y| < C|y|/t$ whenever $t > K$.

From the condition of Theorem 2 and (99) we have

$$0 < \boldsymbol{m}(\alpha y)R(t) < m(t) < \boldsymbol{M}(\alpha y)R(t),\ \boldsymbol{m}(\alpha y)\boldsymbol{m}(R)(t\text{-}a(t)) < R(t) < \boldsymbol{M}(\alpha y)\boldsymbol{M}(R)(t\text{-}a(t)), \tag{100}$$

and hence

$$0 < \boldsymbol{m}(R)/[\boldsymbol{M}(\alpha y)\boldsymbol{M}(R)] < t-a(t) < \boldsymbol{M}(R)/[\boldsymbol{m}(\alpha y)\boldsymbol{m}(R)]. \tag{101}$$

From (99) - (101) it follows that

$$m(t) \cong \alpha yR,\ c(t) \cong \beta(1-y)R,\ t-a(t) \cong 1/\alpha y. \tag{102}$$

We assume, in addition, that the input data fulfil the condition of Theorem 3 and have the *asymptotic properties*

$$\alpha(t,s)y(s) \cong C_1/t,\ \beta(t,s)[1-y(s)] \cong C_2/t,\ f(t) \cong C_3/t, \tag{103}$$

where C_i, $i = 1, 2, 3$, are positive constants. From Theorem 3 and (103) we have

$$0 < C_4/tR(t) < m(t) < C_5/tR(t),\ 0 < C_6/tR(t) < f(t) < C_7/R(t)t,\ C_8/t < f(t) < C_9/t, \tag{104}$$

where C_i, $i = 4, \ldots, 9$, are certain positive constants, and hence

$$0 < C_8/C_7 < R(t) < C_9/C_6,\ C_4C_8/(C_7t) < m(t) < C_5C_9/(C_6t),$$

$$t\text{-}a(t) = t[1-a(t)/t],\ D_1 < \ln\,[t/a(t)] < D_2,\ D_1 = C_6C_8/(C_5C_7C_9),$$

$$D_2 = C_7C_9/(C_4C_6C_8). \tag{105}$$

From (103)-(105) it follows that

$$m \cong C_1C_3/[(C_1+C_2)t],\ c \cong C_2C_3/[(C_1+C_2)t],\ R \cong C_3/(C_1+C_2),\ t-a \cong t(1-\exp(-1/C_1)). \quad (106)$$

In a similar way, we can obtain the asymptotic properties of the solutions in the case of the system (9, Ch. 5) under the conditions of Theorems 4 and 5.

For the system (46, Ch. 5) we assume that the input data fulfil the condition of Theorem 6 and have the following asymptotic properties $x(s) \cong x$, $y(s) \cong y$, $R(t) \cong R$, where x, y, and R are positive constants. Then in the case of $\nu = 1$, $\mu = 1$, $c_1 = 0$ we have

$$\alpha \cong \alpha\text{^}(0) + x/yM,\ M'' \cong x[M - M(a)]M',\ M - M(a) \cong R, \quad (107)$$

and

$$M \cong \exp(xRt)/xR,\ \alpha \cong x/y\exp(xRt)/xR,\ a \cong t,\ m \cong \exp(xRt). \quad (108)$$

For arbitrary $R(t)$, $R'(t) > 0$, and positive constants x, y, using the relations $M - M(a) = R$, $m - m(a)a' = R'$, we reduce the case $\nu = 1$, $\mu = 1$, $c_1 = 0$ to the equation of the form (10), asymptotic properties of which have been investigated in J. Murray [142]. In the other cases as well as for the system (49-51, Ch. 5) under the conditions of Theorems 7-10, we can obtain similar relations.

We can summarize the results of this section as the following:

Theorem 14. *In the class of MM considered we have the asymptotic properties of solutions expressed by the relations of* (99) - (108)*-types.*

The result (106) seems to be unexpected because we can hold almost a constant for the functioning biomass R in spite of the fact that α, β, m, c, $f \to 0$ for $t \to \infty$.

7. Some Additions and Remarks

It is natural that with the new class of MM under consideration, many problems of investigation remain open. Among them are the problems of existence and uniqueness solutions for the system of the equations (17-19, Ch. 5) and for the equations of a continuous MM of (98)-type. The respective existence and uniqueness theorems under more general assumptions in the sense of C. Caratheodory [31] can be found in V. Ivanov [91]. It seems interesting to consider the models of (46 and 49-51, Ch. 5)-type in the cases $\nu > 1$ and, for example, in the case $\nu = -1$, which means $\alpha^{-1} = \int_0^t \alpha(s)ds$.

All Theorems 1-10 can be generalized in the cases when the functions of α, β-types depend on the other elements of the models. Let us dwell on the respective generalization of Theorem 2 when instead of MM (5, 6, Ch. 5) we have

$$m(t) = \int_{a(t)}^{t} \alpha[t,s,m(s),c(s)]y(s)m(s)ds,\ c(t) = \int_{a(t)}^{t} \beta[t,s,m(s),c(s)][1-y(s)]m(s)ds, \quad (109)$$

where α, y, β are given and m, a, and c are sought for.

Theorem 2'. *Let in addition to the assumptions of Theorem* 2 α *and* β *in* (109) *comply with the conditions of Lipschitz over m, c for which it is sufficient that* $|\partial\alpha/\partial m|$, $|\partial\alpha/\partial c|$, $|\partial\beta/\partial m|$, $|\partial\beta/\partial c|$ *be bounded. Then the system* (109) *has the unique nonnegative solution m, a, c* $\in C^1([t^\wedge, T])$, *which can be found by the iterative process*

$$m_{i+1}(t) = \int_{a(t)}^{t} \alpha[t,s,m_i(s),c_i(s)]y(s)m_i(s)ds,\ R(t) = \int_{a(t)}^{t} m_i(s)ds,\ c_{i+1}(t) =$$

$$\int_{a,i(t)}^{t} \beta[t,s,m_i(s),c_i(s)][1-y(s)]m_i(s)ds,\ i = 1, 2, \dots,\ m_1(t) = \mathbf{m}(\alpha y,t)R(t),\ t^\wedge \le t \le T, \qquad (110)$$

where $c_1(t)$ *is an arbitrary integrable function,* $t \in [t^\wedge, T]$, *which converges at the rate exceeding that of geometrical progressions.*

Let us consider more the *problem of* necessary and sufficient conditions for the *solution solvability* on the example of the system (5), (6), Ch. 5. It is easy to see that the condition

$$0 \le R(t) \le M(t^\wedge) + \int_{t^\wedge}^{t} m^+(s)ds,\ m^+(t) = m(t^\wedge) + \int_{t^\wedge}^{t} \alpha(t,s)y(s)m^+(s)ds,\ t^\wedge \le t, \qquad (111)$$

is necessary for solvability.

Theorem 2". *Under the condition of Theorem* 2 *and the additional conditions*

$$D > \alpha(t,s)y(s) > d > 0,\ t^\wedge > \mathbf{M}(R)/[d\mathbf{m}(R)] > d^* > 0, \qquad (112)$$

the system (5), (6), Ch. 5 *is solvable, iff the condition* (111) *holds.*

The proof can be based on the relation of (11)-type:

$$m(t) = u(t) + \int_{t^\wedge}^{t} g(t,s)u(s)ds,\ g(t,s) = \sum_{k=0}^{\infty} g_{k+1}(t,s),\ g_1(t,s) = \alpha(t,s)y(s),$$

$$g_{k+1}(t,s) = \int_{s}^{t} g_1(t,z)g_k(z,s)dz,\ k = 1, 2, \dots;\ u(t) = \int_{a(t)}^{t^\wedge} g_1(t,s)m^\wedge(s)ds, \qquad (113)$$

and

$$R(t) = \int_{a(t)}^{t^\wedge} m^\wedge(s)ds + \int_{t^\wedge}^{t} u(s)[1 + \int_{s}^{t} g(s,r))dr]ds,\ t^\wedge_1 \ge t \ge t^\wedge,\ a(t^\wedge_1) = t^\wedge. \qquad (114)$$

On the strength of (112), this last equation has the unique solution $a(t)$, $0 \le a(t) \le t^\wedge$. Later, when $t^\wedge < a(t)$, the role $t^\wedge_1$ plays $t^\wedge_2$, $a(t^\wedge_2) = t^\wedge_1$, and so on. Since we have

$$t^\wedge_r - t^\wedge_{r-1} = t^\wedge_r - a(t^\wedge_r) > \mathbf{M}(R)/[D\mathbf{m}(R)] > (D/d)d^* > 0 \qquad (115)$$

independently of t, any time-interval can be passed. It should be noted that without the restriction on the duration of the prehistory we could weaken the condition (112).

Chapter 15

INVESTIGATION OF OPTIMIZATION PROBLEMS

Preliminary knowledge into the general theory of control and the examples of analytic solutions for some optimization problems are given. The conditions of existence of solutions are investigated. Stability of solutions and sequence of formulae for determining gradients of specific functionals are established. The structure of solutions for different optimization problems is considered in detail. The conditions of uniqueness of solutions and the example of non-uniqueness of solutions are adduced. Asymptotic properties of some solutions are derived.

1. Preliminary Concepts and Results

In a rather important case, MM (5-10-types, Ch. 5) are described by the following system of equations and inequalities:

$$x_i(t) = \sum_{j=1}^{n} \int_{z_j(t)}^{t} k_{ij}(t,s)y_{ij}(s)x_j(s)ds,\ i = 1,\dots, n; \quad (1)$$

$$p_i(t) = \sum_{j=1}^{n} \int_{z_j(t)}^{t} k_{ij}(t,s)y_{ij}(s)x_j(s)ds,\ i = n+1,\dots, 2n; \quad (2)$$

$$0 \le y_{ij} \le 1,\ 0 \le z_j(t) \le t,\ 0 \le s \le t,\ 0 < t^\wedge \le t \le T; \quad (3)$$

with the initial conditions

$$x_j(s) = x_j^\wedge(s),\ y_{ij}(s) = y_{ij}^\wedge(s),\ s \in [0, t^\wedge],\ z_j(t^\wedge) = z_j^\wedge,\ j = 1, \dots, n,\ i = 1, \dots, 2n. \quad (4)$$

We consider the functions k_{ij} and p_i to be given, y_{ij} to be control functions, and x_j and z_j to be phase variables. Note that some of y_{ij} can be linearly dependent. Then the control y is a collection $\{y_{ij}\}$ after elimination of linearly dependant elements.

Let us consider the *problem of optimal control*: to determine the control $y(t)$, $t \in [t^\wedge, T]$, with the restrictions (1)-(4) and the respective variables $x = (x_1, \dots, x_n)$ and $z = (z_1, \dots, z_n)$ for which the functional

$$I = \int_{t^\wedge}^{T} \Phi(x,y,z,t)dt = \min, \quad (5)$$

where Φ is a differentiable function of x, y, z.

We will assume that the given functions k_{ij} and p_i are continuous and the control y is sought in the class of piecewise continuous functions with the restrictions (3, 4): $y \in U$.

If the system (1, 2) is solvable, then its solution x is continuous. If the class U of controls turns out to be too restricted, we can assume that the vector function y is measurable, $y_{ij} \in L_\infty([0, T])$, or relaxed (generalized) (L. Pontryagin et al., [153]).

Let us introduce the *function of LaGrange*

$$L(x,y,z,\varphi) = \int_{t^\wedge}^{T} \{\Phi(x,y,z,t) + \sum_{i=1}^{n} \varphi_i(t)[x_i(t) - \sum_{j=1}^{n} \int_{z,(t)}^{t} k_{ij}(t,s)y_{ij}(s)x_j(s)ds] +$$

$$\sum_{i=n+1}^{2n} \varphi_i(t)[p_i(t) - \sum_{j=1}^{n} \int_{z,(t)}^{t} k_{ij}(t,s)y_{ij}(s)x_j(s)ds]\}dt, \tag{6}$$

where $\varphi(t) = \{\varphi_1(t), \ldots, \varphi_{2n}(t)\}$ is the so-called LaGrange's coefficients. Under the assumptions accepted, there exists the *derivative L' by Fresher* of x, y, z determined by the formula

$$\delta L = (L',(\delta x,\delta y,\delta z)) = \int_{t^\wedge}^{T} [\sum_{j=1}^{n} \partial L/\partial x_j \delta x_j(t) +$$

$$\sum_{i=1}^{2n} \sum_{j=1}^{n} \partial L/\partial y_{ij} \delta y_{ij}(t) + \sum_{j=1}^{n} \partial L/\partial z,j \delta z,j(t)]dt, \tag{7}$$

where δu means a variation for any u and

$$\partial L/\partial x_j = \partial\Phi/\partial x_j(t) + \varphi_j(t) - \sum_{i=1}^{n} y_{ij}(t)\int_{J,j(t)} k_{ij}(s,t)n_i(s)ds,\ \partial L/\partial y_{ij} = \partial\Phi/\partial y_{ij}(t) -$$

$$x_j(t)\int_{J,j(t)} k_{ij}(s,t)\varphi_i(s)ds, \partial L/\partial z,j = \partial\Phi/\partial z,j(t) + \sum_{i=1}^{2n} k_{ij}(t,z,j(t))y_{ij}(z,j(t))x_j(z,j(t))n_i(t). \tag{8}$$

The formulae (7), (8) are obtained by means of L increment evaluation after changing integration order. The $J,j(t) \subset [t, T]$ is determined in the following way. We divide $[t^\wedge, T]$ on the intervals of monotonousness of the function $z,j(t)$ and denote $z,j^{(1)}(t)$, $z,j^{(3)}(t)$, ... as the increasing branches of $z,j(t)$ and $z,j^{(2)}(t)$, $z,j^{(4)}(t)$, ... as its decreasing branches. After this we evaluate the reciprocal functions

$$z,j^{(k)\,-1}(t) = z,j^{(k)\,-1}(d_j^{k-}),\ t^\wedge \le t \le d_j^{k};\ z,j^{(k)\,-1}(t) = z,j^{(k)\,-1}(t),\ d_j^{k-} = \min z,j^{(k)} \le t \le d_j^{k+} = \max z,j^{(k)},$$

$$z,j^{(k)\,-1}(t) = z,j^{(k)\,-1}(d_j^{k+}),\ d_j^{k+} \le t \le T. \tag{9}$$

Then

$$J,j(t) = [t, z,j^{(1)\,-1}(t)] \cup [z,j^{(2)\,-1}(t), z,j^{(3)\,-1}(t)] \cup \ldots. \tag{10}$$

Note that if $z,j'(t) \ge 0$, then the formulae (9, 10) can be considerably simplified:

$$J,j(t) = [t, Z,j^{-1}(t)],\ Z,j^{-1}(t) = z,j^{-1}(t),\ t^\wedge \le t \le z,j(T),\ Z,j^{-1}(t) = T,\ z,j(T) \le t \le T. \tag{11}$$

We assume that the phase variables x and z pertain to the open domain. Then we obtain the *conjugate equations*

$$\partial L/\partial x_j = 0, \partial L/\partial z_j = 0, j = 1, ..., n, \tag{12}$$

which will have the unique solution with respect to $\varphi_1, ..., \varphi_{2n}$, $t \in [t^\wedge, T]$, if the respective determinant does not equal to 0 under any y_{ij} and z_j from (3), (4). Then there exists the *gradient of the functional* (5) over y:

$$I'(y) = \{\partial L/\partial y_{ij}\}_{j=1,...,n}^{i=1,...,2n} \tag{13}$$

The necessary condition of optimum (minimum) has the form

$$(I'(y^*), (y-y^*)) \geq 0, y \in U, \tag{14}$$

where (u, v) means the scalar product in Hilbert space of functions u and v, y^* is any solution of the problem (1)-(5). For the intrinsic solutions, (14) is identical to

$$I'(y^*) = 0. \tag{15}$$

It should be noted that MM (51, 53-55, Ch. 5) do not belong to the class of MM (1)-(4), but the above technique of the construction of the relations (6)-(15) remains quite similar there.

Let us introduce now the main additional concepts and results following the monograph J. Warga [203]. A *general problem of optimization* is one in which some general sets W, X_1, and $C_1 \subset X_1$ and functions g_0: $W \to R$ and g_1: $W \to X_1$ are given, and it is desired to minimize g_0 on the set $\{w \in W \mid g_1(w) \in C_1\}$. W is usually a subset of the Cartesian product of three sets: the *sets Y of states*, whose elements are usually functions with values in some vector space, the *set U of control functions* on some measure space T into some space R, and the set B *of control parameters* (endowed with some structure).

Let us restrict attention to W such that points $(y, u, b) \in Y \times U \times B$ satisfy the *equation of motion* $y = F(y,u,b)$ and, among such points, consider those that "come close" to satisfying the restriction $g_1(y,u,b) \in C_1$ and, in so doing, tend to minimize $g_0(y,u,b)$. Later we reduce the problem to the case when the equation of motion is a *functional-integral equation* and $C_1 = \{0\}$.

Now the problem is to find

$$\inf g_0(y, u, b) \tag{16}$$

on the set

$$H(U) = \{(y, u, b) \in Y \times U \times B \mid y = F(y, u, b)\}, \tag{17}$$

where the equation of motion or state equation

$$y(t) = F(y, u, b) = \int f[t, s, \xi(y)(s), u(s), b(t)]ds, t \in T = [0, T], \tag{18}$$

under the restriction

$$g_1(y, u, b) = 0, \tag{19}$$

where y is a state variable, u is a control function, and b is a control parameter (y, u, and b may be vectors).

Let us put the set

$$A(U) = \{(y, u, b) \in Y \times U \times \mathrm{B} \mid y = F(y,u,b),\ g_1(y,u,b) = 0\} \tag{20}$$

and let $Y = C(T, R^n)$. As a *control space U* we consider the space U of ordinary (measurable by Lebesgue) functions and the space S of *generalized* or *relax* [203] *functions* representing functions of the time with the values on the *set of measures*. Let $R^*(t)$ be the *admissible set* of control values and

$$U^* = \{u \in U \mid u(t) \in R^*(t)\},\ S^* = \{\sigma(t,r) \in S \mid \textstyle\int_{R^*(t)} \sigma(t,r)dr = 1\}. \tag{21}$$

If the measure σ is concentrated on the finite number of points r_i, $i = 1, \ldots, n$, (what is typical in our case), then

$$\sigma(t,r) = \textstyle\sum_{i=1}^{n} p_i(t)\delta(r - r_i),\ \sum_{i=1}^{n} p_i(t) = 1, \tag{22}$$

where δ is delta-function of Dirac, $p_i(t)$ are probabilities that a control u accepts the value r_i at the instant t.

Following [203], we call:

i) a point $(y, u, b) \in A(U^*)$ is a *minimizing U-solution* of the problem if it minimizes the function g_0 on the set $A(U^*)$;

ii) a sequence (y_j, u_j, b_j) in $H(U^*)$ is an *approximate U-solution* if, for every neighborhood G of 0 in R^1, $g_1(y_j,u_j,b_j) \in G$ for all sufficiently large integers j. An approximate U-solution $(y_j\hat{},\ u_j\hat{},\ b_j\hat{})$ is a *minimizing approximate U-solution* if $\lim_j g_0(y_j\hat{},u_j\hat{},b_j\hat{})$ exists in R^1 and $\lim_j g_0(y_j\hat{},u_j\hat{},b_j\hat{}) \le \lim_j \inf g_0(y_j,u_j,b_j)$ for every approximate U-solution;

iii) a point $(y, \sigma, b) \in A(S^*)$ is a *minimizing relaxed* (generalized) *solution* if it minimizes the function g_0 on the set $A(S^*)$.

In the last case the state equation (18) is substituted by

$$y(t) = \textstyle\int\int_{R^*(s)} f[t,s,\xi(y)(s),r,b(t)]\sigma(s,r)drds = \int f[t,s,\xi(y)(s),\sigma(s),b(t)]\mu(ds). \tag{23}$$

The physical sense of the substitution consists in that for every t, s, y, and b we do not calculate a function f itself but its mean value by $r \in R^*(t)$ and here $\sigma(s,r)$ decides what kinds of value r and weight are involved in averaging.

Let us introduce some additional notions from [203], which are necessary for understanding the subsequent theorems. Let R^1 be the *metric space* of real numbers or real measures, $C(T, X)$ the *metric space* of all bounded continuous functions on T to X, T a topological space, X a Banach space; (T, Σ, μ) a positive finite *measure space* (μ is finite if its values are in R^1), (T, Σ, μ, X) the *metric space*, the set of all μ-measurable functions on T to X; $B(T, \Sigma, \mu, S; X)$ [or $B(T, S; X)$ or B) the vector space of functions φ: $T \times S \to X$ such that, for all $(t, s) \in T \times S$, $\varphi(t, .) \in C(S, X)$, $\varphi(.,s)$ is μ-measurable and there exists a μ-integrable ψ_φ: $T \to R^1$ with $|\varphi(t,.)|_{\sup} \le \psi_\varphi(t)$. The elements of B are *Caratheodory functions*, the function $|\varphi|_B = \varphi(t,.)_{\sup}\, \mu(dt)$: $B \to R$ is a norm on B. The function $t \to \varphi(t,\xi(t))$: $T \to X$ is μ-integrable for every choice of $\varphi \in B$ and μ-measurable ξ: $T \to S$. Let more *frm*(T) be the vector space of all Radon measures in T (Radon measure is the finite regular Borel measure, in particular, Dirac measure), *frm*$^+(S)$ the set of all positive measure in S, *rpm*(S) the set of all Radon probability measure in S. A compact metric space R (or T) can be embedded in *frm*(R) by identifying each $r \in R$ with the Dirac measure δ_r at r (or identifying r with the element l_r of functionals $C(R)^*$) such that $l_r(c) = c(r)$ for all $c \in C(R)$, $r \in R$. Let finally the sequentially closure of a set A (*seq cl*(A)) be the collection of all limits of convergent sequences in A, *seq cl*$(A) \subset A$^, which is the intersection of all the closed sets that contain A.

Theorem 1. *Under the relations* (16)-(23), *given* k, n, $V \in R^k$, $W \in R^n$, *and functions* f: $T \times T \times V \times R \times B \to R^n$ *and* ξ: $C(T, W) \to L_\infty(T, \Sigma, \mu, V)$, *if* i) *$V$ and W are closed*; ii) *B has a compact topology*; iii) *the function* $g = (g_0, g_1)$ *seq cl*$(A(S^*))$ *is continuous*; iv) $A(S^*) \ne \varnothing$; v) $f \in C(T, B(T, V, R, B; R^n))$ *and ξ is continuous, then the problem has the minimizing relaxed solution* $(y, \sigma, b) \in V \times S^* \times B$.

Theorem 2. *Assume that the conditions of Theorem* 1 *are satisfied with* $A(S^*)$ *replaced by* $H(S^*)$. *Let* (y^, σ^, b^) *be a minimizing relaxed solution, U an abundant subset of* $R^\#$, *and assume, furthermore, that* y^ *is the unique solution of the equation* $y = F(y,\sigma$^$,b$^$)$, *and the equation* $y = F(y,u,b$^$)$ *has a solution y for each* $u \in U^*$ *and sufficiently close to* σ^ *(in the metric space S). Then there exists a sequence* (y_j, u_j) *in* $Y \times U^*$ *such that* $(y_j, u_j, b$^$)$ *is a minimizing approximate U-solution and* $\lim_j g(y_j,u_j,b$^$) = g(y$^$,\omega$^$,b$^$)$.

Theorem 3. *Let the conditions of Theorem* 1 *be satisfied and* (y^, σ^, b^) *be any minimizing relaxed solution. If* i) $g(y,\sigma,b) = g$^(y,b) *for all* $(y, \sigma, b) \in Y \times S^* \times B$; ii) *the set* $R^*(s)$ *is closed for all* $s \in T$; iii) $f(.\ .,v,r,b)$ *is* $\Sigma \times \Sigma$*-measurable for all* $(v, r, b) \in V \times R \times B$; *and that there exists* $\nu \in$ *frm*$^+(T)$ *such that* iv) $\nu(A) > 0$ *for any open* $A \subset T$; *and* v) $\{f(.,s,v,r',b\} \mid r' \in R^*(s)\}$ *is a convex subset of* (T, Σ, v, R^n) *for all* $(s, v, b) \in T \times V \times B$, *then the problem has the minimizing U-solution.*

Theorem 4. *Let the conditions of Theorem* 1 *be satisfied, and assume furthermore that, for all* $(t, s, v, r, b, y, \sigma) \in T \times T \times V \times R \times B \times Y \times S^*$, *we have* $g(y,\sigma,b) = g$^(y,b), $R^*(s)$ *is closed, and*

$$f(t,s,v,r,b) = \sum_{j=1}^{l} \alpha^{j}(t,s,v,b)\beta^{j}(s,v,r,b), \tag{24}$$

where α^j: $T \times T \times V \times B \rightarrow R^1$, β^j: $T \times V \times R \times B \rightarrow R^n$, $\beta^j(s,.,.,.)$ *are continuous, and* $\beta^j(.,v,r,b)$ μ*-measurable. Then there exists a minimizing relaxed solution* (y^, σ^, b^) *such that, for all* $s \in T$, σ^(s) *is a measure supported on a set of at most* $ln + 1$ *points of* $R^*(s)$.

These Theorems 1-4 are proved in [203, pp. 408-413] (under some more general conditions and objects).

2. Examples of Analytical Solutions

Let $a^{-1}(t\text{^}) - t\text{^} \geq T - t\text{^} = 1/(\alpha y\text{^})$, $y \leq y$^. Then

$$I(y) = \int_{t\text{^}}^{T} c(t)dt = \beta\text{^}(1-y\text{^})R(T-t\text{^}) + \int_{t\text{^}}^{T} [\beta(1-y) - \beta\text{^}(1-y\text{^})]m(t)dt$$

and

$$I'(y) = \int_{t\text{^}}^{T} (T-t)\{-\beta + \alpha(t-t\text{^})[\beta(1-y) - \beta\text{^}(1-y\text{^})]\}m(t)dt.$$

If $\beta = \beta$^, then $I'(y) < 0$ and hence

$$\max I(y)\ (0 \leq y \leq 1) = I(0) = \beta\text{^}(1-y\text{^})R(T-t\text{^}) + \beta\text{^}y\text{^}\int_{t\text{^}}^{T} (T-t)m(t)dt.$$

Let $\beta = \beta\text{^}\exp\,[c_\beta(t-t\text{^})]$, $d = c_\beta - \alpha y$^. Then under large d with respect to the main terms

$$I'(0) = (y\text{^}^{-1} - 1)/d^2 \exp\,[d(T-t\text{^})] > 0,\ I'(1) < 0.$$

So, $\max I(y)$ $(0 \leq y \leq 1) = I(y^*)$, $y^* > 0$.

We also consider the problem: $\max(T-t\text{^})$ $(0 \leq y \leq 1)$ under the restriction $c(t) \geq c^0(t)$. Let $c^0(t) = c^0 = \beta(1-y)R(t)$ and α, β, y, and R be constants. On the strength of (5, 6, Ch. 5)

$$c(t) = c^0,\ R(t) = R,\ m(t) = \alpha yR = m\text{^},\ a(t) = t - t\text{^}$$

and therefore

$$\max I_7(y)\ (0 \leq y(t) \leq 1) = \max\,(T-t\text{^})\ (0 \leq y(t) \leq 1) = T(y) - t\text{^} = \infty.$$

Of course, this conclusion is valid under the condition that any process of disintegration is absent here. Indeed, let α, y, and R be the same, but $\beta = \beta\text{^}\exp\,(-dt)$, where β^ and d are positive constants. Then

$$m(t) = \alpha yP,\ c(t) = \alpha\beta\text{^}Ry(1-y)\exp\,[-d(t-\text{^})][1-\exp\,(-dt\text{^})]/d$$

and therefore under $c(t) \geq c^0 = \text{constant}$,

$$t - t\text{^} \le \ln \{\alpha\beta\text{^}Ry(1-y)[1-\exp(-dt\text{^})]/[dc^0]\}/d,$$

from where

$$\max (T\text{-}t\text{^})\ (0 \le y \le 1) = \ln \{1/4\alpha\beta\text{^}R[1-\exp(-dt\text{^})]/[dc^0]/d.$$

Let us consider one more example when R, and $m\text{^}$ are positive constants, $y\text{^} = 1, \alpha = \beta = xs, xt\text{^}^2 = 2$. We will find y, m, c, and a for which $I_1(y) = \max$. In our case

$$I_1 = I = I(y) = \int_{t\text{^}}^{T} c(t)dt = \int_{t\text{^}}^{T} [\int_{a(t)}^{t} \alpha(s)m(s)ds - m(t)]dt.$$

So, regarding the infinitesimal of the first order, the variations

$$dR = 0 = \int_{c(t)+da(t)}^{t} (m+dm)ds - \int_{a(t)}^{t} mds,$$

$$dI = \int_{t\text{^}}^{T} \{\int_{a(t)}^{t} [\alpha(s)-\alpha(a)]dm(s)ds - dm(t)\}dt = \int_{t\text{^}}^{T} \{\int_{a\text{^}(t)}^{t} [\alpha(t)-\alpha(a(s))]ds - 1\}dm(t)dt,$$

where $\text{^}a(t) = a^{-1}(t)$, $t \in [t\text{^}, a(T)]$, $\text{^}a(t) = T$, $t \in [a(T), T]$. Note that the last formula is obtained on the basis of rearrangement of the integration order.

Let us put

$$\int_{t}^{a\text{^}(t)} [\alpha(t)-\alpha(a(s))]ds - 1 = 0,\ t \in [t\text{^}, a(T)]\ (\alpha(t) = xt).$$

It is not difficult to verify that the solution of this equation is $a(t) = t - t\text{^}$. Therefore,

$$m(t) = m\text{^} = R/t\text{^},\ y(t) = (t-t\text{^})y(t-t\text{^})/t = (t-kt\text{^})/t,\ t \in [kt\text{^}, (k+1)t\text{^}],$$

$$k = 1, 2, \ldots,\ c(t) = xm\text{^}t\text{^}(t-t\text{^}),\ t \in [t\text{^}, T-t\text{^}].$$

With regard to the infinitesimal of the second order in the vicinity of the y obtained on the segment $[t\text{^}, T-t\text{^}]$, the variation

$$dI = \int_{t\text{^}}^{T-t\text{^}} \{\int_{a(t)+da(t)}^{a(t)} [\alpha(s)-\alpha(a(t))]dsm\text{^}dt < 0.$$

Thus, the m, y, c are optimal on $[t\text{^}, T-t\text{^}]$. On the segment $[T-t\text{^}, T]$, we have

$$dI = \int_{T-t\text{^}}^{T} \{\int_{t}^{T} [\alpha(t)-\alpha(s-t\text{^})]ds - 1\}dm(t)dt,$$

$$dm = \int_{t-t\text{^}}^{t} [\alpha y - \alpha(t-t\text{^})y(t-t\text{^})]dmds + \int_{t-t\text{^}}^{t} \alpha mdyds,$$

where $\int_t^T [\alpha(t)-\alpha(s-t\text{^})]ds < 1$. We assume that

$$T = (P+1)t\text{^},\ P > 1;\ dy \leq 0,\ s \in [Pt\text{^}, (P+1)t\text{^}],\ dy = 0,\ s \in [t\text{^}, Pt\text{^}],$$

and show that $dm \leq 0$. In our case, $\alpha(t-t\text{^})y(t-t\text{^}) = x(t-Pt\text{^})$ and therefore

$$dm + x(t-Pt\text{^})\int_{Pt\text{^}}^{t} dmds = \int_{Pt\text{^}}^{t} \alpha ydmds + \int_{Pt\text{^}}^{t} \alpha mdyds$$

from which

$$dm = \int_{Pt\text{^}}^{t} (\alpha y\, dm + \alpha mdy)\{1 - x(t-Pt\text{^})(t-s)\exp[-x(t-s)(t+s-2Pt\text{^})/2]\}ds,$$

where expression in { } > 0 on the strength of the inequality $\exp(u) > 2u$, $u > 0$. Therefore $dm \leq 0$ and, hence, $dI \geq 0$. It means that the optimal

$$y = 0,\ s \in [Pt\text{^}, (P+1)t\text{^}],\ m = \int_{t-t\text{^}}^{Pt\text{^}} xsy(s)m\text{^}ds =$$

$$xm\text{^}[(P+1)t\text{^}-t][t-(P-1)t\text{^}]/2,\ a'(t) = m/m\text{^},\ a(Pt\text{^}) = (P-1)t\text{^},$$

$$c(t) = xm\text{^}\int_{t-t\text{^}}^{Pt\text{^}} sds - m + x\int_{Pt\text{^}}^{t} sm(s)ds,\ t \in [Pt\text{^}, (P+1)t\text{^}].$$

The optimal (maximal) value of the functional I on the segment $[t\text{^}, (P+1)t\text{^}]$ is

$$I^* = xm\text{^}t\text{^}\int_{t\text{^}}^{Pt\text{^}} (t-t\text{^})dt + \int_{Pt\text{^}}^{(P+1)t\text{^}} c(t)dt,$$

One can see that $I^* \to \infty$ with the speed P^2 in order for $P \to \infty$, even though R is a constant.

3. Existence and Uniqueness of Solutions

Theorem 5. *Under the assumptions of Theorems* 2-5, Ch. 14, *the problems* (C.1-C.7, Ch. 5) *have the minimizing generalized solutions, the minimizing approximate U-solutions, and the minimizing U-solutions (i.e., the ordinary solutions y^* on the set of measurable functions U^*).*

Proof. We give the proof for the problem (C.1, Ch. 5) by analogy with Yu. Yatsenko [213]. In the other cases the proof is similar. Let in notations of [203]

$$y = x = (m, c),\ u = y,\ b = a,$$

$$g_0(y,u,b) = -\int_{t\text{^}}^{t} c(t)dt,\ g_1(y,u,b) = \int_{a(t)}^{t} m(s)ds - R(t),\ 0 \leq a(t) \leq t, \quad (25)$$

and the state equation (18) have the form

$$m(t) = \int_0^t f_1 ds,\ c(t) = \int_0^t f_2 ds,\ f_1 = f_2 = 0,\ 0 \le s \le a(t);$$

$$f_1 = \alpha(t,s)y\hat{}(s)m\hat{}(s),\ a(t) \le s \le t\hat{},\ f_2 = \beta(t,s)[1-y\hat{}(s)]m\hat{}(s),\ a(t) \le s \le t\hat{},$$

$$f_1 = \alpha(t,s)y(s)m(s),\ t\hat{} \le s \le T;\ f_2 = \beta(t,s)[1-y(s)]m(s),\ t\hat{} \le s \le T, = 0,\ 0 \le s \le a(t). \quad (26)$$

Let us accept

$$Y = C([t\hat{},\ T]) \times C([t\hat{},\ T]),\ B = \{a(t) \in C^1([t\hat{},\ T]),\ 0 \le a(t) \le T\}. \quad (27)$$

As a control space U we consider the space ordinary controlling functions y: $[t\hat{}, T] \to R^1$ and the space Σ of generalized controlling functions σ representing functions of the time with the values on the set of measures. Respectively, let us determine the admissible set of control values

$$R^*(t) = [0,\ 1] \subset R^1, \quad (28)$$

the control functions

$$U^* = \{y \in U \mid y(t) \in R^*(t)\} \quad (29)$$

and the generalized control functions

$$\Sigma^* = \{\sigma(t,r) \in \Sigma \mid \int_{R^*(t)} \sigma(t,r)dr = 1\}. \quad (30)$$

Let V be the set of values for the state variable $x = (m,\ c)$ under $y \in R^*$ and $a \in B$, $V \subset R^1 \times R^1$. These are the following facts:

(i) on the strength of continuity and monotonicity of the operator in the state equation (127), the set V is bounded and closed;

(ii) on the strength of the assumptions of Theorem 2, the set B is compact;

(iii) the function $g = (g_1,\ g_2)$ is continuous;

(iv) the set $A(\Sigma^*) \ne \varnothing$.

Furthermore, the following properties of the function $f(f_1, f_2)$, f: $[t\hat{},\ T] \times [0,\ T] \times V \times R^* \times B \to R^1 \times R^1$, are true:

(i) f is continuous over t under $(s,\ x,\ y,\ a) \in [0,\ T] \times V \times R^* \times B$;

(ii) f is continuous over $(x,\ y,\ a)$ under $(t,\ s) \in [t\hat{},\ T] \times [0,\ T]$;

(iii) f has a discontinuity under $s = a(t)$ but it is measurable by s under $(t,\ x,\ y,\ a) \in [t\hat{},\ T] \times V \times R^* \times B$;

(iv) there exists an integrable function $\psi : [0,\ T] \to R^1$ such that

$$\sup |f(.,s,.,.,.)| ([t\hat{},\ T] \times V \times R^* \times B) \le \psi(s),\ s \in [0,\ T]. \quad (31)$$

Thus, all the conditions of Theorem 1 have been realized and therefore the problem C.1

(Ch. 5) has the minimizing solution $(m, c, \sigma, a) \in V \times \Sigma^* \times B$.

Let us prove now the existence of the approximate U-solution. Let $(x^\wedge, \sigma^\wedge, a^\wedge)$ be the minimizing relaxed solution. It is easy to see that $x^\wedge$ is the unique solution of the Volterra-type equation $x = F(x,\sigma^\wedge,a^\wedge)$ and the equation $x = F(x,y,a^\wedge)$ has a solution for each $y \in U^*$ and is sufficiently close to $\sigma^\wedge$ in the metric space Σ. It means that the conditions of Theorem 2 are valid.

Note that according to [203] the sequence y_j will converge to the measure σ if

$$\text{Lim} \int_0^T \varphi(s,y_j(s))ds \; (j \to \infty) = \int_0^T ds \int_0^1 \varphi(s,r)\sigma(r)dr \tag{32}$$

for arbitrary $\varphi \in L([0, T] \times [0, 1])$.

Let us prove, at last, the existence of the ordinary solution y on the set of measurable functions. Indeed, there are the following properties of the problem:

i) $g(x,\sigma,a) = g^\wedge(x,a)$, $g^\wedge(x,a)\colon X \times B \to R^1 \times R^1$ for all $(x, \sigma, a) \in X \times \Sigma^* \times B$;

the condition ii) - iv) of Theorem 3 are obviously valid;

v) the set of functions $f^\wedge(r) = f(.,s,x,r,a)$, $f^\wedge\colon R \to L([t^\wedge, T])$, makes up a convex subset $\{f^\wedge(r) \mid r \in R^*(s)\}$ in the metric space of integrable functions $[t^\wedge, T] \to R^1 \times R^1$ (it follows from the formula (25)-(29)) for all $(s, x, a) \in [0, T] \times V \times B$.

Thus, for the problem under consideration, Theorem 3 is valid. On the strength of Theorem 3, our problem has the minimizing U-solution (x, y, a).

It should be noted that the set B in [203] is usually the convex subset of a finite vector space. But in Ch. V, VI, and VIII of [203] there are remarks about the validity of the theorems when B is, in particular, the convex compact subset of a space of continuous functions.

At any case (and to weaken the conditions of Theorem 5), we can give another proof based on reducing the equation of motion to the form

$$m(t) = \int_{a(t)}^{t^\wedge} \alpha(t,s)y(s)m(s)ds + \int_{t^\wedge}^{t} \alpha(t,s)y(s)m(s)\}ds,$$

$$M(t) = \int_0^t m(s)ds,\; a(t) = M^{-1}[M(t) - R(t)],\; t \in [t^\wedge, T], \tag{33}$$

i.e., to the case of $n = 3$ in Theorem 1. If $a(t) \le t^\wedge$ in (33), then we can establish easily the validity of all the conditions of Theorem 1. Otherwise, we have to consider (33) under $t \in [t^\wedge, t_1]$, $a(t) \le t^\wedge$, then under $t \in [t_1, t_2]$, $a(t) \le t_1$, and so on until we pass T.

And also we have to prove that the operator F corresponding to the system (33) is continuous.

This last property of F is almost obvious with regard to the natural restriction $\alpha(t,s)y^\wedge(s)m^\wedge(s) > 0$, $s \in [0, t^\wedge]$.

Theorem 6. *Under the assumptions of Theorems* 6, 7, Ch. 14, *the problems* C.1-C.7, Ch. 5, *have the minimizing U-solutions* x^*, y^*.

Proof. It is similar to the one above, but the role of a state equation (18) will play the system of the equations in (54), Ch. 5, with respect to m, c and α.

Theorem 7. *Under the assumptions of Theorem* 10, *Ch.* 14, *the problems* C.1-C.7, Ch. 5, *have the minimizing U-solutions* $x^*, y^*, z^*, e^*, g^*, r^*, w^*$.

Proof. It is similar to the one above.

Theorem 8. *In the particular case when* $\beta(t,s) = k(t)\alpha(t,s)$, *the problem* C.1 *has a minimizing relaxed solution* (m, c, σ, a) *such that for all* $s \in [t\hat{}, T]$ $\sigma(s)$ *is a measure supported on a set of at most* 2 *points of* $[0, 1]$.

Proof. We can rewrite the relations (25), (26) in the form

$$g_0(y,u,b) = -\int_{t\hat{}}^{t} c(t)dt,\ c(t) = k(t)\int_{a(t)}^{t} \alpha(t,s)m(s)ds - k(t)m(t),$$

$$g_1(y,u,b) = \int_{a(t)}^{t} m(s)ds - R(t),\ 0 \le a(t) \le t,\ m(t) = \int_0^t f[t,s,m(s),y(s),a(t)]ds,$$

$$f = 0,\ \ 0 \le s \le a(t);\ = \alpha(t,s)y\hat{}(s)m\hat{}(s),\ a(t) \le s \le t\hat{},\ = \alpha(t,s)y(s)m(s),\ t\hat{} \le s \le T. \tag{34}$$

It means that for our problem Theorem 4 takes place and also $l = n = 1$.

Later we will show that in our case the measure σ is supported on two points, $r = y_{\min}(t)$ (in particular, $r = 0$) and $r = 1$, on the boundary of $R^*(t)$. Then we have

$$\sigma(s,r) = p(s)\delta(r\text{-}1) + [1-p(s)]\delta[r-y_{\min}(s)], \tag{35}$$

where δ is Dirac measure, and therefore

$$x(t) = F(x,c,b) = \int_0^t \int_{R^*(s)} f[t,s,m(s),r,b(t)]\sigma(s,r)drds =$$

$$\int_0^t \{f[t,s,m(s),1,b(t)]\,p(s) + f[t,s,m(s),y_{\min}(s),b][1-p(s)]\}ds. \tag{36}$$

From (36), (16), and Theorem 5 it is easy to conclude that

$$y^*(t) = p(t) - y_{\min}(t)[1-p(t)] \tag{37}$$

is the desired ordinary minimized solution. In the case $R^*(t) = [0, 1]$ the optimal control $y^*(t) = p(t)$, which is the probability that the control will be equal to the value $r = 1$.

Let $g_0{}'_u(y,u,b) = g_0{}'(t)$ be the gradient of $g_0(y, u, b)$ and let $g_0{}'(t)$ be positive, negative for t belonging to the segments T_1, T_2 correspondingly for all admissible u. Then in (35) and (37) we have $p(t) = 1, t \in T_1$, and $p(t) = 0, t \in T_2$. Thus, for determination of the structure of the solutions, we need to investigate the gradients of the respective specific functionals.

The similar theorems can be also established for other optimization problems.

4. On Stability of Solutions

Theorem 9. *Under the assumption that the initial functions are continuously differentiable, the remaining term R in the formula of the finite increment*

$$L(x+\delta x, y+\delta y, z+\delta z, t) - L(x,y,z,t) = \delta L(x,y,z,t) + R, \tag{38}$$

where L and δL are given by the relations (6)-(11), *satisfies the condition*

$$R = o(|\delta y|_C) \text{ or (what is the same) } R/|\delta y|_C \to 0 \text{ for } |\delta y|_C \to 0. \tag{39}$$

Besides, regarding the main terms, the variation of the functional I (5) *is* $\delta I = I'(y)\delta y$, *where* $I'(y)$ *is the respective gradient or derivative.*

Proof. Under the assumption of s. 1, there exist finite values $|k|_C$, $|\Phi|_C$, $|p|_C$, $s \in [0, T]$, $t \in [t^\wedge, T]$, and also norms $|x|_C$ and $|\varphi|_C$ of the desired solutions $x(t)$ and $\varphi(t)$, $t \in [t^\wedge, T]$, under any controls y, $y \in U$.

On the strength of continuity of the relation $x(t)$ of $y(t)$ and finiteness of the segment $[t^\wedge, T]$, the equality $|\delta x_C| = O(|\delta y_L|)$ or (what is the same) $|\delta x_C|/|\delta y_L| < c$, where c is a certain constant, is valid.

The remainder term R contains an analytical expression over products of variations δy and δx and depends on the given and desired functions (F. Vasil'ev [192]), and also on $T-t^\wedge$. Thus, we have $R = o(|\delta y|_C)$. The last conclusion of Theorem 9 is trivial.

This theorem means that in the case of MM (5-9, Ch. 5) and under the assumptions considered, we have the property of *technical stability* of solutions for the optimization problem C.1-C.7, Ch. 5. Let us dwell on some of the problems of C.1-C.7, Ch. 5 in more detail.

Lemma 1. *The gradient of the functional I, with regard to MM* 5, 6, Ch. 5 *under the condition that R is given, has the form*

$$I'(y) = -m(t)\int_t^{A^\wedge(t)} [\alpha(s,t)\psi(s)+\beta(s,t)]ds,$$

$$A^\wedge(t) = a^{-1}(t),\ t \in [t^\wedge, a(T)];\ T,\ t \in [a(T), T], \tag{40}$$

where $\psi(t)$, $t \in [t^\wedge, T]$, *is determined uniquely under any given* $y(t)$ *with the help of the equations* (5, 6, Ch. 5) *and the conjugate equation*

$$\psi(t) = \int_t^{A^\wedge(t)} [\alpha(s,t)y(t) - \alpha(s,a(s))y(a(s))]\psi(s)ds - \int_t^{A^\wedge(t)} \{\beta(s,t)[1-y(t)]$$

$$-\beta(s,a(s))[1-y(a(s))]\}ds. \tag{41}$$

Lemma 2. *The gradient of the functional I, with regard to MM* (5, 6, Ch. 5) where *f is given, has the form*

$$I'(y) = -m(t)\int_t^{A^\wedge(t)} \{[\alpha(s,t)\psi_1(s) + \beta(s,t)] + [\alpha(s,t) - \beta(s,t)]\psi_2(s)\}ds, \quad (42)$$

where ψ_i, $i = 1, 2$, *are determined with the conjugate equations*

$$\psi_1(t) = \int_t^{A^\wedge(t)} \{\alpha(s,t)y(t)\psi_1(s) - \beta(s,t)[1-y(t)] + [\alpha(s,t)y(t) + \beta(s,t)(1-y(t))]\psi_2(s)\}ds,$$

$$\psi_2(t)[\alpha(t,a(t))y(a(t))+\beta(t,a(t))(1-y(a(t))] = -\psi_1(t)\alpha(t,a(t))y(a(t))+\beta(t,a(t))(1-y(a(t)). \quad (43)$$

Lemma 3. *The gradient of the functional*

$$I_1 = \int_{t^\wedge}^{T} \int_{b(t)}^{t} \beta(t,s)[1-y(s)]m(s)dsdt \quad (44)$$

with regard to MM (9, Ch. 5), R_I *and* R_2 *are given, has the form*

$$I_1'(y) = -m(t)\{\int_t^{A^\wedge(t)} [\alpha(s,t) - \alpha(s,a(s))]\psi(s)ds + \int_t^{B^\wedge(t)} [\beta(s,t) - \beta(s,b(s))]ds\},$$

$$B^\wedge(t) = b^{-1}(t),\ t \in [t^\wedge, b(T)];\ T,\ t \in [b(T), T], \quad (45)$$

the function ψ *being defined by*

$$\psi(t) = y(t)\int_t^{A^\wedge(t)} [\alpha(s,t) - \alpha(s,a(s))]\psi(s)ds - (1-y(t))\int_t^{B^\wedge(t)} [\beta(s,t)-\beta(s,b(s))]ds, \quad (46)$$

where a, b, and m are the solutions of the system considered (*see Theorem* 4).

Lemma 4. *The gradient of the functional* (44), *with regard to MM* 9, Ch. 5, where *f and R are given, has the form*

$$I_1' = -m(t)\{\int_t^{A^\wedge(t)} [\alpha(s,t)(\psi_1(s)+\psi_3(s))+\psi_2(s)]ds - \int_t^{B^\wedge(t)} [\beta(s,t)(\psi_3(s)-1)+\psi_2(s)]ds\}, \quad (47)$$

the functions ψ_i, $i = 1, 2, 3$, *being defined by the equations*

$$\psi_1(t) = y(t)\int_t^{A^\wedge(t)} \{\alpha(s,t)[\psi_1(s) + \psi_3(s)] + \psi_2(s)\}ds + [1-y(t)]\int_t^{B^\wedge(t)} \{\beta(s,t)[\psi_3(s)-1] +$$

$$\psi_2(s)\}]ds,\ [\psi_1(t) + \psi_3(t)]\alpha(t,a(t)) + \psi_2(t) = 0,\ [\psi_3(t)-1]\beta(t,b(t)) + \psi_2(t) = 0, \quad (48)$$

where a, b, and m are the solutions of the system considered (*see Theorem* 5).

Lemma 5. *The gradient of the functional* I_3, *with regard to MM* (5, 6, Ch. 5) *where c is given, has the form*

$$I_3'(y) = -m(t)\int_t^{A^\wedge(t)} [\alpha(s,t)\psi_1(s) - \beta(s,t)\psi_2(s)]ds, \tag{49}$$

where ψ_i, $i = 1, 2$, *are determined with the conjugate equations*

$$\psi_1(t) = \int_t^{A^\wedge(t)} \{\alpha(s,t)y(t)\psi_1(s) + \beta(s,t)[1-y(t)]\psi_2(s) - 1\}ds,$$

$$\psi_1(t)\alpha(t,a(t))y(a(t)) + \psi_2(t)\beta(t,a(t))[1-y(a(t))] = 1. \tag{50}$$

The similar lemmas are also valid for all the functionals I_1--I_7 (Ch. 5).

Lemma 6. *The gradient of the functional I* ($I = \int_{t^\wedge}^{T} \int_{a(t)}^{t} \beta(t,s)[1-x(s)-y(s)]m(s)dsdt$), *with regard to MM* (46, Ch. 5), $\nu = \mu = 1$, $c_1 = 0$, *R is given*, $\alpha_1 = \alpha(s) = \beta(t,s)$, *has the form*

$$I'(x,y) = (I'_x, I'_y), I'_x = m(t)\alpha(t)\int_t^{A^\wedge(t)} [\psi_1(s)+1]ds, I'_y = -m(t)\alpha(t)\int_t^{A^\wedge(t)} [\psi_2(s)+1]ds, \tag{51}$$

where the conjugate variables ψ_i, $i = 1, 2, 3$, *are determined with the relations*

$$\psi_1(t)+\int_t^T m(u)\int_u^{A^\wedge(u)} [1-x(u)(\psi_1(s)+1)-y(u)(\psi_2(s)+1)]ds\}du = 0,$$

$$\psi_2(t)+\int_t^{A^\wedge(t)} \{\alpha(t)[1-x(t)(\psi_1(s)+1)-y(t)(\psi_2(s)+1)]-\psi_3(s)\}ds=0,$$

$$\alpha(a(t))[1-x(a(t))(\psi_1(t)+1)-y(a(t))(\psi_2(t)+1)]-\psi_3(t) = 0, \tag{52}$$

and also m, a, and α *are the solutions of the system* 46, Ch. 5.

Lemma 7. *The gradient of the functional I* ($I = \int_{t^\wedge}^{T}\int_{a(t)}^{t} \beta(t,s)[1-x(s)-y(s)-z(s)-e(s)-g(s)-r(s)-w(s)]m(s)dsdt$), *with regard to MM* (51, Ch. 5), $\nu = 1$, $\mu = 1$, $c_1 = d_1 = e_1 = r_1 = 0$, $a,i(t) = a(t)$, $i = 1, \ldots, 8$, *where R is given, has the form*

$$I'(x,y,z,e,g,r,w) = (I'_x, I'_y, I'_z, I'_e, I'_g, I'_r, I'_w),$$

$$I'_x = -m(t)\int_t^{A^\wedge(t)} [\beta(t)+\alpha(t)\psi_1(s)]ds, I'_y = -m(t)\int_t^{A^\wedge(t)} [\beta(t)+\alpha(t)\psi_2(s)]ds,$$

$$I'_z = -m(t)\int_t^{A^\wedge(t)} [\beta(t)+\alpha(t)\psi_3(s)]ds, I'_e = I'_g = I'_r = I'_w = -m(t)\beta(t)\int_t^{A^\wedge(t)} ds, \tag{53}$$

where

$$\psi_1(t) - \int_t^T \int_u^{A^\wedge(u)} m(u)d(u,s)dsdu = 0, d(u,s) = x(u)\psi_1(s) + y(u)\psi_2(s) + z(u)\psi_3(s),$$

$$\psi_2(t) - \int_t^{A\hat{}(t)} \{[\beta(1-x-y-z-e-g-r-w)]_t - \alpha(t)d(t,s) - \psi_4(s)\}ds = 0,\ \psi_3(t) + \int_t^{A\hat{}(t)} [(1-x-y-z-e-g-r-w)m]_t\, ds = 0,\ -[\beta(1-x-y-z-e-g-r-w)]_{a(t)} + \alpha(a(t))d(a(t),t) + \psi_4(t) = 0. \quad (54)$$

The proof of Lemmas 1-7 is almost trivial according to the general theory of control. It should be noted, however, that for the proof of Lemmas 6, 7, we need to use, in addition, the formulae of the type

$$\int_{t\hat{}}^T \psi_1(t)\int_{a(t)}^t \alpha(s)x(s)m(s)dsdt = \int_{t\hat{}}^T [\alpha(t\hat{}) + \int_{t\hat{}}^t \alpha'(u)du]x(t)m(t)\int_t^{A\hat{}(t)} \psi_1(s)dsdt =$$

$$\int_{t\hat{}}^T \int_u^T x(t)m(t)\int_t^{A\hat{}(t)} \psi_1(s)dsdt\alpha'(u)du + \int_{t\hat{}}^T \alpha(t\hat{})x(t)m(t)\int_t^{A\hat{}(t)} \psi_1(s)dsdt. \quad (55)$$

<u>Theorem 10.</u> *With regard to the main terms, the variations of the functionals I_1 and I_3 for different cases of MM* (5-9, 46, 51, Ch. 5) *are given by the formulae $\delta I = I'(y)\delta y$, where $I'(y)$ are given by the respective formulae* (40)-(54).

<u>Proof</u>. It is obvious from Lemmas 1-7 and the proof of Theorem 9.

5. On Investigation of Solution Structure

The first essential result on the properties of solutions of the problem C.1 (see Ch. 5) has been obtained by V. Glushkov and V. Ivanov [62] (also see Introduction). The *structure of solution* for the problem C.1 and other optimization problems has been investigated in detail mainly by Yu. Yatsenko [213].

Let us refer to the appropriate results below without detailed proofs (these proofs can be found in [213]). Some of those results will follow from Theorem 10 and also from the asymptotic properties of the solutions for some of the optimization problems.

<u>Lemma 8.</u> *If $\alpha(t,s)$ and $R'(t)$ belong to the class of Lipschitz over t, $t \in [t\hat{}, T]$, then on the segment $[t\hat{}, T]$, there exists the unique integrable function $y_{min}(t) = 0$, $t \in \Delta \subset [t\hat{}, T]$: $a'_{min}(t) > 0$; $y_{min}(t) = y^{\cdot}(t) \geq 0$, $t \in [t\hat{}, T]\backslash\Delta$: $a'_{min}(t) = 0$, where $a_{min}(t)$ is determined uniquely from the equation* 5, 6, Ch. 5, *under $y = y_{min}$. And also $0 < y_{min}(t) < 1$, $t \in [t\hat{}, T]$, if*

$$R''(t) < \alpha(t,t)R'(t) + \int_{t\hat{}}^t \alpha'_t(t,s)R'(s)ds + \int_0^{t\hat{}} \alpha'_t(t,s)y\hat{}(s)m\hat{}(s)\mathrm{d}s, \quad (56)$$

and $y_{min}(t) = 0$, if $R'(t) \leq 0$ or

$$R'(t) \leq \int_{z(t)}^{t\hat{}} \alpha(t,s)y\hat{}(s)m\hat{}(s)ds,\ m\hat{}(z(t))z'(t) - \int_{z(t)}^{t\hat{}} \alpha(t,s)y\hat{}(s)m\hat{}(s)ds = -R'(t),\ z(t\hat{}) = 0. \quad (57)$$

We will assume later that the condition $0 \leq y_{min}(t) \leq 1$ is valid.

Theorem 11. *There exists the instant* θ, $t^\wedge \le \theta < T$, *such that* $I'(y_{min},t) < 0$ *under* $t \in (\theta, T)$ *and* $y^*(t) = y_{min}(t)$. If $\theta > t^\wedge$, *then the segment* $[t^\wedge, \theta]$ *consists of subintervals, where either* $I'(y^*,t) > 0$ *and* $y^*(t) = 1$, *or* $I'(y^*,t) < 0$ *and* $y^*(t) = y_{min}(t)$, *or* $I'(y^*,t) = 0$ *and* $y_{min}(t) \le y^*(t) \le 1$, *and here, all these cases are possible depending on the given functions of the problem.*

The more refined results can be obtained in the special cases

$$\text{A.}\quad \beta(t,s) = k(t)\alpha(t,s),$$

$$\text{B.}\quad a(T) \le t^\wedge \text{ (then } a^{-1}(t) = T \text{ and } y(a(s)) = y^\wedge(a(s)). \tag{58}$$

Lemma 9. *Let* A. *or* B. *be valid and* $I'_y(a^*,s) = 0$, $s \in [t, A^\wedge(t)]$. *If* $a(t) > a^*(t)$, $a(u) \ge a^*(u)$, $u \in (t, T]$, *then* $I'_y(a(t),t) < 0$, *and if* $a(t) < a^*(t)$, $a(u) \le a^*(u)$, $u \in (t, T]$, *then* $I'_y(a(t),t) > 0$.

We give a sketch of the proof in the case of both conditions *A*. and *B*.
By Lemma 1, we have in the case of A.

$$\psi(t) = -k(t), \tag{59}$$

and in the case of B.

$$\psi(t) = -\int_t^T [\beta(s,t) - \beta(s,a(s))]ds = \int_t^T [\alpha(t,s) - \alpha(s,a(s))]\psi(s)ds. \tag{60}$$

So, in both cases we have

$$I(a) = \int_{t^\wedge}^T k(t)[\int_{a(t)}^{t^\wedge} \alpha(t,s)m^\wedge(s)ds + \int_{t^\wedge}^t \alpha(t,s)m(s)ds - m(t)]dt,$$

$$dI = I(a+da) - I(a) = \int_{t^\wedge}^T k(t)[\int_{(t)+da(t)}^{a(t)} \alpha(t,s)m^\wedge(s)ds +$$

$$\int_{t^\wedge}^t \alpha(t,s)dm(s)ds - dm(t)]dt,\ R(t) = \int_{a(t)}^{t^\wedge} m^\wedge(s)ds + \int_{t^\wedge}^t m(s)ds,$$

$$dR = \int_{(t)+da(t)}^{a(t)} m^\wedge(s)ds + \int_{t^\wedge}^t dm(s)ds = 0. \tag{61}$$

Using (59) - (61) and

$$\int_{t^\wedge}^t \alpha(t,s)dm(s)ds = \alpha(t,t)\int_{t^\wedge}^t dm(v)dv - \int_{t^\wedge}^t \alpha'_s(t,s)\int_{t^\wedge}^s dm(v)dvds,$$

$$\int_{t^\wedge}^T k(t)\int_{t^\wedge}^t \alpha'_s(t,s)\int_{t^\wedge}^s dm(v)dvdsdt = \int_{t^\wedge}^T \int_t^T k(s)\alpha'_t(s,t)\int_{t^\wedge}^t dm(v)dvdsdt,$$

$$-\int_{t^\wedge}^T k'(t)\int_{t^\wedge}^t dm(s)dsdt = k(T)\int_{(t)+da(t)}^{a(t)} m^\wedge(s)ds + \int_{t^\wedge}^T k(t)dm(t)dt \tag{62}$$

obtained by integration by parts and inversion of the order of integration, we have

$$dI = \int_{t^\wedge}^{T} k(t)\int_{(t)+da(t)}^{a(t)} [\alpha(t,s) - \alpha(s,a(t))]m^\wedge(s)dsdt + k(T)\int_{a(T)+da(T)}^{a(T)} m^\wedge(s)ds, \qquad (63)$$

The desired properties follow from (63) immediately.

Theorem 12. *For the problem* C.1 *in the cases* (33) *and* (58), *there exists the "best" function* $a^*(t)$, $t \in [t^\wedge, T]$, *such that* $I'(a^*,t) = 0$, $t \in [t^\wedge, \theta]$, $t^\wedge \le \theta < T$, *and* $I'(a^*,t) < 0$, $t \in (\theta, T)$. *Depending on the values of a disagreement*

$$\Delta_0 = |a^*(t^\wedge) - a(t^\wedge)|,\ \Delta_1 = |a^{*\prime}(t^\wedge) - a'(t^\wedge)| \qquad (64)$$

and a length of the planning interval $T - t^\wedge$ *it is possible that the optimal solutions* $a(t)$ *and* $y(t)$ *have the following behavior variants:*

(*i*) $y(t) = y_{\min}(t)$, $t \in [t^\wedge, T]$;
(*ii*) $y(t) = 1$, $t \in (t^\wedge, s_1)$, $a(t)$ *does not intersect* $a^*(t)$, $y(t) = y_{\min}(t)$, $t \in [s_1, T]$;
(*iii*) $y(t) = 1$, $t \in (s_{i-1}, s_i)$, $y(t) = y_{\min}(t)$, $t \in [s_i, s_{i+1}]$, $i = 1, \ldots, N-1$, $s_0 = t^\wedge$, $s_N = T$, $a(t)$ *intersects* $a^*(t)$ N *times*, $N \ge 1$. (65)

The case of unlimited quantity of switching is possible. In this case under $s_i \to s^\wedge$, *where* $s^\wedge$ *is a condensation point, the values* $|s_i - s_{i-1}|$ *are decreasing geometrically with a ratio* $q < 1$.

Theorem 13. *For MM* (46, Ch. 5), $\nu = 1$, $\mu = 1$, $z = c_1 = 0$, $\alpha_1 = \beta_1 = \alpha(s)$, *if*

$$m^\wedge > 0,\ 0 < R'(t),\ R'_{\max} = \int_{a(t)}^{t^\wedge} \alpha^\wedge(s)y^\wedge(s)m^\wedge(s)ds,$$

$$m^\wedge(a(t))a'(t) - \int_{a(t)}^{t^\wedge} \alpha^\wedge(s)y^\wedge(s)m^\wedge(s)ds = -R(t),\ t \in [t^\wedge, T], \qquad (66)$$

and the segment $[t^\wedge, T]$ *is sufficiently large, then the optimal*

$$x^*(t) = 1,\ t \in [t^\wedge, t_1],\ y^*(t) = 1,\ t \in [t_1, t_2],\ x^*(t) = y^*(t) = 0,\ t \in [t_2, T], \qquad (67)$$

and

$$t_1 - t^\wedge = \Theta(T - t^\wedge),\ t_2 - t_1 = \Theta[(T - t^\wedge)^{1/2}],\ T - t_2 = \Theta(1). \qquad (68)$$

One of the interpretations of Theorem 27 is the following.

To achieve the maximum of the summarized output of the external product for the sufficient large period $T - t^\wedge$, it is necessary that all new active centers be directed on the largest initial part $[t^\wedge, t_1]$ of $[t^\wedge, T]$ to the Genome (G), after this on some internal part $[t_1, t_2]$ of $[t^\wedge, T]$ to the recreation of means of recreation (A), and at last only on the end period $T - t_2$ to the creation of the product (B).

Reduction of the maximum possible period of the creation of the external product from $T - t^\wedge$ to $T - t_2$ is compensated by essential growth of new AC efficiency.

Theorem 14. *If the condition* A. (58) *is fulfilled;* $y_{min}(t) < 1$, $t \in [t\hat{}, T]$; $\beta(t,s)$ *is a monotone increasing function of* s; α *and R are slowly varying functions* (*i.e.*, α'_s, *and* $R'(t) \ll 1$) *and the solution* y^* *of the problem* C.1, Ch. 5 *has a finite number of switches, then this solution is unique.*

Theorem 15. *If the conditions* A. *and* B. *are fulfilled;* $\beta(t,s)[1-y\hat{}(s)]$ *and* $\alpha(t,s)y\hat{}(s)$ *are monotone increasing functions of s, then the problem* C.1, Ch. 5 *has no more than one solution.*

Theorem 16 (on the attracting locus [213]). *Let the following conditions*

$$i)\ \beta(s) = b^s,\ b > 1, \text{ or } \beta(s) = s^k,\ k > 0;\ ii)\ \alpha(t,s) = k(t)\beta(s), \tag{69}$$

where k is rapidly enough decreasing function of t, be fulfilled. Then for the problem C.1 *there exists the "best" function* $a^0(t)$ *("turnpike") such that* $I'_y(a^0) = 0$, *the solution of the problem* C.1 $a^*(t) \to a^0(t)$ *under* $t, T \to \infty$, *and also the behavior of* $a^0(t)$ *under* $t \to \infty$ *does not depend on the initial values* $m\hat{}(t)$, $y\hat{}(t)$.

6. Case of Non-Uniqueness of Solutions

Let us consider now a case of *non-uniqueness of solutions* for optimization problems on the example of the same problem C.1, Ch. 5.

Let α, $m\hat{}$, $y\hat{}$, R' be constants, $\beta = \beta(s)$. Then on the base of (40) and (41), supposing $a(t) \le t\hat{}$, $t \in [t\hat{}, T]$, from $I' = I'(y,s) = 0$, $t \in [t\hat{}, \theta]$, $t\hat{} < \theta < T$, it follows that

$$\alpha\int_s^T \psi(t)dt = -\beta(s)(T-s),\ \psi(s) = (1-y\hat{})\int_s^T [\beta(a(t)) - \beta(s)]dt,\ t \in [t\hat{}, \theta];$$

$$\psi(t) = \alpha[y(t)-y\hat{}]\int_t^T \psi(s)ds - \beta(t)(T-t)[1-y(t)] + (1-y\hat{})\int_t^T \beta(a(s))ds,\ t \in [\theta, T];$$

$$m' = \alpha(y-y\hat{})m + R'\alpha y\hat{}\ ,\ m\hat{} = \alpha y\hat{}m\hat{}t\hat{},\ [t\hat{}-a(t)]m\hat{} = R - \int_{t\hat{}}^t m(s)ds,\ t \in [t\hat{}, T];$$

$$I(y) = \int_{t\hat{}}^T \{m\hat{}(1-y\hat{})\int_{a(t)}^{t\hat{}} \beta(s)ds + \int_{t\hat{}}^t [1-y(s)]\beta(s)m(s)ds\}dt. \tag{70}$$

Note that any equation of the form $f(t) = 0$, $t \in [t\hat{}, \theta]$, where f is a differentiable function, is identical to the equation $f'(t) = 0, f(t\hat{}) = 0$ or $f'(t) = 0, f(\theta) = 0$. Therefore under $t\hat{} = 1$, $\alpha y\hat{} = 1$, $y\hat{} = 1/2$, the relation (70) is identical to

$$\beta(a(s)) = -u'' - u',\ u = \beta(s)(T-s),\ u'(1) + u(1) = \int_1^T \beta(a(s))ds;\ m(t) = R' + m\hat{}a'(t),$$

$$y(t) = 1/2 + [m'(t)-R']/[2m(t)],\ t \in [1, \theta],\ a(1) = 0,\ a'(1) = [m\hat{}-R']/m\hat{} \ge 0, \tag{71}$$

where the value θ is determined by the condition $I'(y,\theta) = 0$, $I(0,t) < 0$, $\theta < t < T$.
We have, besides (71),

$$2\int_{\theta}^{T} \psi(s)ds = -u(\theta),\ \psi(t) = -\int_{t}^{T} \psi(s)ds - \beta(t)(T-t) + 1/2\int_{t}^{T} \beta(a(s))ds,\ \theta < t < T. \quad (72)$$

Let us put, for example,

$$\beta(s) = \beta(\theta)[1+3(s-\theta)],\ s \in [\theta, T];\ \beta(s) = s,\ s \in [0, 1/2];$$

$$\beta(s) = s-1/2,\ s \in [1/2, 1];\ \beta(a(s)) = -u'' - u' = 1/2(1-e^{-(t-1)}),$$

$$u(s) = \beta(s)(T-s),\ u'(1) + u(1) = \int_{1}^{T} \beta(a(s))ds,\ s \in [1, \theta];\ \beta(a(s)) = 1/2(1-e^{-(\theta-1)}). \quad (73)$$

It is not difficult to verify that the equation (73) with respect to $a(s)$ has the following two solutions

$$a_1(s) = 1/2(1-e^{-(s-1)}),\ a_2(s) = 1-1/2e^{-(s-1)},\ s \in [1, \theta], \quad (74)$$

and, besides,

$$u(t) = 1/2(T-t)(1-e^{-(t-1)}) + C(e^{-(t-1)} - e^{-(T-1)}),\ t \in [1, \theta],\ C = 1/2(e^{T-1} - e^{T-\theta}) + 1/2(T-\theta-1)e^{T-\theta},$$

$$\psi(t) = [1/4(1-e^{-(\theta-1)}) - \beta(\theta)(1+3(T-\theta+2)](1-e^{t-T}) + 6\beta(\theta)(t-T), \quad (75)$$

where θ is the solution of the equation

$$f(\theta) = \beta(\theta)[1+6(T-\theta)]+[1/2(1-e^{-(\theta-1)})-2\beta(\theta)(1+3(T-\theta+2)](T-\theta+e^{-(T-\theta)}-1)/(T-\theta) = 0. \quad (76)$$

It is easy to see that $f(T) = \beta(T) > 0$, $f(T-1) < 0$. It means that the desired θ meets the relation $T-1 < \theta < T$. We show that $a_1(t)$ and $a_2\hat{}(t) = a_m(t)$, $t \in [1, \theta\hat{}]$; $a_2\hat{}(t) = a_2(t)$, $t \in [\theta\hat{}, T]$, where $a_m(t)$ corresponds to $y_{max}(t) = 1$, $t \in [1, \theta\hat{}]$, deliver local maximums to the functional I. The function $a_m(t)$ and value $\theta\hat{}$ can be determined by the formulae:

$$a_m(t) = (2-1/[3(T-1)])(e^{t-1}-1)-2(1-1/[3(T-1)])(t-1),\ a_m(\theta\hat{}) = a_2(\theta\hat{}),\ a_m'(\theta\hat{}) = a_2'(\theta\hat{}).$$

It is not difficult to see that under large T we have $1 < \theta\hat{} < 2$. Then, we have

$$\delta I = \int_{\theta}^{T} \{-m(t)\int_{t}^{T}[2\psi(s)+\beta(t)]ds\}dt - \int_{1}^{T} [\beta'(a(t))(\delta a(t))^2/2 - \int_{1}^{t} \delta y(s)\delta m(s)\beta(s)ds]dt,$$

$$\delta m(t) = 2\int_{1}^{t} (y(s)-1/2)\delta m(s)ds + 2\int_{1}^{t} m(s)\delta y(s)ds. \quad (77)$$

Therefore

$$\delta m(t) = 2\int^t_1 m(s)\exp\{2\int^t_s [y(u)-1/2]du\}\delta y(s)ds \quad (78)$$

and also

$$\int^t_1 \delta y(s)\delta m(s)\beta(s)ds \geq \mathbf{m}(\beta)\mathbf{m}(m)(\int^t_1 \delta y(v)dv)^2/2 \geq 0. \quad (79)$$

Since $\int^T_t [2\psi(s) + \beta(t)]ds \geq 0$, $t \in [\theta, T]$, and besides $\beta'(a_i(t)) > 0$, $i = 1, 2$, we have

$$\delta I(y_1), \delta I(y_2) \leq 0, y_1(t) = 0, t \in [0, T]; y_2(t) = 1, t \in [1, \theta\text{^}]; y_2(t) = 0, t \in [\theta\text{^}, T]. \quad (80)$$

It means that $y_1(t)$, $y_2(t)$ and the respective $a_1(t)$, $a_2\text{^}(t)$ deliver local maximums to I. We can also obtain the case when $I(y_1) = I(y_2)$ (V. Glushkov, V. Ivanov, Yu. Yatsenko [70, 71]).

7. Asymptotic Properties of Solutions

Let us consider some *asymptotic properties* of the relations (40)--(54) as $t \to T$ or $t, T \to \infty$. If $t \to T$, then $\text{^}A(t) = T$ for $a(T) \leq t \leq T$ and $\psi(t)$ in (41) tends to 0. So, in (40) under $\alpha(s,t)\psi(s) > -\beta(s,t)$, $s \in [t, T]$, $I_1'(y) < 0$. It means that the desired y that maximizes $I = I1$ has to be minimum possible in the end of $[t\text{^}, T]$ or on the whole segment $[t\text{^}, T]$ if $T-t\text{^}$ is sufficiently small. The same property takes place for the relations (42)--(46).

Under the condition of Lemma 5 and $t \to T$, we have

$$I_3'(y) \approx m(t)\int_t^{A\text{^}(t)} \beta(s,t)/\{\beta(s,a(s))[1-y(a(s))]\}ds > 0, \quad (81)$$

that is, the desired y that minimizes I_3 also has to be a possible minimum in the end of $[t\text{^}, T]$ or on the whole $[t\text{^}, T]$, if $T-t\text{^}$ is sufficiently small. In a similar way, under the condition of Lemma 6 and $t \to T$, we have $\psi_1(t)$, $\psi_2(t) \to 0$ and therefore $I_1{}'_x$, $I_1{}'_y < 0$ in the end of $[t\text{^}, T]$. It means that the desired x and y which maximize I_1 have to be minimum in the end of $[t\text{^}, T]$.

At last, under the conditions of Lemma 7 and $t \to T$, we have $\psi_i(t) = 0$, $i = 1, 2, 3, 4$, and hence $I_1{}'_j < 0$, $j = x, y, z, e$, in the end of $[t\text{^}, T]$, that is the desired x, y, z, and e have to be minimum in the end of $[t\text{^}, T]$.

Let us show that $y(t) = 0$ cannot be the solution of the problem C.1 under certain conditions. Let $y(t) = 0$, $R(t) \neq 0$, $t \in [t\text{^}, T]$, in MM (5, 6, Ch. 5). Then $a(t) < t\text{^}$. Indeed, if $a(t) \geq t\text{^}$, then $m(t) = 0$ and $R(t) = \int^t_{a(t)} m(s)ds = 0$, but it is impossible. Let

$$0 < y\text{^}(s), \alpha(t,s) = \Theta(\alpha\exp[-c_1(t-s)]), \beta(t,s) = \Theta(\beta\exp[-c_2(t-s)]),$$

$$c_1, c_2 > 0, \alpha > c_1, t \to \infty. \quad (82)$$

From (82), (40), and (41) it follows that under t, $T-t \to \infty$ ($y(t) = 0$, $t \in [t\text{^}, T]$)

$$-\psi(t) >\sim -\alpha m(y\text{^})\int^T_t \exp(-c_1 s)[-\psi(s)]ds + \beta/c_2, \quad (83)$$

from which

$$-\psi(t) >\sim \beta/c_2,\ I'(y) = \Theta(m(t)\int_t^T \{\alpha\exp[-c_1(s-t)][-\psi(s)]-\beta\exp[-c_2(s-t)]\}ds) >\sim$$

$$m(t)\beta/c_2(\alpha/c_1-1) > 0. \tag{84}$$

It means that $y(t) = 0$, $t \in [t^\wedge, T]$, cannot be the solution of the problem C.1.

Let more $x(t) = y(t) = 0$, $t \in [t^\wedge, T]$, in MM (46, Ch. 5), $\alpha = \beta$, $\mu = 1$, $\nu = 1$, $c_1 = 0$. Then $a(t) < t^\wedge$ and under t, $T-t \to \infty$ from (51) and (52) it follows that

$$I'_x = -m(t)\alpha(t)\int_t^T [\psi_1(s)+1]ds,\ I'_y = -m(t)\alpha(t)\int_t^T [\psi_2(s)+1]ds,$$

$$\psi_1(t) = -\int_t^T (T-s)m(s)ds,\ \psi_2(t) = \int_t^T \psi_3(s)ds-\alpha(t)(T-t),$$

$$\psi_3(t) = -\alpha(a)x(a)[\psi_1(t)+1] - \alpha(a)y(a)[\psi_2(t)+1] + \alpha(a), \tag{85}$$

from which

$$I'_x \geq m(t)\alpha(t)\int_t^T [m(m)(T-s)^2/2-1]ds >\sim 0. \tag{86}$$

Apparently, $I'_x(r) = 0$ implies $r = T$.

Under

$$0 < x^\wedge(t),\ y^\wedge(t) \leq 1,\ \alpha^\wedge(t) > 0,\ t \in [0, t^\wedge], \tag{87}$$

we have

$$\psi_2(t) = -\int_t^T \alpha^\wedge(a(s))y^\wedge(a(s))\psi_2(s)ds + f(t),\ f(t) = f_1(t) - f_2(t),\ f_2(t) = \alpha(t)(T-t),$$

$$f_1(t) = -\int_t^T \{x^\wedge(a(s))\alpha^\wedge(a(s))[\psi_1(s)+1] + \alpha^\wedge(a(s))[1-y^\wedge(a(s))]\}ds,. \tag{88}$$

Under $T-t$, $t \to \infty$,

$$f_1(t) = \Theta[(T-t)^3],\ f_2(t) = \Theta[(T-t)(t-t^\wedge)]. \tag{89}$$

This relation follows from

$$\alpha'(t) = \int_{a(t)}^{t^\wedge} \alpha^\wedge(s)x^\wedge(s)m^\wedge(s)ds < \text{constant}. \tag{90}$$

So, on the greater part of the segment $[t, T]$, the asymptotic behavior of $f(t)$ is determined by the function $f_1(t)$. Besides,

$$\psi_1(t) = \Theta[-(T-t)^2],\ I'_x > 0, \tag{91}$$

and

$$\psi_2(t) = -\int^T_t f'(s)\exp\,[-\int^s_t \alpha\text{^}(a(r))y\text{^}(a(r))dr]ds > 0,\ I'_y < 0. \tag{92}$$

Let $(t-t\text{^}) = \Theta[(T-t)^2]$. Then the mutual influence of $f_1(t)$ and $f_2(t)$ is compensated. And what is more, on the segment $[T\text{^}, T]$, $T\text{^} = T-(T-t\text{^})^{1/2}$, the asymptotic behavior of $f(t)$ is determined by the function $f_2(t)$, and also

$$-\psi_2(t) = \Theta[(t-t\text{^})(1-e^{-(T-t)})] > 0,\ I'_y > 0,\ t \in (T\text{^}, T\text{^^}) \subset [t\text{^}, T]. \tag{93}$$

It means that $x^*(t)$ for the problem C.1 is more than 0 on $[t\text{^}, r]$, $r = \Theta(T)$, and $y^*(t)$ is more than $y(t) = 0$ on $(T\text{^}, T\text{^^})$, where (see Theorem 13)

$$T\text{^^} = \Theta(T),\ T\text{^^} - T\text{^} = \Theta[(T-t\text{^})^{1/2}]. \tag{94}$$

Thus, we have the following:

Theorem 17. *Under the conditions of theorems above and* (82), *the asymptotic properties of solutions are expressed by the formulae* (83), (84), (93), and (94).

8. Additions and Remarks

The similar problems of investigation for multivariate MM still remain open.

We dwell here on two problems: optimization by using a certain other control functions and property of the functionals in the vicinity of the optimal internal control functions. The first problem is especially important for CAO since the distributions *x*, *y*, *z*- types can be pre-assigned and we have to use other control functions such as, for example, the functions *u*, *v*-type described in Ch. 5 (see (52), (53), Ch. 5). With regard to *u*, we can consider the same problems as above with the same results if we have, for example,

$$m(t) = \int^t_{a(t)} \alpha(t,s)u(s)y(s)m(s)ds,\ c(t) = \int^t_{a(t)} \beta(t,s)[U-u(s)][1-y(s)]m(s)ds,\ 0 < u \le U. \tag{95}$$

The reason for such kinds of relations can be scantiness of external resources $f(t)$. Otherwise, we will have simpler MM than (94), namely

$$m(t) = \int^t_{a(t)} \alpha(t,s)u_1(s)y(s)m(s)ds,\ c(t) = \int^t_{a(t)} \beta(t,s)u_2(s)[1-y(s)]m(s)ds,$$

$$0 \le u_i \le U_i,\ i = 1,\ 2. \tag{96}$$

In (96) the control functions u_1 and u_2 can be created independently for optimization of functionals that depend on *m*, *c*. As to the function *v*, with the replacement of *m* by *m*/*v* we will actually have the same MM and optimization problems as above.

One more possibility for optimization of CAO is creating their "prehistory" in such a manner that their following behavior would be pre-assigned. If in MM (96) α, β, and uy are given, and $m(t)$, $a(t)$ are pre-assigned for $t \in [t^\wedge, T]$, $a(t) < t^\wedge$, then we can determine the prehistory $m^\wedge(t)$, $t \in [0, t^\wedge]$, solving the equation

$$-\alpha(t,a)u(a)y(a)m^\wedge(a)a' + \int_a^{t^\wedge} \alpha'_t(t,s)u(s)y(s)m^\wedge(s)ds = [m(t) - \int_{t^\wedge}^{t} \alpha(t,s)u(s)y(s)m(s)ds]'. \quad (97)$$

The second problem is important for the numerical solution of the optimization problems. If *functional is strongly convex* in the vicinity of that t for which the gradient of functional equals to zero, then the well-known numerical methods for functional optimization will be convergent in the same vicinity by the rate of geometric progression.

We show here that the typical functional above

$$I = \int_{t^\wedge}^{T} \int_{a(t)}^{t} \beta(t,s)m^\wedge(s)dsdt \quad (98)$$

is strongly convex in the vicinity of internal optimal values of the control function y under both conditions A. and B. (58) and under the usual condition $\alpha'_s(t,s)m^\wedge(s) > 0$.

Indeed, let $I'(y,t) = 0$, $t \in [d, e] \subset [t^\wedge, T]$. Then we have by analogy with (61)--(63)

$$dI(y) = I(y+dy) - I(y) = (I''(y)dy, dy) = \int_d^e k(t) \int_{a(t)+da(t)}^{a(t)} [\alpha(t,s) - \alpha(s,a(t))]m^\wedge(s)dsdt =$$

$$-\int_d^e k(t)\alpha'_s(t,a(t))m^\wedge(a(t))(da)^2/2dt < 0 \quad (99)$$

to within the second order of da and hence dy, since da is an increment corresponding to dy by MM (5, 6, Ch. 5).

Similar results are also derived from some other optimization problems [213].

Chapter 16

NUMERICAL METHODS AND SOFTWARE

A *complete* or total absolute *error* Δ for the approximate solution of Volterra integral equations of the second and the first kind is considered. This error is less or equal to the sum of errors of three types due to input data inaccuracy and incompleteness (*inherent error*), approximate algorithms (*method error*), and realization of algorithms on computers (*round-off error*). The *optimal numerical methods* with *pre-assigned accuracy* allowing us to obtain $\Delta < \varepsilon$, where $\varepsilon > 0$ and pre-assigned, are determined. The minimal in order numbers of necessary basic computers' operations under realization of the methods and the respective algorithms are estimated. The equations and the methods for which the numbers of operations are admissible in practice under rather small ε are investigated.

It is shown that many problems of applied mathematics can be reduced to the solutions of Volterra-type integral and integro-differential equations. So, the numerical methods for V.i.e. can be used for the numerical solution of wide class of problems. In addition, a general procedure for the optimal by accuracy finite approximation of initial data of a given class is considered.

1. On the Solution with the Pre-Assigned Accuracy

1. 1. SECOND KIND OF VOLTERRA EQUATIONS

Let us consider *Volterra integral equations* (V.i.e.) *of the second kind*:

$$x(t) = \int_0^t f(t,u,x(u))du,\ t \in [0, T],\ x(0) = 0, \tag{1}$$

where $f(t)$ is a function of the class F (considered below) and $x(t)$ is the desired solution.

1.1.1. *Inherent errors*

Let X be a linear normalized space of functions determined on $[0, T]$ with the norm $\|\cdot\|_t$, $0 < t \le T$, $\|\cdot\|_T = \|\cdot\|$, such that

$$\|x_1x_2\|_t \le \text{vrai max } |x_1(u)|\,\|x_2\|_t\ (0 \le u \le t),\quad \left\|\int_0^t x(u)du\right\| \le \int_0^t \|x\|du,\ x_1, x_2 \in X. \tag{2}$$

Lemma. *In the cases of* $X = L_p = L_p([0, T])$, $1 \le p < \infty$, *and* $X = C([0, T])$, *the properties* (2) *are valid.*

Proof. Validity of the first one is obvious. The proof of the second one in the case $X = C$ is almost trivial. In the case $X = L_p$ the second inequality is the consequence of the generalized inequality of Minkowski (see G. Hardy et al. [76]):

$$\{\int_0^t |\int_0^t x(v,u)du|^p dv\}^{1/p} \le \int_0^t \{\int_0^t |x(v,u)|^p dv\}^{1/p} du,\ 1 \le p < \infty, \tag{3}$$

in which we need to put $x(v,u) = x(v\text{-}u)$, $x(v\text{-}u) = 0$ for $u > v$.

We assume that the equation (1) has the unique solution $x \in X$ for each $f \in F$ and that $F \subset X$. When f runs F, the respective solution of equation (1) runs the determined set $X^\wedge$. Namely, the whole $X^\wedge$ is the approximate solution of the equation (1) without loss of information. However, for easing of the subsequent applications the approximate solution is usually the determined $x^\wedge \in X^\wedge$ corresponding to the determined $f^\wedge \in F$. Then the *inherent* or hereditary *error* sought for is

$$\Delta_1 = \sup \|x - x^\wedge\|_t\ (x \in X^\wedge),\ 0 < t \le T. \tag{4}$$

Denoting

$$\Delta f = \int_0^t [f(t,u,x(u)) - f^\wedge(t,u,x^\wedge(u))]du,\ \Delta x = x(t) - x^\wedge(t), \tag{5}$$

we easily find

$$\Delta x = \int_0^t [f(t,u,x(u)) - f(t,u,x^\wedge(u))]du + \Delta f, \tag{6}$$

from which under the condition that f belongs to the class of *Lipschitz functions* over x:

$$\|f(.,x_1) - f(.,x_2)\| \le L\|x_1 - x_2\|, \tag{7}$$

on the strength of the Lemma and the inequality by Chaplygin we have

$$\|\Delta x\|_t \le L\int_0^t \|\Delta x\|_u du + \|\Delta f\|, \tag{8}$$

and also

$$\|\Delta x\|_t \le L\int_0^t \|\Delta f\|_u \exp[L(t-u)]du + \|\Delta f\|_t \le \max \|\Delta f\|_u\ (0 < u < t)e^{Lt},$$

$$0 < t \le T,\ \Delta_1 \le \sup \|\Delta f\|\ (f \in F)e^{LT}. \tag{9}$$

Thus, we have proved the following.

Theorem 1. *Under the condition that the function f in* (1) *complies with* (7) *and under the condition* (2), *for the inherent error* Δ_1 *the estimates* (4) *and* (9) *are valid.*

Note that these estimates are well known in the space *C*.
Let the condition

$$L'(x_1 - x_2) < f(.,x_1) - f(.,x_2) < L''(x_1 - x_2),\ -L < L',\ L'' < L, \tag{7'}$$

be given instead of (7). Then

$$L'\int_0^t \Delta x du + \Delta f < \Delta x < L''\int_0^t \Delta x du + \Delta f,\ L'\int_0^t \Delta f e^{L'(t-u)} du + \Delta f < \Delta x < L''\int_0^t \Delta f e^{L''(t-u)} du + \Delta f,$$

$$|\Delta x| < \max\left[|L'|\int_0^t |\Delta f| e^{L'(t-u)} du,\ |L''|\int_0^t |\Delta f|\, e^{L''(t-u)} du\right] + |\Delta f| \tag{8'}$$

from which

$$\Delta_1 \leq \max\left(2, \exp\left[\max(L', L'')T\right]\right)\max|\Delta f|\ (0 < u < T). \tag{9'}$$

In particular,

$$\Delta_1 \leq 2\max|\Delta f|\ (0 < u < T),\ \max(L', L'') \leq 0. \tag{9''}$$

Thus, we also have

Theorem 1'. *Under the condition of Theorem* 1 *with the replacement of* (7) *by* (7') *for the inherent absolute error* (4), *the estimates* (9') *and* (9") *are valid.*

It should be noted that under (7')

$$L'\int_0^t x(u)du + f_1(t) \leq x(t) \leq L''\int_0^t x(u)du + f_1(t),\ f_1(t) = \int_0^t f(t,u,0)du, \tag{10}$$

from which we can estimate $|x(t)|$ from both above and below in several cases. This can give more precise estimations of the hereditary relative error regarding (9') and (9").
Let the constants a and b be determined such that

$$f^+(t,u,x) = \sup f(t,u,x)\ (f \in F) > ax + b,\ a > 0, \tag{11}$$

and, besides,

$$L^+\delta x < f(.,x) - f(.,x^+) < L^{+\prime}\delta x,\ \delta x = x - x^+,\ x^+(t) = \int_0^t f^+(t,u,x^+(u))du. \tag{7''}$$

Then

$$x^+(t) > a\int_0^t x^+(u)du + bt,\ \delta f = \int_0^t [f(t,u,x^+(u)) - f^+(t,u,x^+(u))]du \leq 0, \tag{12}$$

and also

$$|x^+(t)| > |b|/a|1 - e^{-at}(1+at)|e^{at},$$

$$L^+\int_0^t \delta f \exp(L^+(t-u))du + \delta f < \delta x < L^{+\prime}\int_0^t \delta f \exp(L^{+\prime}(t-u))du + \delta f, \tag{13}$$

from which

$$|\delta x| < \max(|L^+|, |L^{+\prime}|)\int_0^t |\delta f| \exp(L^+(t-u))du + |\delta f|,$$

$$\|x\|_t < \max 2, \exp(L^+t))\max \|\delta f\|_u\ (0 < u < t), \tag{14}$$

and more

$$\|x/x^+\|_t < a/|b|e^{-at}1/|1 - e^{-at}(1+at)|\max(2, \exp(L^+t))\max \|f\|_u\ (0 < u < t) \tag{15}$$

Thus, we have

Theorem 2. *Under the condition* (2), (7"), *and* (11), *for the inherent relative error, the estimate* (15) *is valid.*

We will show, at last, how the least possible value

$$\Delta^* = r^*_x = \inf[\sup \|x - x\hat{}\|\ (x \in X\hat{})]\ (x\hat{} \in X\hat{}) \tag{16}$$

can be estimated. We assume that $f \in F$ has everywhere a continuously differentiable f'_x and that the derivative f'_x runs F' while f runs F.

For an arbitrary set F with the norm $\|\cdot\|$ we denote by f^* *Chebyshev center* and by r^*_f *Chebyshev radius* of F:

$$r^*_f = \inf \sup \|f_1 - f\|\ (f \in F) = \sup \|f^* - f\|\ (f \in F), \tag{17}$$

where lower bound means over all the elements f_1 for which $\|f_1 - f\|$ make sense.

Let f^*, $f'_x{}^*$, and x^* be Chebyshev centers for F, F', and $X\hat{}$, respectively.

Theorem 3. *Under the condition that*

$$\|f^* - f_*\| = o(\sup \|f - f^*\|\ (f \in F)),\ f_*{}'_x = f'_x{}^*, \tag{18}$$

and F is symmetrical relative to its Chebyshev center f^*, the solution x_* of the equation

$$x(t) = \int_0^t f^*(t,u,x(u))du \tag{19}$$

is asymptotically optimal by accuracy, i.e.,

$$\|x^* - x_*\| = o(\sup \|x - x^*\| \ (x \in X^\wedge)) = o(r^*_x), \ \ r^*_x \to 0, \tag{20}$$

Proof. We have

$$x - x_* = \int_0^t [f(t,u,x) - f^*(t,u,x_*)]du =$$

$$\int_0^t \{[f_*(t,u,x) - f_*(t,u,x_*)] + [f^*(t,u,x) - f_*(t,u,x)] + [f(t,u,x) - f^*(t,u,x)]\}du$$

$$= \int_0^t f_*{}'_x(x_*)(x - x_*)du + \int_0^t [f(t,u,x) - f^*(t,u,x)]du + o(\|x - x_*\|), \tag{21}$$

after which the matter is reduced to the linear case that is trivial.

Consequently, from Theorem 3 it follows that any approximate *solution* of the equation (19) with higher order accuracy than r^*_f will be asymptotically optimal by accuracy. It will be the same result if f^* is given with the sufficient accuracy.

For the methods, algorithms, and software for determinations of optimal by accuracy approximations of f^* from several classes see V. Ivanov et al. [93].

1.1.2. *Method errors*

To begin with, we will solve the equation (1) on the segment $[0, \Delta T]$, $\Delta T < T$. We consider two methods of the *approximately iterative* (AI)-type (V. Dzyadyk [42]): the first one (M_1) for the case when f does not have smoothness of high degree and the second one (M_2) when f is an analytical function.

Let $f(t,u,x)$ apply with (7) and have ith, $i = 0, 1, 2$, derivatives over all the arguments that are less by modules than the respective given constants $Li = Li(f)$ (in a domain when the solution exists). Then, as the approximate solution of the equation (1), we take the approach to

$$x_N(t) = \Sigma_{\nu=1}^{N-1} (x_{\nu+1} - x_\nu); \ x_{\nu+1} = \int_0^t f(t,u,x_\nu(u))du, \ x_0 = 0, \tag{22}$$

where the integrals are computed by the formula of the trapezium.

Applying the well-known steps, we have on $[0, \Delta T]$

$$\|x_1 - x_0\|_t = \|f_0\|_t, \ f_0 = \int_0^t f(t,u,0)du, \ \|x_2 - x_1\|_t \le Lt\max\|f_0\|_u \ (0 < u < t), \ \ldots,$$

$$\|x_{\nu+1} - x_\nu\|_t \le (Lt)^\nu/\nu! \max \|f_0\|_u \ (0 < u < t), \tag{23}$$

from which the error

$$\Delta_2(M_1,t) = \|x-x_N\|_t \leq \sum_{\nu=1}^{\infty} \|x_{\nu+1}-x_\nu\|_t \leq$$

$$\max \|f_0\|_u (0 < u < t)(Lt)^N/N![1+ Lt/(N+1) + (Lt)^2/(N+1)(N+2)+...] <$$

$$\max \|f_0\|_u(0 < u < t)(2\pi N)^{-1/2}(Lte/N)^N[1+Lt/(N+1)+(Lt)^2/(N+1)(N+2)+...], 0 \leq t \leq \Delta T. \quad (24)$$

Let us divide the segment [0, ΔT] into M parts by the points

$$t_j = j\Delta t,\ \Delta t = \Delta T/M,\ j = 1, ..., M, \quad (25)$$

and put

$$x\hat{}_{\nu+1,j} = \Delta t\{[f(t_j,0,0)+f(t_j,t_j,x\hat{}_{\nu j})]/2+\Sigma_{s=1}^{j-1} f(t_j,t_s,x\hat{}_{\nu s})\} = \int_0^t {'f}(t_j,u,x\hat{}_\nu(u))du, \quad (26)$$

where for an arbitrary function f a function $'f$ denotes the polygon:

$$'f(u) = [f(t_s)(t_{s+1} - u) + f(t_{s+1})(u - t_s)]/\Delta t,\ u \in [t_s, t_{s+1}], \quad (27)$$

where $x\hat{}_\nu(u)$ is a step function with the values $x\hat{}_{\nu s}$ for $u \in [t_s, t_{s+1}]$. It is well known (see, for example, N. Bakhvalov [16]) that

$$\int |f - 'f|du\ (u \in [0, t_j]) < L_2(f)/12(\Delta t)^2 t_j. \quad (28)$$

Therefore,

$$|x\hat{}_{\nu+1,j} - x_{\nu+1}(t_j)| \leq |x_{\nu+1}(t_j) - \int_0^t {'f}(t_j,u,x_\nu(u))du| + |\Sigma^{j-1}_{s=0}\int_{t_s}^{t_{s+1}} ['f(t_j,u,x_\nu(u)) - 'f(t_j,u,x\hat{}_\nu(u))]du| <$$

$$2(f(.,x_\nu))/12(\Delta t)^2 t_j + Lt_j \max |x\hat{}_{\nu,s} - x_{\nu,s}|\ (1 \leq s \leq j). \quad (29)$$

It is not difficult to show that

$$df(.,.,x_\nu(u))/du = \partial f/\partial u + \partial f/\partial x dx_\nu/du;\ d^2f(.,.,x_\nu(u))/du^2 =$$

$$\partial^2 f/\partial u^2 + \partial^2 f/\partial x^2(dx_\nu/du)^2 + \partial f/\partial x d^2x_\nu/du^2 + 2\partial^2 f/[\partial u \partial x]dx_\nu/du. \quad (30)$$

So, we should estimate $(x_\nu)'$ and $(x_\nu)''$ for $\nu = 1,..., N$-1. We have

$$(x_{\nu+1})' = f(t,t,x_\nu(t)) + \int_0^t \partial f/\partial t(t,u,x_\nu(u))du,\ (x_{\nu+1})'' =$$

$$\partial f/\partial t + \partial f/\partial u(u=t) + \partial f/\partial x\,(x_\nu)' + \partial f/\partial t(t,t,x_\nu(t)) + \int_0^t \partial^2 f/\partial t^2(t,u,x_\nu(u))du, \quad (31)$$

and therefore,

$$(x_{\nu+1})' < L_0 + L_1\Delta T,\ (x_{\nu+1})'' < L_1(3 + L_0 + L_1\Delta T) + L_2\Delta T =$$

$$3L_1 + L_0L_1 + (L_1^2 + L_2)\Delta T. \qquad (32)$$

Combining the obtained estimations, we easily find $L_2(f(.,x_\nu)) < d$,

$$d = L_2 + L_2L_0^2 + 2L_0L_2 + 3L_1^2 + L_0L_1^2 +$$

$$(2L_0L_1L_2 + L_1^3 + 3L_1L_2)\Delta T + L_2L_1^2\Delta T \qquad (33)$$

and also

$$\Delta x_{\nu+1} < dt_j/12(\Delta t)^2 + Lt_j\Delta x_\nu;\ \Delta x_\nu = \max |x\wedge_{\nu,s} - x_{\nu,s}|\ (1 \le s \le j), \qquad (34)$$

from which

$$\Delta x_\nu < dt_j/12(\Delta t)^2(1-(Lt_j)^\nu)/(1-Lt_j) < dt_j/6(\Delta t)^2,\ Lt_j < L\Delta T < 1/2. \qquad (35)$$

Now regarding (24), it is not difficult to obtain

$$\Delta_2(M_1,t_j) = |x(t_j) - x\wedge_{Nj}| \le |x(t_j) - x_N(t_j)| + |x_N(t_j) - x\wedge_{Nj}| \le$$

$$L_0t_j(2\pi N)^{-1/2}(e/2N)^N\{1+1/[2(N+1)] + \dots\} + dt_j/6(\Delta t)^2,\ Lt_j \le 1/2,\ 1 \le j \le M. \qquad (36)$$

As a result we have the following.

<u>Theorem 4.</u> *The estimate* (33) *and* (36) *give the error of the method* (22)-(27). *In addition, the number of f-computation operations does not exceed* $(M+1)(M+2)N/2$.

It should be noted that in a similar way we can estimate the error and the number of operations if we apply the more precise formula by Simpson (involving the 4th derivation of f). It is convenient to have the approximate solution as the respective splice-functions (see, for example, V. Ivanov [89]) instead of the array of $x\wedge_{Nj}$. Under the condition of smaller smoothness of f, one can create the method of numerical realization of (22) based on the rational approximation of f [89] or the *method of Monte-Carlo* (G. Mikhaylov [138]).

Let $x\wedge(t)$ be the approximate solution of the equation (1) and let us consider the case $t \in [k\Delta T, (k+1)\Delta T]$, $k = 1, \dots, R-1$, $R\Delta T = T$. Let us also put

$$x_k(t) = \int_0^{k\Delta T} f(t,u,x\wedge(u))du + \int_{k\Delta T}^{t} f(t,u,x_k(u))]du,$$

$$x_{k,v}(t) = \int_0^{k\Delta T} f(t,u,x\hat{}(u))du + \int_{k\Delta T}^{t} f(t,u,x_{k,v-1}(u))du,\ x_{k,0}(t) = \int_0^{k\Delta T} f(t,u,x\hat{}(u))du, \quad (37)$$

and

$$x\hat{}_{k,v,j} = x\hat{}_{k,0,j} + \int {}'f(t_{k,j},u,x\hat{}_{k,v-1}(u))du\ (u \in [k\Delta T, t_j]),$$

$$x\hat{}_{k,0,j} = \int_0^{k\Delta T} {}'f(t_{kj},u,x\hat{}(u))du,\ t_{kj} = k\Delta T + j\Delta t,\ j = 1, \ldots, M,\ \Delta t = \Delta T/M,\ v = 1, \ldots, N, \quad (38)$$

where $x\hat{}_{k,v}(u)$ is the stepwise function which is equal to $x\hat{}_{k,\,Ns}$ on the segment $[t_{k,s}, t_{k,s+1}]$ and $x\hat{}(t)$ is the respective polygon of t, $t \in [t_j, t_{j+1}]$.

By analogy with (33) and (36), from (37) and (38) it follows

$$\Delta x = x(t) - x\hat{}(t) = x(t) - x_k(t) + x_k(t) - x_{kN}(t) + x_{kN}(t) - x\hat{}(t) = \int_0^t [f(t,u,x(u)) - f(t,u,x\hat{}(u))]du + r_k(t),$$

$$r_k(t) = \int_{k\Delta T}^{t} [f(t,u,x\hat{}(u)) - f(t,u,x_{k,N}(u))]du + \int_{k\Delta T}^{t} [f(t,u,x_{kN}(u)) - f(t,u,x_k(u))]du +$$

$$x_k(t) - x_{k,N}(t) + x_{kN}(t) - x\hat{}(t), \quad (39)$$

from which by the virtue of (7') and (9"), where $r_k(t)$ plays the role of Δf, we have

$$\Delta_2(M_1) = \max|x(t) - x\hat{}(t)|\ (t \in [0, T]) \le 3[kL_0\Delta T/N! + kd\Delta T/6(\Delta t)^2] + |x''(t)|/8(\Delta t)^2 \le$$

$$3L_0T/N! + dT/2(\Delta t)^2 + 1/8[3L_1 + L_0L_1 + (L_1^2 + L_2)\Delta T](\Delta t)^2,$$

$$j = 1, \ldots, M;\ k = 1, \ldots, R-1;\ R\Delta T = T;\ \Delta t = \Delta T/M. \quad (40)$$

Thus, it establishes the following:

Theorem 4'. *The method* (37), (38) *error is given by the estimate* (33) *and* (40). *In addition, the number of f-computations does not exceed* $(M+1)(M+2)NR/2$.

Let us consider now the second case when $f(z,v,w)$ is an analytical function of 3 complex-valued arguments z, v, w in the domain

$$D = E_r \times E_r \times D(H), \quad (41)$$

where E_r is the closed domain bounded by the *ellipse of Zhukovsky*

$$\partial E_r = \{(x,y)\,|\,x = \Delta T/2 + a_r\cos t,\ y = b_r \sin t,\ t \in [-\pi, \pi]\},$$

$$a_r = \Delta T(r + r^{-1})/4,\ b_r = \Delta T(r - r^{-1})/4,\ r > 1, \quad (42)$$

and the domain $D(H)$ is the closed circle with the radius H and the center 0.

Let us consider another *approximately iterative* (*AI*) *method* [42], setting

$$w(z) = \int_0^z f(z,s,w(s))ds,\ w_v(n;z) = \int_0^z L_n(f(.,w_{v-1}(n;.),s))ds = \sum_{i=0}^n f(z,x_i,w_{v-1}(n;x_i)) \int_0^z p_i(s)ds,$$

$$w_v(n;x_j) = w_{vj} = \Delta T/2 \sum_{i=0}^n a_{ij} f(x_j,x_i,w_{(v-1)i});\ j = 0, ..., n;\ v = 0, ..., N;$$

$$a_{ij} = e_i/n[1 - c_j + c_i/2(1-c_{2j}) + \sum_{v=2}^n e_v c_{iv}(c_{j(v-1)}/(v-1) - c_{j(v+1)}/(v+1) - 2/(v^2-1))];$$

$$e_0 = e_n = ½;\ e_j = 1, j = 1,..., n-1;\ c_k = \cos(k\pi/n);\ x_j = x_j(n) = \Delta T/2(1 - \cos(j\pi/n)). \tag{43}$$

Theorem 5. *Let in the equation*

$$w(z) = \int_0^z f(z,s,w(s))ds, \tag{44}$$

the function f be analytical in the domain $D^\wedge = E_r^\wedge \times E_r^\wedge \times D(H^\wedge)$, $\Delta T^\wedge > 0$, $H^\wedge > 0$, $r^\wedge > 1$.

Then if under $r \in [1, r^\wedge]$ *and natural numbers n and N we take the number* $H \le H^\wedge$, $\Delta T \le \Delta T^\wedge$, *such that the inequality*

$$H > (a_r + \Delta T/2)(\|f\|_{C(D^\wedge)} + L\Delta_{2,\Delta T}),\ a_r = a_r(\Delta T) = \Delta T/4(r + r^{-1}),$$

$$D = E_r \times E_r \times D(H), \tag{45}$$

is fulfilled, where $\Delta_{2,\Delta T}$ *denotes the right part of the inequality*

$$|w(z) - w_N(n;z)| \le L_0(f)\{(\|L_n\|+1)r(1+r^{-2n-2})/[(1-q)(r^\wedge - r)](r/r^\wedge)^n(a_r + \Delta T/2) +$$

$$L^{-1}e^q q^{N+1}/(N+1)!\},\ \|L_n\| < 2/\pi \ln n + 1,\ q = L(a_r + \Delta T/2),\ L_0(f) = \|f\|_{C(D^\wedge)}, \tag{46}$$

L is Lipschitz constant for f over the argument w (in *D*), *then the polynomial* $w_N(n;z)$ *created by the AI-algorithm* (43) *approximates the solution* w(z) *of the equation* (44) *in the domain* E_r *such that the inequality* (46) *is fulfilled.*

The proof is similar to that of Theorem 2 in the articles [42]; V. Dzyadyk, V. Ivanov [43].

It follows from Theorem 5 and the proof of Theorem 4' the following:

Theorem 5'. *If the segment* [0, *T*] *is divided on the sub-segments* $[(k-1)\Delta T, k\Delta T]$, $L\Delta T < 1/2$, $k = 1,..., R$, *and for every of these sub-segments the condition of Theorem* 5 *takes place, then the error* $\Delta_2(M_2)$ *for* [0, *T*] *under the piecewise polynomial approximation* (43)*-type is estimated by the relation*

$$\Delta_2(M_2) < \exp(LT) 4\Delta_{2,\Delta T},\ L\Delta T < 1/2. \tag{47}$$

In addition, the total number of f-computations does not exceed $RN(n+1)^2$.

Let us explain the origin of the constant 4. Under $L\Delta T = 1/2$, $r_k < k\Delta_{2,\Delta T}$, we have

$$\Delta_2(M_2) < \Delta_{2,\Delta T}\Sigma_{k=1}^{R-1} \exp((R-k)L\Delta T)k < e^{LT}\Delta_{2,\Delta T}\int_0^\infty xe^{-x/2}dx = 4e^{LT}\Delta_{2,\Delta T}.$$

1.1.3. *Round-off errors*

Let us denote f_l for f computed on a computer, where l is the number of binary digits in an adder of the computer. Not infrequently the estimate

$$\Delta_3(f) = |f - f_l| < C_f(|f|+1)2^{-l} \leq C_f(L0+1)2^{-l}, \tag{48}$$

is valid [16], where the constant C_f is not very large. Let us see the additional error of the solution that can result in the error of (48)-type.

Theorem 6. *Under the condition* (7'), (9"), *and* (48) *the round-off error for the approximate solution* (38) (*in the floating point system*)

$$\Delta_3(M_1) < \{2.12L_0[(M+1)R+2]^2 + 4C_f(L_0+1)(M+1)R\}2^{-l},\ 2^{-l}(M+1)R < 0.1. \tag{49}$$

Proof. Let us apply to the scheme (38) the well-known estimate (J. Wilkinson [210]) of the type

$$|\Sigma_{k=1}^{m} [(a_k)_l - a_k]| < \Sigma_{k=1}^{m} |a_k||E_k| < 1.06\cdot 2^{-l}\max|a_k|\ (1 \leq k \leq m)(m+2)^2/2,\ 2m < 0.1. \tag{50}$$

As a result, the maximal round-off error along one step (see also (26)) of an iteration

$$\Delta_{3,1} < 1.062^{-l}L_0[(M+1)R+2]^2/2 + \Delta_3(f)(M+1)R,\ 2^{-l}(M+1)R < 0.1. \tag{51}$$

By the virtue of convergence of the iterative process (38) over v with the similarity ratio $L\Delta T < 1/2$, the summarizing error along one step does not exceed $2\Delta_{3,1}$ [89]. This error along subsequent steps will be as a hereditary error.

So, by the virtue of (9"), the accumulated error does not exceed $4\Delta_{3,1}$.

Theorem 7. *The round-off error for the approximate solution* (43)

$$\Delta_3(M_2) < 1/(1-q)[n+3)^2/2L_0 \max|a_{ij}|1.06\cdot 2^{-l} + \Delta_3(a_{ij})L_0 + \Delta_3(f)\max|a_{ij}|(n+1)],$$

$$(n+2)2^{-l} < 0.1;\ \Delta_3(a_{ij}) < (n+7)\ln n/n 1.06\cdot 2^{-l}\Delta_3(\cos)\ln n;$$

$$\Delta_3(\cos) < 2^{-l+1};\ |a_{ij}| < 3\ln n/n;\ q = L(a_r+\Delta T/2) < 1. \tag{52}$$

The proof can be obtained by analogy with the proof of Theorem 6 on the strength of (43) and (49) and the well-known estimates (J. Wilkinson [210]) of the round-off error for any arithmetical operation *:

$$(a*b)_l = (a*b)(1+e_x),\ |e_x| < 2^{-l}. \tag{53}$$

From Theorem 7 and the mode of the proof of Theorem 6, the next also follows:

Theorem 7'. *Under the condition of* (7) *and Theorem* 5' *the round-off error for the piecewise polynomial approximation* (*with regard only to the main terms*)

$$\Delta_3(M_2,T) < e^{LT}4\Delta_3(M_2,\Delta T) = L_0e^{LT}2^{-l+3}4n\ln n/(1-q),$$

$$(n+2)R2^{-l} < 0.1,\ L\Delta T \leq 1/2. \tag{54}$$

1.1.4. *Analysis of the total errors*

Connecting the estimates (9"), (40), and (49) we obtain the estimation of the complete absolute error (in Chebyshev metric) for the approximate solution of V. i. e. (1) by the method M_1 (37), (38):

$$\Delta(M_1) < 2\max|\Delta f|\ (0 < u < T) + 3TL/N! + dT/2(\Delta t)^2 +$$

$$1/8[3L_1+L_0L_1+(L_1^2+L_2)\Delta T](\Delta t)^2 + \{2.12L_0[(M+1)R+2]^2 +$$

$$4C_f(L_0+1)(M+1)R\}2^{-l},\ 2^{-l}(M+1)R < 0.1. \tag{55}$$

In a similar way, connecting the estimates (9), (46), (47), and (54), we obtain the estimation of the *complete error* for the solution of V. i. e. (1) by the AI-method M_2 (43)-type (with regard to the main terms and $q \leq 1/2$, $r/r\hat{} \leq 1/2$):

$$\Delta(M_2) < \exp(LT)\{\max|\Delta f|\ (0 < u < T) + 4L_0[\pi\ln n2^{-n} + eL^{-1}/(2^N N!)]$$

$$-\ 2^{-l+4}L_04n\ln n\},\ 2^{-l}(n+2)R < 0.1. \tag{56}$$

From (55), one can see that in order that $\Delta(M_1)$ has the pre-assigned order of decreasing ε, it is sufficient that $|\Delta f|$, R/M^{2}, R/N^{N}, and $(MR)^2 2^{-l}$ have the order of ε. The number operations M^2NR of the first method will be logarithmically equivalent to the number of input data M^2R for $\varepsilon \to 0$, $1 > \varepsilon > 0$, if, for example, $N = \ln(1/\varepsilon)$, $M^2R = 1/\varepsilon$. In addition, if $\ln R$ has the order of $\ln(1/\varepsilon)$, then l also has the order of $\ln(1/\varepsilon)$. Thus, both hypotheses (N. Bakhvalov [15]) will be fulfilled for the first method. However, it is clear that the conditions of this method efficiency, especially the conditions (9"), are very restrictive.

From (56), it follows that in order for $\Delta(M_2)$ to have the pre-assigned order of decreasing ε, it is sufficient that $e^R|\Delta f|$, $e^R ln\ n/2^n$, $e^R/2^N N!$, and $e^R n\ln n/2^l$ have the order of ε. Under fixed R the number of digits l also has the order of ln $(1/\varepsilon)$ (i.e., the second method will be also stable according to the hypotheses [15]). However, it is clear that in practice we have to use additional R digits (i.e., we have to compute under increased digitalization). The hypothesis about logarithmic equivalence between the numbers of input data and the operations does not take place for the second method. This fact prompts us that there is the determined reserve for improvement of the second method.

Indeed, applying, under the same input data, instead of the simple iteration the method that combines the simple iteration and the *method of Newton* [16, 89], in the respective estimation of the complete error, we will have the term of the form $1/2^{2N}$ instead of $1/(2^N N!)$, after which the property of logarithmic equivalence will take place.

It should be noted that any *method* and the respective *algorithm* for which the hypotheses [15] are valid may naturally be called *optimal* by accuracy and number of operations. We also note the certain *reserves* for raising *accuracy*.

A decrease of *inherent errors* can be achieved at the expense of the restoration of measuring system input data problem solutions that are also reduced to the solutions of Volterra-type and Wiener-Hopph-type equations [89], V. Ivanov et al. [95].

A decrease of *method errors* can be achieved [16] at the expense of accounting of different values of the estimating constants $L_i(f)$, $i = 0, 1, 2$, on each sub-segment $[k\Delta, (k+1)\Delta T]$, $k = 0,\ldots, R-1$, and the special dividing of these sub-segments. Applying instead of Δt the steps Δt_k such that the estimations of the errors for different sub-segments were identical, we can minimize the method error (under a given number of f-computations).

A decrease of *round-off errors* can be achieved ([16]; [95]; V. Voevodin [196]) at the expense of the special techniques of computation. For example, the summation should be realized by way of pair-wise addition of the results of addition. Besides, a probability character of round-off error [196, 210] should be considered.

1.2. FIRST KIND OF VOLTERRA EQUATIONS

1.2.1. *Two main approaches*

Let us consider *V. i. e. of the first kind*

$$Ax = \int_0^t a(t,u,x(u))du = f(t),\ t \in [0, T]. \tag{57}$$

The first approach consists in reducing (57) to the V. i. e. of the second kind

$$x(t) = a^{-1}[-\int_0^t a'_t(t,u,x(u))du + f'(t)],\quad a(x) = a(t,t,x(t)), \tag{58}$$

by way of formal differentiation under the condition of the constant sign of $a'_x(t,t,x(t))$. It can be shown that all the methods of s.1 can be applied to (58) without principal variations.

If, as usually occurs, instead of $a(t,u,x(u))$ and $f(t)$ the respective classes A and F of their approximate value $a^\wedge(t,u,x(u))$ and $f^\wedge(t)$ are given, then under the transition from (57) to (58) the important problem of numerical differentiation or determination of the respective classes A' and F' is aroused (for several new results about the solutions of this problem see [89, 95]). For our purposes it is sufficient to note that we can determine the classes A and F such that A' and F' have been investigated. It is clear that the method of reducing (57) to (58) is not universal and it is still necessary to have an approach for the direct solution of (57) as V. i. e. of the first kind, which is an *ill-posed problem* (A. Tikhonov, V. Arsenin [184]).

1.2.2. *Quasi-solution method*

Let us consider in detail the well-known *method of quasi-solution* [89] in connection with (57) if a priori we know the set M of the solution correctness such that on this set the operator A has the *reciprocal operator* A^{-1} for which the estimate of the *continuity module*

$$w(A^{-1},d) = \sup \|A^{-1}(f_1) - A^{-1}(f_2)\| \; (\|f_1 - f_2\| < d, f_1, f_2 \in A(M) \tag{59}$$

is given, where $\|.\|$ is the sign for the norm of linear normalized Banach or Hilbert space.

The general scheme now consists in solving the problem

$$I(x) = \|Ax - f\| = \min \; (x \in M), A \in \mathrm{A}, f \in F, \tag{60}$$

that is reduced to the convex programming for the convex set M and convexity of $I(x)$ in the domain $D(x_0)$: $I(x) < I(x_0)$, where x_0 is the given initial approximation for the solution of the problem (60). If x^*_M is the exact solution of the problem (60), x^* is the desired solution, $x^* \in M$, and $x_{M\,d}$ is the approximate solution of the problem (60), i.e.,

$$\|Ax_{M\,d} - f\| < \|Ax^*_M - f\| + d, \; d > 0, \; x_{M,\,d} \in M, \tag{61}$$

where d, in a general case, is the *numerical error* for the solution of the problem (60) (does not exceed the sum of the method error and round-off error), then

$$\|Ax^* - Ax_{M\,d}\| < \|Ax^* - f\| + \|Ax^*_M - f\| + d < 2\|Ax^* - f\| + d, \tag{62}$$

and hence

$$\|x^* - x_{M\,d}\| < w(A^{-1}, 2\|Ax^* - f\| + d). \tag{63}$$

But

$$\|Ax^* - f\| \leq d(A,M) + d(F), \tag{64}$$

where

$$d(A,M) = \sup\ [\sup\ \|A_1x\text{-}A_2x\|\ (A_1,\ A_2 \in A)]\ (x \in M),$$

$$d(F) = \sup\ \|f_1\text{-}f_2\|\ (f_1,\ f_2 \in F). \tag{65}$$

By substituting (64), (65) in (63), we have the following:

Theorem 8. *The complete error of the quasi-solution method for the approximation solution* (61) *of the problem* (60)

$$\Delta = \|x^* - x_{M\,d}\| < w(A^{-1}, 2[d(A,M)+d(F)]+ d). \tag{66}$$

Note that a similar estimate (without regarding the round-off error) is well known to specialists. Determined value d can be obtained from the finite-dimensional approximation of the solution of the problem (60) [89, 95].

1.2.3. *Estimate of continuity modules for reciprocal operators*

It is necessary to know the function $w(A^{-1},d)$ for the effective application of (66). There are plenty of scientific works devoted to the estimation of $w(A^{-1},d)$ (see, for example, [96, 184]). We dwell here on a simple estimate that is well known (V. Ivanov [88]) for a linear operator.

Let the *derivative by Fresher* of the operator A in the point $x^\wedge$, $x^\wedge \in D(x_0)$,

$$A'x = \int^{t}_{0} a'(t,u)x(u)du,\ a'(t,u) = a'_x(t,u,x^\wedge), \tag{67}$$

have the property

$$\|Ax - Ax^\wedge - A'x\| < L_2\|x - x^\wedge\|, \tag{68}$$

where L_2 is a given constant, and introduce the equation

$$A_1x_1 = (A')^*A'x_1 = \int^{T}_{0} k(t,u)x_1(u)du = (A')^*f = f_1(t),$$

$$k(t,u) = \int^{T}_{0} a'(v,u)a'(v,t)dv = \int^{u}_{t} a'(v,u)a'(v,t)dv\ (u < t),$$

$$-\int^{T}_{u} a'(v,u)a'(v,t)dv\ (u > t),\ (A')^*f = \int^{t}_{0} a'(u,t)f(u)du = f_1(t). \tag{69}$$

Besides, let the *set of correctness*

$$M(x^\wedge) = \{x \,|\, x = A_1v,\ \|v\| \le R\} \tag{70}$$

Then [88]

$$w(A_1^{-1},d) < (2Rd)^{1/2}. \tag{71}$$

Assuming $\|Ax_1 - Ax_2\| < d$, under condition that $x_1, x_2 \in M \subset S(x^\wedge, r)$, where S is the sphere in Hilbert space H with center $x\hat{}$ and radius r, on the strength of (68) and (71) we have

$$\|A'x_1 - A'x_2\| = \|A'x_1 + Ax\hat{} - Ax_1 - (A'x_2 + Ax\hat{} - Ax_2) + Ax_1 - Ax_2\| < d + r/L_2,$$

$$\|A1(x_1 - x_2)\| \le \|A'\|\ (d + r/L2),\ \|x_1 - x_2\| < (2R\|A'\|\ (d + r/L_2))^{1/2} \qquad (72)$$

By virtue of arbitrariness of r, from here and from (70) it follows that

$$w(A^{-1}, d) < (2\sup \|A'\| Rd\ (x\hat{} \in D(x_0))^{1/2} < (2LRd)^{1/2} \qquad (73)$$

on the set of correctness

$$M = \cap M(x\hat{})\ (x\hat{} \in D(x_0)). \qquad (74)$$

Thus, we have the following:

Theorem 9. *The continuity module $w(A^{-1}, d)$ satisfies the inequality* (73) *in the domain $D(x_0)$ under the condition that the relations'* (67), (69), (70), *and* (74) *determine the set of correctness for the operator A.*

Note that $M \neq \varnothing$ under the sufficient large R and small $d(D(x_0))$, where $d(.)$ is the diameter of domain.

1.2.4. *On regularization for linear equations*
Let us consider the linear equation:

$$A' = A,\ A_1 = A^*A,\ f_1 = A^*f,\ x_1 = x,\ M(x\hat{}) = M, \qquad (75)$$

and introduce the equation (see V. Ivanov [88])

$$A_a x\hat{}_a = (aI + A\hat{}_1)x\hat{}_a = f\hat{}_1,\ x\hat{}_a = 1/af\hat{}_1 + 1/a\int_0^t k(t,u)x\hat{}_a(u)du,\ a > 0, \qquad (76)$$

where I is the unit operator, $A\hat{}_1$ and $f\hat{}_1$ are given estimates of A_1 and f_1. We solve (76) by the *least square method*, i.e., we take as the approximate solution

$$x_n = x_{na} = \Sigma_{k=0}^{n-1} a_k T_k(2/Tt - 1),\ T_k(u) = \cos(k\arccos u), \qquad (77)$$

and we find coefficients a_k by the solution of the linear algebraic system

$$\Sigma_{k=0}^{n-1} a_k(A_c T_k, T_j) = (f\hat{}_1, T_j),\ j = 0, \ldots, n-1, \qquad (78)$$

where for any f_1 and f_2

$$(f_1, f_2) = 1/T\int^T_0 f_1(t)f_2(t)dt. \qquad (79)$$

It follows from the results [88] that the *inherent error* for the solution of (57) or (75)

$$\Delta_1 = \|x\text{-}x\hat{}_a\| < aR+dI(\|f\|+3\|A\|)(R\|A1\|+aR)/a = (a+C_1dI/a)R, dI = d(A)+d(F); \qquad (80)$$

the least square *method error*

$$\Delta_2 = \|x\hat{}_a - x_{n\,a}\| < (a+\|A_1\|)/a\ d_n(x\hat{}_a), \qquad (81)$$

where $d_n(x\hat{}_a)$ is the value of the best approximation for $x\hat{}_a$ by algebraic polynomials in the power n - 1; the *round-off error* under realization of the least square method on a computer

$$\Delta_3 = \|x_{na} - x\hat{}_{nal}\| < (a+\|A_1\|)/a\ n2^{-l}\|x_{n\,a\,l}\|, \qquad (82)$$

where l is the number of digits in a computer's summarizer (we assume here that the elements of (79)-type are computed exactly, otherwise, additional errors as hereditary ones can be included in the estimation (80)).

Noting that $d_n(x\hat{}_a)$ in the presence of rth derivatives of $f_l(t)$ and $k(t,u)$ (by t) can be estimated in the form $d_n(x\hat{}_a) < C_2R/(an^r)$, where C_2 can be found effectively by the virtue of (76), and connecting the estimations (80)-(82), we obtain (with the accuracy of the main terms) the *complete error*

$$\Delta < R(a+C_1dI/a+C_2/(an^r)+C_3/an2^{-l}). \qquad (83)$$

Let us put $C_2/(an^r) = C_3n2^{-l} = C_1dI$. Then (83) can be in the form

$$\Delta < R(a+3C_1dI/a). \qquad (84)$$

Let the parameter $a = (3C_1dI)^{1/2}$, which minimizes the estimate Δ (84). Then

$$\Delta < 2R(3C_1dI)^{1/2};\ C_2/n^r = 3^{1/2}(C_1dI)^{3/2},\ 2^{-l} = C_1/(C_3n)dI. \qquad (85)$$

We summarize the results in the following:

Theorem 10. *The complete error of the method* (75)-(79) *is estimated by the relations* (85), *which allow us to determine necessary parameters of the method and the computer for the solution of the linear equation* (69), (75) *with pre-assigned accuracy.*

It should be noted that for the method under consideration, the hypotheses [15] will also be fulfilled if the algebraic system (78) is solved by the special algorithm [89] combining the iterative process with the approximation of the matrix of this system by a band matrix with the center along the main diagonal.

In conclusion, we note that the considered results without principal difficulties can be transferred to the systems of V. i. e. of the second and first kind. For a few different results of other mathematicians see P. Linz [121].

2. Reduction to Volterra-Type Equations

2.1. MATHEMATICAL PROGRAMMING PROBLEMS

Highly general and important problems are the *mathematical programming problems* of the form

$$f(x) = \inf\,(x \in X),\ X = \{x\colon e_i(x) = 0,\ i = 1, \ldots, l;\ p_j(x) \le 0,\ j = 1, \ldots, m\}, \tag{86}$$

where the functions f, e_i, and p_j are continuously differentiable in R^n (R^n is n-dimensional Euclidean space).

Problems of the solution of linear or nonlinear operator equations of the form $Ax = y$, where the operator A and element y are given, are easily reduced to (86) by means of the solution of the minimization problem for the norm of the residual $f(x) = \|Ax\text{-}y\|$ and a certain finite-dimensional approximation of the desired element x. In a similar way, the more general problems of the solution of equations regarding several restrictions (see, for example, the method of quasi-solution above) are also easily reduced to (86).

We consider how (86) is reduced to the problem of minimization of functions without restrictions, i.e., in all the space R^n (see [89], V. Ivanov, V. Lyudvichenko [100]), and how the last problem is reduced to the solution of V. i. e.

Let

$$F(x,q) = [f(x) - q]_+^{s+1} + g(x), \tag{87}$$

where q is a real parameter; s is an integer; $y_+ = \max\,\{0, y\}$ for arbitrary y; and g is a function with the property

$$g(x) = 0\ (x \in X),\ > 0\ (x \notin X). \tag{88}$$

The examples of g(x) are

$$g(x) = \Sigma^{l}_{i=1}\ |e_i(x)| + \Sigma^{m}_{j=1}\ [p_j(x)]_+ \tag{89}$$

and

$$g(x) = \Sigma^{l}_{i=1}\ e_i^2(x)\ + \Sigma^{m}_{j=1}\ [p_j(x)]_+^{s}. \tag{90}$$

Let the values of $f(x)$ belong to the segment $[q^-, q^+]$, $q^- = \inf f(x)\ (x \in X) < \infty$.

Lemma 1. *If a settled value $q^\wedge \geq q^-$ ($q^\wedge < q^-$), then the value of the global minimum of $F(x,q)$ in R^n is equal to* 0 (> 0).

Lemma 2. *Let the function f, e_i, and p_j be convex. Then the function* (87), *where g is given by* (89) *or* (90) *is also convex.*

Thus, in the condition of this Lemma we have the well-known *convex programming problem.*

Lemma 3. *The function* (87), *where g given by* (90), *will be s-time differentiable if each of the functions f, e_i, and p_j has the same property.*

The proof of Lemmas 1-3 is obvious.

For analytical functions instead of (87) one can introduce, for example,

$$F(x,q,r) = [f(x) - q]^2 + \Sigma^{l}_{i=1} e_i^2(x) + \Sigma^{m}_{j=1} [p_j(x)+r_j^2]^2,\ r = (r_1, \ldots, r_m). \qquad (91)$$

Lemma 4. *The function $F(x,q,r)$ is analytical over all the arguments and the desired solution $x^* = x(q^*)$ of the problem* (86) responds to the least $q = q^*$ *for which*

$$F(x^*,q^*,r^*) = \inf F(x,q^*,r)\ (x,\ r \in R^{n+m}) = 0. \qquad (92)$$

The proof of Lemma 4 is also obvious.

Let us consider the following algorithm:

i) create point $x_m = x(q_m)$ for which $|F(x_m,q_m) - \inf F(x,q_m)\ (x \in R^n)| < \varepsilon$, $m = 0, 1, \ldots$;
ii) if $F(x_m,q_m) > \varepsilon$, then $q_{m+1} = q_0 + h2^m$, where h and q_0 are arbitrary numbers, $h > 0$, for example, $h = 1$ and $q_0 = 0$;
iii) if $0 \leq F(x_m,q_m) \leq \varepsilon$, then $q_{m+1} = (q_{m-1}+q_m)/2$;
iv) if $F(x_{m+1},q_{m+1}) > \varepsilon$, then $q_{m+2} = (q_{m+1}+q_m)/2$ and if $F(x_{m+1},q_{m+1}) < \varepsilon$, then $q_{m+2} = (q_{m+1} + q_{m-1})/2$;
v) stop if $q_m - q_{m-1} \leq \varepsilon$.

Theorem 11. *Let the conditions*

$$|f(x) - f(x^\wedge(x))| \leq M(f')|x - x^\wedge(x)|,\ |g(x) - g(x^\wedge(x))| \geq m(g')|x - x^\wedge(x)|,$$

$$x \in \{x: 0 < g(x) < 2\varepsilon\},\ M(f') = \max|f'|,\ m(g') = \min|g'|, \qquad (93)$$

take place, where $x^\wedge(x)$ is the nearest point to the x point of the set $\{x: g(x) = 0\}$, *and let us consider for simplicity the case* (89) *and s* = 0.

Then after the algorithm's stopping, we have

$$|q_m - \inf f(x)\ (x \in X)| \le 2\varepsilon(1 + M(f')/m(g')). \quad (94)$$

Besides, the number of the value $f(x)$ computation does not exceed

$$N = 2\max\,(0, \log_2 [|q^{\cdot} - q_m|/h]) + \log_2 h/\varepsilon. \quad (95)$$

Proof. The relation (95) is obvious. If $q_m \ge q^{\cdot}$, then $q_m - q^{\cdot} \le q_m - q_{m\text{-}1} \le \varepsilon$. If $q_m \le q^{\cdot}$, then on the strength of (93)

$$q^{\cdot} - q_m = q^{\cdot} - f(x^\wedge(x)) + f(x^\wedge(x)) - f(x_m) + f(x_m) - q_m < 2\varepsilon(1 + M(f')/m(g')). \quad (96)$$

In a similar way we can receive the estimates in the other cases of the functions f and g under the more general condition $|g(x) - g(x^\wedge(x))| > v|x - x^\wedge(x)|$, $v > 0$, $x \in \{x: 0 < g(x) < 2\varepsilon\}$.

Thus, we have the reduction of the initial problem (86)-type to the finite sequence of the minimization problems in R^n.

Let us consider the reduction of the last problem to the solution of V.i.e.

Lemma 5. *Any finite solution of the systems of V.i.e. of the form*

$$x(t) = x_0 - \int_0^t f'(x(u))du,\ f'(x) = \operatorname{grad} f(x),\ x(0) = x_0,\ t \ge 0, \quad (97)$$

has the property $f'(x(\infty)) = 0$ (i.e., $x(\infty)$ is a local minimum of $f(x)$).

The proof is obvious.

Theorem 12. *Let $x(t)$ be the solution of (97) and*

$$\|x_0 - x(t)\| \le d,\ \ t \ge 0,\ \cos\,(F'(x), x_0 - x(\infty)) \ge c > 0,\ F(x) - F(x^*) \le \|F'(x)\|\,\|x - x^*\|. \quad (98)$$

Then

$$F(x(T)) - F(x(\infty)) \le \varepsilon,\ T \ge d^2/(c\varepsilon). \quad (99)$$

Proof. On the strength of (98) we have

$$d \ge \|x_0 - x(t)\| \ \ge \int_0^t \|F'(x(u))\cos\,(F'(x(u), x_0 - x(\infty))\|du \ge c\int_0^t \|F'(x(u))\|du, \quad (100)$$

from which

$$\int_0^t \|F'(x(u))\|du \le d/c. \quad (101)$$

Therefore, there exists the least $t^\wedge$, $t^\wedge \leq d/(c\varepsilon)$, such that $\|F'(x(t^\wedge))\| \leq \varepsilon/d$. Indeed, otherwise

$$\int^{t^\wedge}_0 \|F'(x(u))\| du > \varepsilon t^\wedge/d \geq d/c. \tag{102}$$

So, $F(x(t^\wedge))-F(x^*) \leq \varepsilon$. It remains to note that $F(x(T))-F(x^*) \leq F(x(t^\wedge))-F(x^*)$ for $T \geq t^\wedge$.

Theorem 13. *Let*

$$\|F'(x)\| \leq M, \ (F'(x), x-x(\infty)) \geq m\|x-x(\infty)\|^2. \tag{103}$$

Then

$$\|x(t)-x^*\| \leq (\varepsilon/M)^{1/2}, \ F(x)-F(x^*) \leq \varepsilon, \ T \geq 1/(2m)\ln[M\|x_0-x^*\|/\varepsilon]. \tag{104}$$

Proof. On the strength of (103) we have

$$1/2d/dt\|x-x^*\|^2 = ((x-x^*)', x-x^*) = (F'(x) - F'(x^*), x-x^*) = -m\|x-x^*\|^2,$$

$$F(x)-F(x^*) \leq M\|x-x^*\|^2, \tag{105}$$

from which (104) follows rather easily.

Thus, the matter reduces to the solution of Volterra-type equations of the form (97) on the finite segment $[0, T]$, which is needed for the above-considered methods of s. 1.

2.2. SOME OTHER PROBLEMS

The problem of the solution of *operational equations* of the form

$$Ax = y, \tag{106}$$

where usually two elements are to be given and third is to be found, reduces to the mathematical programming of the problem of minimizing the residual:

$$\|Ax - y\| = \min\ (x, y, A \in X \times Y \times \mathrm{A}), \tag{107}$$

where X, Y, and A are given sets: a finite approximation of all the elements in (106) (see s. 3.1 below). In particular, by means of (106) and (107) the solution of the boundary problems of mathematical physics can be reduced to the mathematical programming and by the same token to the solution of V.i.e.

The important problem is the selection of the *base system of* the *elements* $e_1, \ldots, e_n$ with the help of which we need to find the solution of the equation $Ax = y$. The "good" system is any one that gives a "good" approximation, for example, for the desired solution x under not "large" n. In practice, the sense of the solution prompts us usually to select the "good" system. In theory, there are many well-known "good" systems for some classes of functions [89] and for some certain new, rather effective ones (see, for example, V. Ivanov [92]).

3. On Optimal Numerical Methods

3.1. APPROXIMATION OF FUNCTIONS

Reduction of many other problems to (86) and hence to (97) is based on the *approximation* of initial data and desired results with the help of objects that are completely determined by finite number of parameters. In practice, we only have finite information about each applied problem of infinite-type. This is the main reason for each problem to be naturally imbedded in the class of problems with the same information about their initial data.

The main *postulate* of the so-called "*Information-based complexity theory*" ([89]; J. Traub, H. Wozniakowski {186]) is the following. If we have any problems with at least two possible solutions under given information then there does not exist an algorithm that allows us to find any solution with confidence. We can find only the approximate solution with accuracy that is equal or greater than 1/2 of the distance between the different solutions. This finite information determines classes of data and results under the same information.

We consider the so-called *method of bound functions* [89] to construct the *optimal by accuracy approximation*. Let class functions F be given. We determine two bound functions:

$$f^{+}(x) = \sup f(x)\ (f \in F), f^{-}(x) = \inf f(x)\ (f \in F). \tag{108}$$

Then the *optimal by accuracy approximation* $f^{*}(x)$ of functions from the class F and the respective error $\rho^{*}(x)$ can be found with the help of the formulae:

$$f^{*}(x) = [f^{+}(x) + f^{-}(x)]/2,\ \rho^{*}(x) = [f^{+}(x) - f^{-}(x)]/2. \tag{109}$$

The functions f^{*} and ρ^{*} are called *Chebyshev center and radius* of F.

As an example of application, we consider the following class of functions:

$$F \equiv C^{n}_{1,L,N,\varepsilon}: |f(x) - f(y)| \le L\|x-y\|_1 = L\max |x_i - y_i|\ (i = 1, \ldots, n),$$

$$x, y \in \pi: [0, 1]^n, |f(x^j) - f^{\wedge}_j| \le \varepsilon_j, x^j = (x_1^{\,j}, \ldots, x_n^{\,j}), j = 1, \ldots, N, \tag{110}$$

where $L, N, n, x_i^{\,j}, f^{\wedge}_j, \varepsilon_j, i = 1, \ldots, n; j = 1, \ldots, N,$ are given real numbers.

Theorem 14. (see [89]). *The optimal by accuracy approximation and the respective error on the class* $C^{n}_{1,L,N,\varepsilon}$ *are given by the formulae* (109), *where*

$$f^{+}(x) = \min \{f^{\wedge}_j + \varepsilon_j + L\|x-x^j\|_1\}\ (j = 1, \ldots, N),$$

$$f^{-}(x) = \max \{f^{\wedge}_j - \varepsilon_j - L\|x-x^j\|_1\}\ (j = 1, \ldots, N), \tag{111}$$

and

$$\rho^*(x) \le \max |\varepsilon_j| \ (j = 1, \ldots, N) + L/N^{1/n}. \tag{112}$$

Note that the error $\rho^*(x)$ depends on $(2 + n)N + 3$ magnitudes here. Examples can be constructed, when $\rho^*(x) = 0$ for arbitrary values of L and ε_j, $j = 1, \ldots, N$.

The similar results can be obtained in many classes of functions [89, 95]. A priori given estimates of L and/or ε_j, $j = 1, \ldots, N$, can be strongly overstated. To improve them, we can solve the problem (the *method of quasi-solution*)

$$L = \max |f(x^j) - f(x^k)| / \|x^j - x^k\|_1 \ (j, k = 1, \ldots, N, j \ne k) = \min \tag{113}$$

under given $f^\wedge_j$, ε_j, $j = 1, \ldots, N$, and observance of the relation (110), or the problem (the *method of residual principle*)

$$\varepsilon = \max |\varepsilon_j| \ (j = 1, \ldots, N) = \min \tag{114}$$

under given L and $f^\wedge_j$, $j = 1, \ldots, N$, and the relation (110). The problem (113) and (114) can be reduced to the *linear programming problem* ([89]; L. Khachiyan [111]).

3.2. ON THE SOLUTION OF OTHER APPLIED PROBLEMS

The method and the results of s. 3.1 can be expanded for many other applied problems: solutions of algebraic, differential and integral equations, minimization of functions and functionals, etc. (see, for example, [89, 92]).

After reduction of an initial problem or from the beginning, when we have a certain finite problem for its solution on a computer, the questions about round-off errors and numbers of computation operations arise. We can answer these questions in a reasonable way following the results from above. We dwell here on some results concerning numbers of the necessary main operations on computers.

For linear algebraic systems in the form

$$\textstyle\sum_{i=1}^{n} x_i(Ae_i, \ Ae_j) = (f, Ae_j), j = 1, \ldots, n, \tag{115}$$

which appear by the well known *method of least square* for the solution of the linear operator equation $Ax = f$, we can economize the number of basic operations at the expense of approximately tape structure of the respective matrices $\{(Ae_i, \ Ae_j)\}$. As a result [8, 12], we can achieve n^2 operations instead of n^3 in order, where n is the number of the equations. The same result can be obtained for many other direct methods of mathematical physics.

The problem of *dynamic programming* is actually a consequence of the following relation

$$\inf f(x_1, \ldots, x_n) \text{ (over } x_1, \ldots, x_n) = \inf [\ldots [\inf f(x_1, \ldots, x_n) \text{ (over } x_n)] \ldots \text{(over } x_1)]. \tag{116}$$

For the problem of quadratic *Boolean programming*, i.e., determination of at least one vector x for which

$$f^* = \sup f(x)\ (x \in B^n), f(x) = \sum_{i=1}^{n} x_i(1 - \sum_{j=i+1}^{n} a_{ij}x_j),\ x = (x_1, \ldots, x_n),$$

$$x_i,\ a_{ij} \in B = \{0, 1\}, j = i+1, \ldots, n;\ i = 1, \ldots, n,\ n > 1, \tag{117}$$

where B^n is the set of all the vertices of n-dimensional cube, it is known that we can achieve n^m operations if $a_{ij} = 0$ for $|i-j| > m$, where m does not depend on n. This result can probably be generalized on any applied mathematical programming problem, for the correlation among variables decreases quickly depending on their distances.

The problem (117) in a general case can be easily reduced (V. Trubin [187]) to the so-called *NP-complete (and NP-hard) problems* (M. Garey, D. Johnson [53]) that are very difficult for the solution under large n. There is the conjecture that this problem does not belong to the class P, i.e., it cannot be solved with the help of a polynomial number of basic operations.

In contrast, it was proved L. Khachiyan [111] that the *problem* of *linear programming* belongs to the class P and the polynomial-time algorithms were constructed for its solution.

Theorem 15 (V. Ivanov [98]). *Let L_n be the set of all 2^n binary words, and let*

$$I_p = \sum Re^{pf(x)}\ (x \in L_n)/\sum e^{pf(x)}\ (x \in L_n). \tag{118}$$

Then

$$I_p \to I = 1/N \sum x\ (x \in L^*),\ p \to \infty;\ 1 \le N \le 2^n, \tag{119}$$

where L^* is the set of all points x^* for the global maximum f^* of the function $f(x)$ on L_n, N is the number of those points. Besides, if $N = 1$, then $x^* = \lim I_p$ $(p \to \infty)$.

Proof. For $x \notin L^*$, we have $f(x) < f^*$. Therefore,

$$I_p = \{\sum x\ (x \in L^*) + \sum xe^{p[f(x) - f^*]}\ (x \notin L^*)\}/\{N + \sum e^{p[f(x) - f^*]}\ (x \notin L^*)\} \to$$

$$1/N \sum x\ (x \in L^*),\ p \to \infty, \tag{120}$$

i.e., (119) is proven.

Note that the formulae (118)-(119) can be generalized on any problem of global optimization, including the problems minimization of a free energy for any macromolecule.

We show now that even stronger result can be obtained under a certain finite value of p.

Theorem 16. *Let $x_t = (x_{1,t}, \ldots, x_{n,t})$; $x^*_t = (x^*_{1,t}, \ldots, x^*_{n,t})$, $t = 1, \ldots, N$, $1 \le N \le 2^n$, and*

$$J_p = \sum x_{1,t}\exp\,[pf(x_t)]\ (t = 1, \ldots, 2^n)/\sum \exp\,[pf(R_t)]\ (t = 1, \ldots, 2^n). \tag{121}$$

Then there exist finite values $p \geq n$ *and* $t^\wedge$, $1 \leq t^\wedge \leq N$, *such that* $|J - J_p| < 1/2, p \geq n,$

$$J = 1/N\sum x^*_{1,t}\ (t = 1, \ldots, N),\ x^*_{1,t^\wedge} = 0\ (0 \leq J_p \leq 1/2),\ 1\ (1 \geq J_p > 1/2). \quad (122)$$

Proof. It follows from $J_p \to J$, when $p \to \infty$, that for a certain finite $p = p^\wedge$, we have $|J - J_{p^\wedge}| < 1/2$. Since all possible values of f^* are $0, 1, \ldots, n-1$ (see V. Trubin [187]), depending on values a_{ij}, we have

$$q_n = \min\ (f^* - f_t)\ (t = N+1, \ldots, 2^n) \geq 1, f_t = f(R_t), \quad (123)$$

Using (121) and (123), we find

$$J - J_p = \{\sum\ [1/N\sum x^*_{1,t}\ (t = 1, \ldots, N) - x_{1,t}]\exp\ [p(f_t - f^*)]\ (t = N+1, \ldots, 2^n)\}/S,$$

$$S = N + \sum \exp\ [p(f_t - f^*)]\ (t = N+1, \ldots, 2^n). \quad (124)$$

Since $|1/N\sum r^*_{1,t}\ (t = 1, \ldots, N) - r_{1,t}| \leq 1$, we have

$$|J - J_p| \leq X/(N + X),\ X = \sum \exp\ [-p(f_t - f^*)]\ (t = N+1, \ldots, 2^n) < e^{-p}(2^n - N). \quad (125)$$

The function $X/(N + X)$ is increasing while X is increasing. So, $|J - J_p| < 1/2$, if $X/(N + X) < 1/2$ or $X < N$. In turn, $X < N$, if $e^{-p} < N/(2^n - N)$, which is valid, if $e^{-p} \leq 2^{-n}$ or $p \geq n\ln 2 > n$. The property $0 \leq J_p \leq 1$ follows from definition of J_p. Since $r^*_{1,t} = 0$ or 1, we have $J \leq 1$. Thus, if $J_p \leq 1/2$, then $J < 1$, and hence at least one of $r^*_{1,t}$ is equal to 0. If $J_p > 1/2$, then $J \neq 0$, and hence at least one of $r^*_{1,t}$ is equal to 1.

Thus, after computing $J_{,p}$ in (122) for $p = n$ (or $p = n\ln 2$) we can find the first digit of one of the solution of the problem (117), i.e., we can reduce this problem from n to n-1. The number of operations of classical computer will be more than any degree of n, in general.

We can still likely determine J_p effectively by a special device with regard to J_p in (118) and (121) are associated with the coordinates of the centers of mass for the system of particles with masses $\exp\ (pf_t)$, $t = 1, \ldots, 2^n$. We can use also another interpretation:

$$P_t = \exp\ (pf_t)/\sum \exp\ (pf_s)\ (s = 1, \ldots, 2^n) = \exp\ [p(f_t - f^*)]/S \leq e^{-p},\ t = N + 1, \ldots, 2^n, \quad (126)$$

are the probabilities of occurrence of the atoms of protein with the free energy $f(x)$ in the state x_t for certain values of p (A. Finkelstein, O. Ptitcyn, [48]).

Accepting the following *nature postulate*: *The nature could not permit events with too low probabilities, and there is no need to measure or compute those probabilities, while it implements them for a process*, we can explain the fast process of protein folding.

And this can serve as a basis for the construction of a non-algorithmic type device for the effective solution of NP-hard problems.

In general, we need to solve an applied problem $P(I)$ with the help of an algorithm $A(X)$ on a computer $C(Y)$, where I, X, and Y are sets of parameters of the problem, algorithm, and computer, respectively. Then, in practice, we have to deal with the so-called complete error E, necessary time T, and memory M of the computer,

$$E = E(I,X,Y),\ T = T(I,X,Y),\ M = M(I,X,Y). \tag{127}$$

The optimal by E algorithm $A(X)$ is important, because with insufficient accuracy, we should try to find reserve in improving the parameters I and/or Y, but not the algorithm.

The important *reserve* of improving the *accuracy* of initial data $I = I(t)$, $t \in [0, R]$, of an applied problem can be realized (V. Ivanov et al. [95]) by solving (again) the Volterra-type equation of the first kind in the form

$$\int_{t-t^\wedge}^{t} K(t,s)I^\wedge(s)ds = I(t),\ t \in [0, R], \tag{128}$$

where $I^\wedge(s)$ is the desired vector of more precise data, $K(t,s)$ is the matrix of impulse transient functions of the respective measuring devices (which is known with the sufficient accuracy), $t^\wedge$ is the estimate from below for the time of end of all the transient processes.

For improving the round-off error we can carry out computations with the additional accuracy (with extra binary digit bits). The usual mode for improving the time $T(I,X,Y)$ is the computation in a parallel of the collection of the independent sub-problems representing the initial problem. Improving $M(I,X,Y)$ usually implies making worse $T(I,X,Y)$ and vice versa.

The solutions of integral equations exist in the so-called classes of generalized functions. To find approximations of generalized solutions on the basis of, for example, the least square method, it is necessary to have the complete systems of generalized and easily computable functions. One of such systems is constructed and justified in V. Ivanov [92].

4. On Algorithms and Software

One mass problem under realization of mathematical modeling, based on MM in question, is the problem of a numerical solution of the system of Volterra-type integral equations.

For the effective numerical solution of this class of the equation with the pre-assigned accuracy and with regard to the errors of three types (at the expense of the initial data inaccuracy, approximation methods and round-off errors under realization of the approximation methods on the computer) see s. 1 and 2. It has been shown that the solution of optimization problems can be also reduced to the solution of the integral equations of Volterra-types (s. 3).

Very sophisticated software, including the apparatus for the application of the new technique under consideration, is contained in V. Ivanov et al. [96, 97]; V. Mikhalevich, V. Ivanov et al. [136]. Peculiarity of the program complex [96] consists of that as a rule the so-called estimating subroutines are contained in it, side by side with the ordinary subroutines resolving principal standard problems of computation mathematics.

The former subroutines estimate the number of the necessary basic computer operations, the required computer memory capacity, and different kind of errors accompanying a process of the solution of applied problems on a computer.

The program complexes [97] and [136] are based on [96] and focused on application of mathematical methods to economy and biology, including the solution of the problems of identification, prediction, and optimization for MM under consideration.

The software [136] consists of the subroutines for the solution of the problems of CAO MM identification, for modeling of the problems of development prediction in general and in the form "if . . . , then . . .", and for modeling some CAO optimization problems.

The appropriate software developed in the USA can be found in J. More and S. Wright [141], and W. Press et al. [154] (see also Ch. 4, s. 5).

Summary

For the first time, we have novel MM that allow us to model and simulate the problem analysis and synthesis from a unique position for a wide class of CAO. One can relate to CAO: the cell as a whole, different parts and particular processes in the cell in the norm and pathology, cell-to-cell interactions, human organisms and their subsystems in the norm and pathology, AIDS, cancers, aging, etc. There is potentially unlimited utilization of these MM because of unlimited quantity of different CAO that can be integrated or differentiated on any desired level of consideration.

We actually already have a new general methodology that can allow everyone to integrate multiple components of systems. This methodology consists, briefly, in that everyone should consider a given system as one of CAO; decide its subsystems A and B and the prehistory of their functioning; construct, using the genome and the approach of the MM individualization, the indices of efficiency α- and β-types; make decisions about the initial data and desired results; and use MM of CAO in question. Also, everyone has at their disposal the multiple components: the constructive magnitudes α, β, $a(t)$, u, v, $y(t)$, f^*-types and the resources or product magnitudes $f(t)$, $m(t)$, $c(t)$, $R(t)$, $G(t)$-types integrated in the unique entire, which is named MM of CAO (or more general of ES) (Introduction and Ch. 5-8).

In the case of an interaction of two or more CAO, a certain combination essentially of the same MM can be used. Under development modeling of different CAO, everyone can use, evidently, the new control factors, such as the resources' distributions between the subsystems A and B-types of CAO and/or temporal bounds for liquidation of the obsolete technology in CAO, to optimize CAO functioning. In addition, we have the very important qualitative results about regularities of CAO optimal functioning (Ch. 9-12).

One of these results shows that CAO optimal functioning consists of their internal functions priority in comparison with their external functions. The highest priority has the constructive magnitudes corresponding to the genome system, the product magnitudes of enzyme-substrate-types follows after the constructive ones, and the external products of CAO follows at the end. However, the total result of these priorities is the best guarantee for fulfilling the external functions of CAO.

The increase of all the resources of an organism toward its internal needs of restoration most likely plays the crucial role in the case of pathology, in serious cases, of the structural values of α, β, a, u, v, y, and f^*-types (see Ch. 13-16).

A doctor's business itself can be considered as an example of CAO, the internal product of which is the new, more effective doctor's technology and the external product is the number of curing persons. The important result shows that for the number of curing persons to be maximized (for a long-term period of the time), almost all resources have to go to the more effective fulfillment of the internal function of a doctor's business. Popularization of this result, we think, is very important for Health Care's development.

Modeling of CAO allows us to describe in more detail their main structure and functions (both normal and pathologic). However, this description, essentially different from the traditional, requires us to revise the whole statement and technology of necessary biomedical experiments. According to MM of CAO, we have to learn to measure the number of *new* molecules of enzyme-substrate complexes or *new* biomass quantities for different CAO that are always non-negative. This does not mean that their behavior is always monotone. Increasing the function of $a(t)$-type can decrease other values. All these factors demand reconstruction of the infrastructure of medicine and science, in general.

It should be noted that the novel technique under consideration is not developed so completely as the traditional, exploiting, mainly, differential equations. There are many open problems in both general theory and applications of the novel technique. Among them, we would like to name the problems of modeling of phenomena, concerning wave solutions, in particular, the so-called solitons, consideration of problems of bifurcation and non-uniqueness of the solutions in great detail, application to modeling of a cancer in great detail, etc.

The problems of design or synthesis of CAO at the expense of not only the control functions of x, y and u, v-types but also of the functions of α and β-types, which were mentioned in connection with the opportunity of the gene therapy and biotech revolution in Ch. 5 and 7, are also still ahead of us.

Everyone should be aware of the great difficulties in obtaining any new theoretical results in this field and any progress to employ them in practice. It is more so that MM in question are nonlinear systems of integral-functional equations in the presence of some natural restrictions. Experience of work with these MM shows that each qualitative or quantitative result about the new property of CAO can be obtained only through hard efforts. But in return, this new property as a rule appears very important and, if we may say, beautiful in a theoretical sense.

The main advantages of these MM are:

1. All elements of MM have profound physiological interpretation associated with genes of the Human Genome.
2. All relations of MM are exact by definition, which allows us to investigate cells and CAO, in particular, by qualitative methods, knowing only qualitative initial data.
3. Considering in detail a given subsystem, we also have MM of object as a whole.
4. The solution of identification problems for given object does not require the use of animals and can be obtained from the known prehistory of object functioning and the relations of MM themselves.
5. The solution of different problems of the type "If ..., then..." is reduced to the solution of Volterra-type equations that can be solved with prearranged accuracy by optimal numerical methods.
6. The solution of different optimization problems is reduced to several problems of mathematical programming that can be also reduced to the solution of Volterra-type equations.

The models in question can be integrated at many different levels of difficulty, and they allow us to make extensive generalizations and applications that seem to be adequate for some of the mathematical challenges in R. Karp [110].

Bibliography

1. A Survey of Models for Tumor-Immune System Dynamics, Birkhäuser, 1997.
2. Abbas A.K., Lichtman A.H., Pober J.S., Cellular and Molecular Immunology, Saunders, 1991.
3. Adam J.A., General Aspects of Modeling Tumor Growth and Immune Response, in the book [1].
4. Adam L. et al., Protein Structure Prediction by Global Optimization of a Potential Energy Function, Proc. Natl. Acad. Sci.USA, vol. 96, 5482-5485.
5. AIDS is the Acquired Immune - Deficiency Syndrome, Kiev: Zdorov'ya, 1988 (in Russian).
6. Akho A., Khopkroft D., Ul'man D., Construction and Analysis of Computing Algorithms, M.: Mir, 1979 (in Russian).
7. Alberts B., Johnson A., Lewis J., Raff M., Roberts K., Walter P., Molecular Biology of the Cell, Garland Science, 2002.
8. Allman E.S.,Rhodes J.A., Mathematical Models in Biology: An Introduction, Cambridge Univ. Press, 2003.
9. Amaratunga D., Cabrera J., Exploration and Analysis of DNA Microarray and Protein Array Data, Wiley-Interscience, 2004.
10. Applied Mathematical Modeling A Multidisciplinary Approach, CRC Press, 2000.
11. Asachenkov A., Marchuk G., Mohler R., Zuev S., Disease Dynamics, Birkhäuser, 1994.
12. Astrom K.J., Eykhoff P., System Identification - A Survey, Automatica, 7, 1971, 123-167.
13. Baglay R.D., Smirnov K.K., Handling Two Dimensional Signals on Computers, ZVMiMF, 15, N 1, 1975, 241-247 (in Russian).
14. Baker C. T. H., The Numerical Treatment of Integral Equations, Clarendon, 1977.
15. Bakhvalov N.S., On Properties of Optimal Methods for the Solution of the Problems of Mathematical Physics, ZVMiMF, 10, N 3, 1970, 555-568 (in Russian).
16. Bakhvalov N.S., Numerical Methods, M.: Nauka, 1973 (in Russian).
17. Baldi P., Hatfield G.W., DNA Microarrays and Gene Expression, Cambridge Univ. Press, 2002.
18. Bartholomay A.F., Chemical Kinetics and Enzyme Kinetics, in the book [46].
19. Becker W.M., Reece J.B., Poenie M.F., The World of the Cell, The Benjaminn/Cummings, 1996.
20. Bellomo N., Preziosi L., Tumor Immune System Interactions: The Kinetic Cellular Theory, in the book [1].
21. Bellomo N., Pulvirenti M., Generalized Kinetic Models in Applied Sciences, in the book: [139].
22. Bellomo N., Stocker S., Development of Boltzmann Models in Mathematical Biology, in the book [139].
23. Bergers G., Javaherian K., Lo K., Folkman J., Hanahan D., Effects of Angiogenesis Inhibitors on Multistage Carcinogenesis in Mice, Science 284, 1999, 808-812.
24. Bethea R.M., Duran B.S., Boullion T.L., Statistical Methods for Engineers and Scientists, Decker, 1975.
25. Bieri O., Kiefhaber T., Kinetic Models in Protein Folding, in the book [132].
26. Bio-energetic of the Cell: Quantitative Aspects, KAP, 1998.
27. Bolotov B. V., I Teach You Not to Be Ill and Not to Grow Old, Kiev: Informatsionno-izdatel'skoe agenstvo "Ukraina", 1992, 48 pp. (in Russian).
28. Boyer P.D., The ATP Synthase – a Splendid Molecular Machine, Annual Review in Biochemistry, 66, 1997, 717-749.
29. Byers T., Mouchawas J., Marks J. et al., The American Cancer Society Goals How Far Can Cancer Rates Decline in the U. S. by the Year 2015?, Cancer 86, 1999, 715-27.
30. Campbell N.A., Mitchell L.G., Reece J.B., Biology. Concepts & Connections, Benjamin/Cummings, 1997.
31. Caratheodory C., Gesammelte Mathematische Schriften, Munhen, 1954-1957, Rd. 1-5.
32. Caret R.L., Denniston K.J., Topping J.J., Principles and Applications of Organic and Biological Chemistry, WCB, 1997.
33. Chaplain M.A.J., From Mutation to Metastasis: The Mathematical Modeling of the Stages of Tumor Development, in the book [1], 187-236.
34. Clark, H. R., The Cure for All Diseases; With Many Case Histories, New Century Press, 1995.
35. Coffin G.M., HIV Population Dynamics in Vivo: Implication for Genetic Variations, Pathogenesis and Therapy, Science 267, 1995, 483-489.
36. Cooper G.M., The Cell A Molecular Approach, ASM Press, 2000.
37. Curtis H., Biology, Worth Publishers, Inc, 1984.
38. Davydov A.C., Biology and Quantum Mechanics, Kiev: Naukova dumka, 1979 (in Russian).
39. Delisi C., Reflection on Mathematical Contributions to Understanding the Molecular Basis of Life: from 1970 to the 21st Century, in the book [51], 2-27.
40. Doering C.R., Gibbon J.D., Applied Analysis of the Navier-Stokes Equations, Cambridge Univ. Press, 1995.

41. Dwyer R. A., Genomic Perl. From Bioinformatics Basic to Working Code, Cambridge Univ. Press, 2003.
42. Dzyadyk V.K., Approximately Iterative Methods of Approximation by Polynomials for the Solution of the Cauchy Problem, ZVMiMF, 26, N. 3, 1986, 357-372 (in Russian).
43. Dzjadyk V.K., Ivanov V.V., On Asymptotic and Estimates for the Uniform Norms of the LaGrange Interpolation Polynomials Corresponding to the Chebyshev Nodal Points, Analisis Mathematica, 9, 1983, 85-97.
44. Element, The Encyclopedia Americana, Intern. Edition, Grolier Inc., 10, 1990, 203- 209.
45. Fall C., Marland E., Wagner J., Tyson J., Computational Cell Biology, Springer, 2002.
46. Faundations of Mathematical Biology, Academic Press, vol. 1-3, 1971-1972.
47. Fersht A.R., Structure and Mechanism in Protein Science: a Guide to Enzyme Catalysis and Protein Folding, Freeman, 1999.
48. Finkelstein A.V., Ptitcyn O.B., Protein Physics: a Course of Lectures, Academic Press, 2002.
49. Fontalin L.N., Pevnitskiy L.A., Immunology Tolerance, M.: Meditsina, 1978 (in Russian).
50. Fromter E., Transport of Matter Through Biological Membranes, in the book Biophysics, Springer-Verlag, 1993.
51. Frontiers in Mathematical Biology, Springer-Verlag, 1994.
52. Gardiner C.W., Handbook of Stochastic Methods for Physics, Chemistry, and Natural Sciences, Springer, 1990.
53. Garey M.R., Johnson D.S., Computers and Intractability. A Guide to the Theory of NP-Completeness, W.H. Freeman and Co., 1979.
54. Gershenfeld N., The Nature of Mathematical Modeling, Cambridge Univ. Press, 1998.
55. Gerstner W., Kistler W., Spiking Neuron Models. Single Neuron, Populations, Placticity, Cambridge Univ. Press, 2002.
56. Girlin S.K., Modeling of the Arising Developing Systems, DAN Ukr. SSR, N 10, 1987, 65-67 (in Russian).
57. Girlin S.K., Ivanov V.V., Modeling of Developing Systems Interaction, DAN Ukr.SSR, N 1, 1986, 58-60 (in Ukrainian).
58. Glushkov V.M., On a Class of Dynamic Macroeconomic Models, Upravlyayushchie. Sistemy and Mashiny, 2, 1977, 3-6 (in Russian).
59. Glushkov V.M., Cancer Problem from a Position of Systems General Theory, Kiev: Inst. of Cybernetics, Reprint 79-26, 1979 (in Russian).
60. Glushkov V.M., Fluctuation Systematization, Kibernetika, N 2, 1979, 114-115 (in Russian).
61. Glushkov V.M., On Possible Peculiarities of Physical Fields of Bio Systems, Kibernetika, N 3, 1981, 105-106.
62. Glushkov V.M., Ivanov V.V., Modeling of Optimization of Work Places Distribution Between the Branches of Production A and B, Kibernetika, N 6, 1977, 117-131 (in Russian).
63. Glushkov V.M., Ivanov V.V., Yanenko V.M., Methodological Problems of Mathematical Methods Application to Biology, Kiev: Inst. of Cybernetics, Reprint 79-60, 1979 (in Russian).
64. Glushkov V.M., Ivanov V.V., Yanenko V.M., On New Class of Dynamic Models and its Applications to Biology, Kibernetika, N4, 1979, 131-139 (in Russian).
65. Glushkov V.M., Ivanov V.V., Yanenko V.M., On New Class of Dynamic Models and its Applications to Biology, Kibernetika, N4, 1980, 109-118 (in Russian).
66. Glushkov V.M., Ivanov V.V., Yanenko V.M., On New Class of Dynamic Models and its Applications to Biology, Kibernetika, N5, 1981, 113-127 (in Russian).
67. Glushkov V.M., Ivanov V.V., Yanenko V.M., One Class of Nonlinear Dynamic Models and Its Applications, Physica, 2D, 1981, 61-72.
68. Glushkov V.M., Ivanov V.V., Yanenko V.M. et al., Thymus Gland Lymphocytes' Adaptive Redistribution and Regenerating Accumulation Process Modeling, Reprint 82-30, Glushkov Inst of Cybernetics, Kiev, 1982.
69. Glushkov V.M., Ivanov V.V., Yanenko V.M., Developing Systems Modeling, M.: Nauka, 1983 (in Russian).
70. Gluskov V.M., Ivanov V.V., Yatsenko Yu.P., Analytical Investigation of One Class of Dynamic Models, I, Cybernetics, Plenum Publish. Corp., 16, N 2, 1980, 164-177.
71. Glushkov V.M., Ivanov V.V., Yatsenko Yu.P., On Uniqueness of Solutions of One Optimization Problem, DAN SSSR, 256, N 4, 1981, 802-806 (in Russian).
72. Gluskov V.M., Ivanov V.V., Yatsenko Yu.P., Analytical Investigation of One Class of Dynamic Models, II, Cybernetics, Plenum Publish. Corp., 18, N 3, 1982, 390-401.
73. Glushkov V.M., Pshenichnyj B.N., On Mathematical Models of Economic Growth, Cybernetics, 13, N 4, 1977, 1-6.

74. Guide on Physiology. Physiology of Endocrine System, L.: Nauka, 1979 (in Russian).
75. Halvorson G.G., Strong Medicine, Random House, 1993.
76. Hardy G.H., Littlewood J.E., Polya G., Inequalities, Cambridge University Press, 1934.
77. Haskell C.M. (with 168 Contributors), Cancer Treatment, W.B. Saunders Co., 2001.
78. Hayflick L., How and Why We Age, Ballantine Books, 1994.
79. Hoffman G.W., On Network Theory and H-2 Restriction, Contemporary topics in immunology, 11, 1980, 186-226.
80. Hoffman G.W., The Application of Stability Criteria in Evaluating Network Regulation Models, in: Regulation of Immune Response Dynamics, CRC Press, 1, 1982, 137-162.
81. Hoffman G.W. and Gran M.D., Hypothesis: AID is Auto-immune Disease Caused by HIV Plus Allogenic Cells, in International Conference on AIDS. The Scientific and Social Challenge. Montreal, Quebec, Canada June 4-9, 1989. Abstracts, 613.
82. Hritonenko N., Yatsenko Yu., Modeling and Optimization of the Lifetime of Technologies, KAP, 1996.
83. Hritonenko N., Yatsenko Yu., Mathematical Modeling in Economics, Ecology and the Environment, KAP, 1999.
84. Immunology. Reference Book, Kiev: Naukova dumka, 1981, 480 pp. (in Russian) (translated from IMMUNOLOGIE ein Nachschlagewerk, Akademic Verlay, Berlin, 1978).
85. Immunology Methods, Vol. III, Academic Press, 1985.
86. Information, The Royal Swedish Academy of Sciences Press Release about awarding the 1997 Nobel Prize in Chemistry to Prof. Boyer P.D. , Dr. Walker J.E., and Prof. Skou J.C., 1997.
87. Inside the Cell, NIH Publication No. 97-1051, 1997.
88. Ivanov V.V., The Theory of Approximate Methods and Their Application to the Numerical Solution of Singular Integral Equations, Nordhoff Int. Publ., 1976.
89. Ivanov V.V., Methods of Computation. Guidance, Kiev: Naukova dumka, 1986 (in Russian).
90. Ivanov V.V., Optimal by Accuracy Approximation of Functions with Unknown in Advance Singular Points, Proceeding of MI AN USSR, 180, 1987, 120-121 (in Russian).
91. Ivanov V.V., Systems Development Simulation Problems and C. Caratheodory's Concepts, in the book: Constantin Caratheodory: An International Tribute, World Science, I, 1991, 501-526.
92. Ivanov V.V., Model Development and Optimization, KAP, 1999.
93. Ivanov V.V. et al.., Methods for Algorithmization of Continuous Industrial Processes, M.: Nauka, 1975 (in Russian).
94. Ivanov V.V. et al.., Modeling of Idio-type Anti-idio-typical Interactions of Immune Network with Regard to Distinctive Sub-Populations Among Limphosytes, In: Mathematical modeling, North-Holland, 1983, 141-149 (extended version in the book Mathematical modeling, M.: Mir, 1986 (in Russian)).
95. Ivanov V.V. et al.., Characteristics of Problems, Algorithms, and Computers in Complex of Programs of Computation Mathematics, Reprint 84-36 Kiev: Glushkov Inst. of Cybernetics, 1984 (in Russian).
96. Ivanov V.V. et al.., Package of the Programs for Economic Systems Development Prognosis with Regard to the Technical Progress, Upravl. sistemy i mashiny, N 5, 1985, 100-102 (in Russian).
97. Ivanov V.V. et al., Complex of Routines POM-1 (Programs of Optimal Methods), Glushkov Inst. of Cybernetics, Kiev: 1985 (Dep. in GOS FAP SSSR, N 50860000156) (in Russian).
98. Ivanov V.V., Explicit Formulae for Protein Folding, submitted to J. math biosciences in July 2004, #04118.
99. Ivanov V.V., Korzhova V.N., Mathematical Model of the Immune Network of the Patient Infected with HIV, M.: VINITI, N 133-B90, 1990 (in Russian).
100. Ivanov V.V., Lyudvichenko V.A., On One Method of Unconditional Optimization for the Solution of Mathematical Programming Problems, Kibernetika, N 52, 1977, 1-8 (in Russian).
101. Ivanov V.V., Patsalyuk S.S., Yanenko V.M., Efficiency of Work Places Functioning Estimation in the Health Care, Kibernetika, N 6, 1986, 110-113 (in Russian).
102. Ivanov V.V., Tvalodze Yu.G., Zhuzhunashvili A.Sh., On the Solution of Volterra Integral Equations with Pre-assigned Accuracy, M.: VINITI, 1886-89, 1989 (in Russian, see also in English UCD/CCM Report No. 102, 1997, 1-12).
103. Ivanov V.V., Yatsenko Yu.P., Adaptive Algorithms of Optimization in the Integral Macro-Economic Models, DAN SSSR, N 5, 1986, 1053-1058 (in Russian).
104. Ivanov V.V., Yatsenko Yu.P., Galiev U.E., Comparison Some Integral Macroeconomic Models, Avtomatica, 4, 1986, 47-53 (in Russian).
105. Ivanitsky G.R., Krinsky V.I., Sel'kov E.E., Mathematical Bio-Physics of Cell, Moscow: Nauka, 1978 (in Russian).

106. Jacquez J.A., Compartmental Analysis in Biology and Medicine, The Univ. of Michigan Press, 1988.
107. Jerne N.K., Toward a Network Theory of the Immune System, Ann. Immunol., Inst. Pasteur, 1256, 1974, 373-389.
108. Kantorovich L.V., Gor'kov L.I., On Functional Equations under Analysis of One-Product Economic Model, DAN SSSR, 129, N 4, 1959, 732-736 (in Russian).
109. Karmanov V.G., Mathematical Programming, M.: Nauka, 1975 (in Russian).
110. Karp R.M., Mathematical Challenges from Genomics and Molecular Biology, Notices of the AMS, vol 49, # 5, 2002, 544-553.
111. Khachiyan L.G., Polynomial Algorithms in Linear Programming, ZVMiMF, 20, 1980, 51-68 (in Russian).
112. Kimmel M., Axelrod D.E., Branching Processes in Biology, Springer, 2002.
113. Kirscchner D., Using Mathematics to Understand HIV Immune Dynamics, Notices of the AMS, 43, N 2, 1996, 191-202.
114. Kofler M., Maple. An Introduction and Reference, Addison-Wesley, 1997.
115. Korn G.A., Korn T.M., Mathematical Handbook for Scientists and Engineers, McGraw-Hill Book, 1968.
116. Kuang Y., Delay Differential Equations with Applications in Population Dynamics, Academic Press, 1993.
117. Kuznetsov V.A., Basic Models of Tumor-Immune System Interactions Identification, Analysis and Predictions, in the book [1], 237-294.
118. Law, Dornald's Illustrated Medical Dictionary, Saunders, 1988, 899-902.
119. Leturcq D., Winslow B., Johnson P., Thornton G., Neutralization of HIV-1 by Anti-Idio-Typical Antibodies, in the same book as in [81], 554.
120. Levine A.J., Viruses, Scientific American Library, 1992.
121. Linz P., Analytical and Numerical Methods for Volterra Equations, SIAM Studies in Appl. Math., 1985.
122. Lynch S., Dynamical Systems with Applications Using MATLAB, Birkhauser, 2004.
123. Mackey M.C., Mathematical Models of Hematopoietic Cell Replication and Control, in the book [146].
124. Maeda K., Kang S.-M.,Onoda N., Ogava M. et al., Vascular Expansion Growth Factor Expression in Preoperative Biopsy Specimens Correlate with Disease Recurrence in Patients with Early Gastric Carcinoma, Cancer 86, 1999, 566-71.
125. Malcomson J.M., Replacement and the Rental Value of Capital Equipment Subject to Obsolescence, Journal of Economic Theory, 10, 1975, 24-41.
126. Managing Planet Earth, Reading from Science Amer. Magazine, 1990.
127. Marchuk G.I., Mathematical Models in Immunology, M.: Nauka, 1985 (translated into English, the same title, Optimization Software, Inc., New York, 1983).
128. Marzec C.J., Spiegel E.A., Ordinary Differential Equations with Strange Attractors, SIAM J. Appl. Math., 38, N 3, 1980, 403-421.
129. Mathematical Model Research -- Intramural, Source: CRISP Format F Last Update 11-01-1999, Microsoft Word Document, NIH, 1999.
130. Mascagni M.V., Sherman A.S., Numerical Methods for Neuronal Modeling, in the book [134].
131. May R.M., Necessity and Chance: Deterministic Chaos in Ecology and Evolution, Bulletin (New Series) of the AMS, Vol. 32, N. 3, 1995, 291-308.
132. Mechanisms of Protein Folding, Frontiers in Molecular Biology, Oxford Univ. Press, 2000.
133. Medina J.J., The Clock of Ages Why We Age—How We Age—Winding Back the Clock, Cambridg Univ. Press, 1996.
134. Methods in Neuronal Modeling From Ions to Network, The MIT Press, 1998.
135. Michelson S. and Leith J.T., Tumor Heterogeneity and Growth Control, in the book [1].
136. Mikhalevich V.S., Ivanov V.V. et al., Complex of Routines MDS-1 (Modeling of Developing Systems), Glushkov Inst. of Cybernetics, Kiev: 1986 (Dep. in GOS FAP SSSR, N 50880000105) (in Russian).
137. Mikhalevich V.S, Ivanov V.V., Yanenko V.M., Gyulling E.V., Integro-functional Model of Blood Regulation System, Kibernetika, N 3, 1986, 69-77, 86 (in Russian).
138. Mikhaylov G.A., Optimization of Monte-Carlo Weight Methods, M.: Nauka, 1987 (in Russian).
139. Modeling in Applied Sciences A Kinetic Theory Approach, Birkhauser, 2000.
140. Mohler R.R., Bruni C. and Gandolfi A., A Systems Approach to Immunology, IEEE Trans. Automat. Contr., AC-25: 1980, 964-990.
141. More J.J., Wright S.J., Optimization Software Guide, SIAM, 1993.
142. Murray J.D., Asymptotic Analysis, Springer, 1992.
143. Murray J.D., Mathematical Biology I: Introduction; II: Spatial Models and Biomedical Applications, Springer, 2003.

144. Nowak M., Mathematical Models of Virus Infection and Evolution, Abstract, Conference on Mathematical Models in Medical and Health Sciences, Vanderbilt University, May 28-31, 1997.
145. Optimization Methods & Software, Ed. – in – Chief Y. Evtushenko, Taylor & Francis, 2004.
146. Othmer H.G., et al., Case Studies in Mathematical Modeling – Ecology, Physiology, and Cell Biology, Prentice Hall, 1997.
147. Perelson A.S., Immune Network Theory, Immunol. Rev., 110, 1989, 5-36.
148. Petrov A.A., Pospelov I.G., Systematic Analysis of Developing Economy, Izv. AN SSSR, Techn. kibernetika, 1979, N 2, 18-27; N 3, 28-38; N 4, 11-23; N 5, 13-24 (in Russian).
149. Petrov R.V., Stepanenko R.N., Auto-regulative Processes in Immune System and Principle of Immunology Mobile, Kibernetika, N 3, 1986, 78-86 (in Russian).
150. Photosynthesis, The Encyclopedia Americana, Intern. Edition, Grolier Inc., 22, 1990, 38-41.
151. Pogozhev I.B., Application of Mathematical Models of Diseases in Clinic Practice, M.: Nauka, 1988 (in Russian).
152. Poluectov R.A., Pykh Yu.A., Shvytov I.A., Dynamic Model of Ecological Systems, L.: Gidrometioizdat, 1980 (in Russian).
153. Pontryagin L.S., Boltyanskii V.G., Gamkrelidze R.V., and Mischance E.F., The Mathematical Theory of Optimal Processes, Wiley (Inter-science), 1962.
154. Press W.H., Vetterling W.T., Teukolsky S., Flannery B.P., Numerical Recipes in C Numerical Recipes in Fortran The Art of Scientific Computing, C Version; Fortran Version; C Example Book; Fortran Example Book, Cambridge Univ. Press, 1992.
155. Press Release, Nobel, 1996.
156. Press Release, Nobel, 2001.
157. Press Release, Nobel, 2002.
158. Rao C.V., Wolf D.M., Arkin A.P., Control, Exploitation and Tolerance of Intracellular Noise, Nature, vol. 420, 14, 2002, 231-237.
159. Rashevsky N., Mathematical Biophysics, I, II, Dover, 1960.
160. Rifkin J., The Biotech Century, Penguin Putnam Inc., 1998.
161. Roberts D.V., Enzyme Kinetics, Cambridge Univ. Press, 1977.
162. Romanovsky Y. M., Stepanova N.V., Chernavsky D.S., Mathematical Biophysics, M.: Nauka, 1984 (in Russian).
163. Rowe G., Theoretical Models in Biology. The Origen of Life, the Immune System, and the Brain, Oxford Univ. Press, 1994.
164. Samarskii A.A., Mikhailov A.P., Principles of Mathematical Modeling Ideas, Methods, Examples, Taylor & Francis, 2002.
165.Schlick T., Molecular Modeling and Simulation: An Interdisciplinary Guide, Springer, 2002.
166. Segel L.A., Modeling Dynamic Phenomena in Molecular and Cellular Biology, Cambridge Univ. Press, 1984.
167. Self-Organized Biological Dynamics and Nonlinear Control, Cambridge Univ. Press, 2000.
168. Sendov B.Kh., Mathematical Models of Processes of Cell Proliferation and Differentiation, M.: Izdatel'stvo MGU, 1976 (in Russian).
169. Sharpe F.R., Lotka A.J., A Problem in Age Distribution, Philos. Mag., 21, 1911, 435-438.
170. Sherman A., Calcium and Membrane Potential Oscillations in Pancreatic β-Cells, in the book [146].
171. Silverberg G., Technical Progress, Capital Accumulation and Effective Demand: A Self-Organizing Model, in the book: Economic Evolution and Structural Adjustment, Springer Verlag, 1987.
172. Smale L.L., Analytic Geometry and Calculus, Appleton-Century-Crofts, Inc., 1953.
173. Smale S., A Mathematical Models of Two Cells Via Turing's Equation, in the book: Marsden J.E., McCracken M., The Hopf Bifurcation and Its Applications, Springer-Verlag, 1980, 274-283 (in Russian).
174. Smith H.L., Monotone Dynamical Systems: An Introduction to the Theory of Competitive and Cooperative Systems, Mathematical Surveys and Monographs, vol. 41, AMS, 1995.
175. Sonka M., Hlavac V., Boyle R., Image Processing, Analysis, and Machine Vision, An Intern. Thomson Publishing. Co., 1998.
176. Standard Mathematical Tables and Formulae, CRC Press, 1995.
177. Svirezhev Yu.M., Passekov V.P., Fundamentals of Mathematical Evolutionary Genetics, Kluwer Academic Publishers, 1990.
178. The Structure of Life Discovering the Molecular Shapes That Determine Health or Disease, NIH Publication No.91-2778, 1991.

179. Theoretical Immunology, The Proceedings of the Theoretical Immunology Workshop, June 1987, Santa Fe, New Mexico, Addison-Weley. Publ. Co., 1987, Part two, Vol.3 (see pp. 50-100).
180. Thermodynamic Principles, McGraw-Hill Encyclopedia of Science & Technology, Vol. 13, 1982, 645-650.
181. Thieme H.R., Mathematics in Population Biology, Princeton Univ. Press, 2003.
182. Thom R., Structural Stability and Morphogenesis: An Outline of a General Theory of Models, Addison-Wesley, 1989.
183. Tikhonov A.N., Systems of Differential Equations with Small Parameters at Derivatives, Math. Transactions, **31**, # 3, 1952, 575-586 (in Russian).
184. Tikhonov A.N., Arsenin V.P., Methods for the Solution of Ill-Posed Problems, M.: Nauka, 1974 (in Russian).
185. Tikhonov A.N., Vasil'eva A.B., Sveshnikov A.G., Differential Equations, M: Nauka, 1980 (in Russian).
186. Traub J.F., Wozniakowski H., Perspective on Information-Based Complexity, Bulletin of the AMS, 26, N 1, 1992, 29-52.
187. Trubin V.A., Universality of one class quadratic integer problems, Kibernetika, 2, 1977, 147 (in Russian).
188. Tyson J.J., What Everyone Should Know About the Belousov-Zhabotinsky Reaction, in the book [51].
189. Tyson J.J., Chen K., Novak B., The Eukaryotic Cell Cycle: Molecules, Mechanisms, and Mathematical Models, in the book [146].
190. Vareba F.J. and Continto A., Second-Generation Immune Networks, Immunology Today, 12, N5, 1991, 159-166.
191. Vasil'ev F.P., Numerical Methods for the Solution of Extreme Problems, M.: Nauka, 1980 (in Russian).
192. Vasil'ev F.P., Extreme Problems Solution Methods, M: Nauka, 1981, 400 pp. (in Russian).
193. Väth M., Volterra and Integral Equations of Vector Functions, Marcel Dekker, 1999.
194. Vernadsky V.I., Living Substance, M.: Nauka, 1978 (in Russian).
195. Vladimirov V.S., Equations of Mathematical Physics, M.: Nauka, 1967.
196. Voevodin V.V. Computational Foundation for Linear Algebra, M.: Nauka, 1977 (in Russian).
197. Voit E.D., Computational Analysis of Biochemical Systems, A Practical Guide for Biochemists and Molecular Biologists, Cambridge Univ. Press, 2000.
198. Volterra V., Theory of Functional and of Integral and Integro-Differential Equations, Dover Publications, 1959.
199. Volterra V., Mathematical Theory of Fighting for Existence, M.: Nauka, 1976 (in Russian).
200. Walleczek J., The Frontier and Challenges of Biodynamics Research, in the book [167];
201. Walleczek, Changing Paradigms in Biomedicine: Implications for Future Research and Clinical Applications, in the book [167].
202. Wang T. et al., Regulation of the Innate and Adaptive Immune Responses by Stat-3 Signaling in Tumor Cells, Nature, vol. 10, N. 1, 2004, 48-54.
203. Warga J., Optimal Control of Differential and Functional Equations, Academic Press, 1972.
204. Waterman M.S., Introduction to Computational Biology, Chapman & Hall, 1995.
205. Waterman M.S., Smith T.F., Rapid Dynamic Programming Algorithms for RNA Secondary Structure, Advances in Applied Mathematics, 7, 1986, 455-464.
206. Watson J.D., Molecular Biology of the Gene, W.A. Benjamin, Inc., 1970.
207. Webb G., Theory of Nonlinear Age Dependant Population, Decker, 1985.
208. What does Science Know about AIDS, Scientific American, vol. 259, 1988.
209. Whitten K.W., Gailey K.D., Davis R.E., General Chemistry with Qualitative Analysis, Saunders College Publishing, 1992.
210. Wilkinson J.H., Algebraic Problem of Eigen-Values, M.: Nauka, 1970, 564 pp. (in Russian).
211. Yakushevich L.V., Nonlinear Physics of DNA, John Wiley and Sons, 1998.
212. Yamada W.M., Koch C., Adams P.R., Multiple Channels and Calcium Dynamics, in the book [134], 137-
213. Yatsenko Yu. P., Integral Models of Systems with Controllable Memory, K.: Naukova dumka, 1991 (in Russian).
214. Yu H., Jove R., The Stats of Cancer – New Molecular Targets Come of Age, Nature, vol. 4, Reviews, 2004, 97-105.
215. Zhabotinskiy A.M., Concentrating Auto-oscillations, M.: Nauka, 1974 (in Russion).
216. Zhao X.-Q., Dynamical Systems in Population Biology, CMS BOOKS IN MATH, Vol. 16, 2003.
217. Zullo A., Caplen N.J., Blaese R.M., Gene Therapy for Cancer, Nato ASI Series, H 105, Gene Therapy, Springer-Verlag, 1998, 133-142.

Web Sites of Interest:

1. aacr.org
2. asco.org
3. bettiegraham@nih.gov
4. biooncology.com
5. cancer.org
6. ceolas.org
7. connscurrenttherapy.com
8. ebi.ac.uk/ebi_home.html
9. genentech.com
10. genome.cbs.dtu.dk/gorm/modelorganism.html
11. germline.net
12. mathematical modeling
13. mathematical models
14. mathnetbase.com
15. mbi.osu.edu
16. mips.biohem.mpg.de
17. mtbio.de
18. nanomachina.com
19. ncbi.n/m.nih.gov
20. NelsonEric.com
21. nobel.se
22. nucleus.argon.missouri.edu/enzyme.html
23. proteinsociety.org
24. rcsb.org/
25. santafe.edu/index.html
26. siam.org
27. Stanford.edu/~mariof/ra/knowledgeUnits/ALm/
28. unaid.org/en/defolt.asp
29. v3.calvino.polito.it/~biomat
30. worlded.org/us/health/docs/comp/
31. worlded.org/us/health/lincs/

New Books of Interest

1. Chabner D.-E., Medical Terminology Elsevier Saunders, 2005
2. Drummond M.F. et al, Methods for the Economic Evaluation of Health Care Programmes, Oxford Univ. Press, 2005
3. Edelstein-Keshet L., 4. Mathematical Models in Biology, SIAM, 2005
4. Handbook of Genome Research, Wiley-VCH, 2005
5. Higham D.J., Higham N.J., MATLAB Guide, SIAM, 2005
6. Molecular Diagnostics For the Clinical Laboratorian, Humana Press, 2006
7. Tramontano A., The Ten Most Wanted Solutions in Protein Bioinformatics, CR C Press, 2005

INDEX

About the Authors

According to Marquis WHO'S WHO IN THE WORLD 1993-1994, 11th Edition, IVANOV VIKTOR VLADIMIROVICH, mathematics educator; b. Samara, USSR, June 8, 1929; s. Vladimir Aleksandrovich and Natal'ya Ivanovna (Rudneva) I.; m. Valentina Nikolaevna Chernogub, Dec. 9 1952; 1 child, Natalya. Degree Ph. D. Moscow U., 1956; D. Sc., Siberian Acad. Sci., 1966. 1958-64; head of dept. Glushkov Inst. Cybernetics, Kiev, 1965-87; Prof. Kiev U., 1969-78; lead of SCI's. Glushkov Inst. Cybernetics, Kiev, 1978-92. Author: The Theory of Approximate Methods ..., 1968; Guide Methods of Computations, 1986; author, editor: Methods for Algorithmization of Continuous Industrial Processes, 1975; co-author: (with V. M. Glushkov and V. M. Yanenko) Developing System Modeling, 1983; mem. editorial bd. Ency. of Cybernetics; contbr. over 300 articles to profl. jours. Mem. Am. Math. Soc., 1972 -. Avocations: traveling, problem solving.

He arrived in the USA as a political refugee in August 1992. Since then he was busy improving his English; looking for a job; trying to solve the well-known problem "NP =? P"; working as a volunteer in Tomlinson Adult Learning Center, St. Petersburg, and in University of South Florida, College of Engineering, Tampa; writing the book "Model Development and Optimization" published by KAP in 1999; and writing this monograph.

IVANOVA NATALYA VIKTOROVNA, b. Kiev, Ukraine, November 5, 1983, graduate of Boston University, biology as a major, Spanish as a minor. Currently she is a student at University of Florida College of Dentistry. She contributes to this book by improving its English, introducing certain biological backgrounds and the respective illusrations, and correcting some controversial issues.

Mathematics in Science and Engineering
Edited by C.K. Chui, Stanford University

Recent titles:
I. Podlubny, *Fractional Differential Equations*
E. Castillo, A. Iglesias, R. Ruíz-Cobo, *Functional Equations in Applied Sciences*
V. Hutson, J.S. Pym, M.J. Cloud, *Applications of Functional Analysis and Operator Theory (Second Edition)*
V. Lakshmikantham and S.K. Sen, *Computational Error and Complexity in Science and Engineering*
T.A. Burton, V*olterra Integral and Differential Equations (Second Edition)*
E.N. Chukwu, *A Mathematical Treatment of Economic Cooperation and Competition Among Nations: with Nigeria, USA, UK, China and Middle East Examples*
Wei-Bin Zhang, *Discrete Dynamical Systems, Bifurcations and Chaos in Economics*
C. De Coster and P. Habets, *Two-Point Boundary Value Problems: Lower and Upper Solutions*